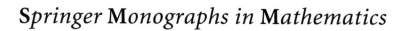

Springer Monographs *in* Mathematics

Stephen Leon Lipscomb

Fractals and Universal
Spaces in Dimension Theory

With 91 Illustrations and 10 Tables

Springer

Stephen Leon Lipscomb
Emeritus Professor of Mathematics
Department of Mathematics
University of Mary Washington
Fredericksburg, VA 22401
USA
slipscomb@umw.edu

ISSN: 1439-7382
ISBN: 978-0-387-85493-9   e-ISBN: 978-0-387-85494-6
DOI 10.1007/978-0-387-85494-6

Library of Congress Control Number: 2008938816

Mathematics Subject Classification (2000): 54xx, 28A80, 57xx

Printed on acid-free paper

springer.com

DEDICATED TO MY WIFE PATTY,
OUR SONS STEPHEN AND DARRIN,

AND

MY MOTHER DEMA ANN (ALKIRE),
AND THE MEMORY OF MY FATHER
DAVID LEON LIPSCOMB

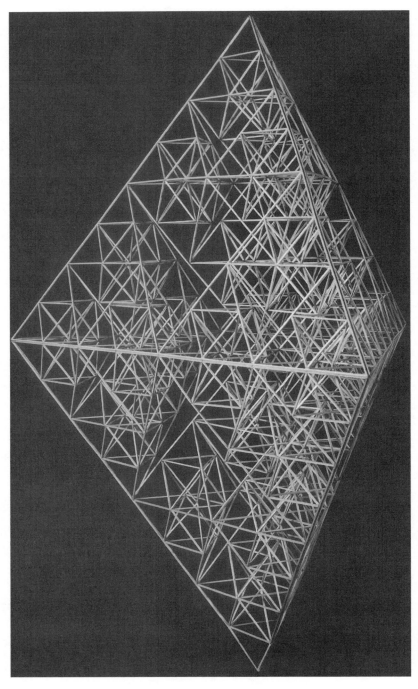

Balsa-wood model of J_5 constructed by Gene Miller.[1]

---

[1]Photograph by Marlin Thomas; graphical adjustments by Bulent Atalay.

Contents

Preface

Mathematics evolves through cycles of expansion, fueled by the analysis of new ideas, and contraction, eventuated by the synthesis of seemingly diverse ideas.

The classical fractals known as Cantor's set, Sierpiński's carpet, and Menger's sponge may be counted among the first examples of universal spaces in dimension theory. Originally, circa 1900, these fractals were constructed by starting with a base-space I, the unit interval, and then recursively cutting holes in the finite-product spaces I^k.

The unit interval as a base space dominated universal space theorems in dimension theory up through 1931 when a subspace of I^{2n+1} was shown to be universal for the class of n-dimensional separable-metric spaces.

For n-dimensional general (not necessarily separable) metric spaces, the unit interval continued to be central through the 1960s. It was used to construct the star space $S(A)$ (hedgehog with $|A|$ prickles — each prickle being a copy of the unit interval); and a subspace of the infinite-product $S(A)^\infty$ was shown to be universal for the class of n-dimensional weight $|A| \geq \aleph_0$ metric spaces.

In the general case, however, the search for a universal space was not over because the exponent of the base-space $S(A)$ is infinite while in the separable case the exponent $(2n + 1)$ is finite.

After nearly half a century (1931–1975) of using the unit interval to construct universal spaces in dimension theory, a new one-dimensional weight $|A|$ base-space J_A (a topological quotient of Baire's space $N(A)$) was introduced, and, a subspace of the finite-product J_A^{n+1} was shown to be the desired universal space.

By 2007 it had been shown that J_A is a *generalized fractal* — an attractor of an infinite iterated function system. So in the beginning classical fractals served as examples of universal spaces, and a century later (1900–2007), we find universal spaces that involve a base space that is one of the first examples of a generalized fractal — to the best of this author's knowledge, the first examples of attractors of infinite iterated function systems were derived from J_A-related research.

The construction of J_A is simply a generalization of the *identify adjacent endpoints in Cantor's set* construction of the unit interval. Indeed, the idea of *endpoints* and *adjacent endpoints in Baire spaces* led naturally to the *identify adjacent endpoints in the Baire space $N(A)$* construction of J_A. As a bonus, the classical ideas of *rationals* and *irrationals* in the unit interval extend to their counterparts in J_A.

To go beyond a superficial understanding of such universal spaces and generalized fractals, one must merge certain aspects of both dimension theory and fractal geometry. This book provides such a development.

For separable metric spaces, the universal spaces and classical fractals are well documented in well-known texts. For general (not necessarily separable) metric spaces, however, most of the research has only appeared in disparate articles. This book unifies the general theory as it currently exists.

Mastery of the mathematics in this book prepares the reader for original research in either dimension theory or fractal geometry. This book contains the motivation and background of several currently open research problems.

In closing this preface, there are those who deserve special thanks. In 1968, when I was employed by the U.S. Navy, I received approval for two years (1968–1970) of Navy-funded Advanced Study at the University of Virginia. For his support, I thank Ray Hughey.

During 1968–1969, Gordon Whyburn was my advisor. In the spring of 1969, based on my idea of "closing the closure," Professor Whyburn said that I would finish my Ph.D. the following year. Sadly, however, in the summer of 1969, Professor Whyburn passed on. For imparting the self-confidence that is required to create original mathematics, I will always remember and value my relationship with Professor Whyburn.

During 1969–1970 Charles Alexander was my advisor. For continuing to impart self-confidence while opening my eyes to dimension theory, and then later approving my dissertation *Imbedding One-Dimensional Metric Spaces*, I greatly appreciate and thank Professor Alexander.

This book would not exist had it not been for the rich history of insights of many mathematicians. Among the many, however, I especially thank a few: Jun-iti Nagata, James Perry, Ivan Ivanšić, and Uroš Milutinović.

Over the last decade, relevant graphics, videos, and models were created. For developing those concrete representations, I give special thanks to two of my former students, Chris Dupilka and Gene Miller.

The theorems and propositions presented in this book rest upon substantive research that spans more than a century. Thus the task of obtaining, evaluating, and organizing these diverse publications into a bibliography was substantial. For her unfailing assistance over two years, I thank in particular one of the librarians at the University of Mary Washington, Carla Bailey.

And to my wife Patty, I give thanks for keeping me healthy and happy.

June 2008   S.L.L.
Spotsylvania, Virginia

Introduction

The writings of Euclid and Aristotle clearly show that the intuitive idea of "dimension" has been around for at least several millennia (Crilly [1999]). By 1810, Bolzano saw the need for a definition, stating, "At the present time there is still lacking a precise definition of the most important concepts: line, surface, solid" (Johnson [1977, page 271]). And circa 1877, Cantor believed that the "coordinate concept of dimension" was basically flawed.

During the early 1900s there was an emergence of a *topological dimension theory* that evolved into an elegant body of mathematics within the context of separable (weight $\leq \aleph_0$) metric spaces (Hurewicz and Wallman [1948]). Almost parallel to the emergence of dimension theory, however, were certain constructions of spaces, now called classical fractals, that mostly served as examples of pathological topological spaces or generalizations of well-known constructions. Among those constructions were *Sierpiński's carpet* and *Menger's sponge* (Figure 49.2). The Carpet and Sponge are, respectively, planar and 3-space generalizations of Cantor's set. In modern terms, each is simultaneously a *fractal* and a *universal space*, the Carpet for planar compact one-dimensional metric spaces, and the Sponge for compact one-dimensional metric spaces.

By the 1940s, an extension of the classical (separable metric) dimension theory to more general spaces seemed improbable. Nevertheless, by the mid-1960s a surprisingly new and natural theory for general (weight $\geq \aleph_0$) metric spaces was rapidly maturing. The extension of the classical theory was initiated by Stone [1948], who recognized a symbiosis between open coverings and metric spaces. This symbiosis was further developed (in the context of general topology) by Bing [1951], Nagata [1950], and Smirnov [1951] in their metrization theorems. And on that foundation, Katětov [1952] and Morita [1954] created a significant and elegant dimension theory for general (weight $\geq \aleph_0$) metric spaces.

One of the remaining problems, however, was the absence of an *analogous* universal space for weight $\geq \aleph_0$ n-dimensional metric spaces (see the quotation from Nagata [1967] on page 9 of this text).

In the classical theory the $(2n + 1)$-dimensional Euclidean cube I^{2n+1} contains the n-dimensional universal space (Nöbeling [1931]). For weight $|A| \geq \aleph_0$ metric spaces, the analogous result appeared in Lipscomb [1975]: A one-dimensional space J_A was obtained by generalizing the *identify adjacent endpoints in Cantor's-set* construction of the unit interval. Indeed, the idea of *endpoints* and *adjacent endpoints in Baire spaces* led naturally to the *identify adjacent-endpoints in the Baire space* $N(A)$ construction of J_A.

The generalization also extends the classical ideas of *rational* and *irrational* to their counterparts in J_A (Definition 3.1). It turned out that the

$(n+1)$-dimensional J_A^{n+1} contains the desired n-dimensional weight $|A| \geq \aleph_0$ universal space.

The method of proof (Lipscomb [1975], and Chapters 7 and 8 in this text) of the J_A Imbedding Theorem was new. And prior to this text, the proof had only appeared in the research literature.

The space J_A was introduced circa 1970 in the context of point-set topology (Chapter 1). For example, the Baire space $N(\{0, 1\})$ is a copy of Cantor's set, and $J_2 = J_{\{0,1\}}$ is a topological copy of the unit interval. And as the *self-similarity* of Cantor's set induces self-similarity of (the unit interval) J_2, the "self-similarity" of $N(A)$ induces "self-similarity" in J_A (§5).

Indeed, J_3 and J_4, respectively, are copies of fractals known as *Sierpiński's triangle* and *cheese*. In fact, J_{n+1} for finite $n \geq 2$ is a topological copy of the n-web fractal ω^n that is the attractor of a finite iterated function system $\mathcal{F}_n$ whose $n + 1$ members are contractions by one-half toward the $n + 1$ vertices of an n-simplex (Chapter 2, §8).

In particular, J_5 lives in 4-space, and had never been viewed in 3-space until Perry and Lipscomb [2003] constructed an isotopy that moves J_5 from 4-space into 3-space with its fractal dimension preserved (Chapter 2 §8,§9,§10; Chapter 12). The existence or non-existence of such isotopies for J_6 in 5-space, J_7 in 6-space, and J_8 in 7-space are open problems.

The term *fractal* was coined by Mandelbrot [1975] the same year that the J_A Imbedding Theorem was introduced. The idea of viewing fractals as attractors of *finite* iterated function systems (IFSs) was introduced in 1981 (Hutchinson [1981]), and then popularized in the late 1980s and early 1990s following the publication of Barnsley's [1988] text.

Also in the early 1990s, Lipscomb and Perry [1992], and independently Milutinović [1992], produced imbeddings of J_A into Hilbert's $l^2(A)$ space. Each imbedding involved an *infinite* IFS. In 1992, however, the IFS theory was limited to IFSs that were *finite*. In 1996, by modifying the topology of J_A, Perry [1996] constructed a subspace ω_c^A of the Tychonoff cube I^A that is *an attractor of an infinite* IFS. Perry also called attention to the open problem of showing that $\omega^A \subset l^2(A)$, a copy of J_A, is the attractor of an infinite IFS (of affine transformations of $l^2(A)$) (§31).

The open problem posed by Perry was solved by Miculescu and Mihail [2008]. Miculescu and Mihail provided the mathematical context with an appropriate Hutchinson operator that had ω^A as its fixed point, i.e., ω^A is indeed the attractor of an infinite IFS (Chapter 5).

Since the introduction of J_A and the J_A Imbedding Theorem in the 1970s, the J_A-related research literature has been growing. In particular, Milutinović's, and, Ivanšić and Milutinović's joint research has been substantial, spanning more than two decades. For example, by modifying the decomposition approach used to prove the J_A^{n+1} Imbedding Theorem, they proved that $(J_3)^{n+1}$ (recall that J_3 is a copy of Sierpiński's triangle) contains a universal space for n-dimensional separable metric spaces (Chapter 11 and the graphic in Figure 55.1).

In addition, analogous to Urysohn's [1925a] Metrization Theorem, which states that *a topological space of weight $|A| \leq \aleph_0$ is metrizable if and only if it is homeomorphic to a subspace of I^∞*, the corresponding J_A^∞ is universal for metrizable spaces of weight $|A| \geq \aleph_0$ (Chapter 10).

Finally, consider *the problem that is inverse to constructing fractals from manifolds*: The emergence of the classical fractals was viewed as one of cutting holes in manifolds. The inverse problem is that of constructing manifolds from fractals. In the context of J_{n+1}, the problem is that of extending the n-web ω^n IFS to an n-simplex IFS. For $n = 2$, Chapter 13 contains the solution; and for $n = 3$, the solution is detailed in Chapter 14. Applications of these two solutions yield new representations of 2-space, 3-space, the 1-sphere, and the 2-sphere.

The *inverse problem is open for $n \geq 4$*. This author believes that the approach used to solve the ω^3 case is general enough to serve as a model for solutions for any n, and the most difficult part of a solution is that of understanding the *hole $\Delta^n \setminus \cup_{w \in \mathcal{F}_n} w(\Delta^n)$* in Δ^n (§97).

Format, Conventions, and Outline

The style of the text is informal, some definitions are neither numbered nor offset. A term defined within a paragraph, however, always appears in italics. In contrast, lemmas, propositions, and theorems always appear in boldface, are always numbered, and always offset.

The sections are numbered sequentially throughout the text, from §**1** in Chapter 1 to §**97** (the last section) in Chapter 14. Then the sections in the Appendices are also sequential, from §**A1** in Appendix 1 to §**A14** in Appendix 3. Each table, lemma, proposition, and theorem is numbered — in §**87** we begin with **87.1 Theorem** and then **Fig. 87.2**, which is followed by **Table 87.3**. The only figures that are not numbered are those of "local interest." The first numbered equation in each chapter has label "(1)" and the following such equations in each chapter are then sequentially numbered.

For specific contents of the chapters, let us consider them individually.

Chapter 1. *Construction of $J_A = J_\alpha$*: Baire's zero-dimensional spaces are illustrated and their relevant properties discussed. The adjacent-endpoint relation is defined and then used to construct J_A. Proofs of Lemmas 3.2 and 3.3 are new and substantially more concise than their original counterparts. The comment section contains an extensive prehistory and history of the mathematics that led to the construction of J_A.

Chapter 2. *Self-similarity and J_{n+1} for Finite n*: The fractal nature of J_A is deduced from that of $N(A)$. Graphic figures of J_{n+1} are provided for small $n \leq 4$. The fractal nature of J_{n+1} is exposed by showing that J_{n+1} is homeomorphic to the attractor ω^n of a finite IFS. The open problems associated with viewing J_{n+1} in 3-space for $n = 5$, 6, and 7 are detailed.

Chapter 3. *No-Carry Property of $\omega^{A'}$*: For some fixed $z \in A$ (z indicates zero), $A' = A \setminus \{z\}$, and a mapping $J_A \to \omega^{A'}$ from J_A into Hilbert space is introduced. (The mapping is shown to be an imbedding in Chapter 4.) The use of star spaces (hedgehogs with $|A|$ prickles) yield the no-carry characterization of $\omega^{A'}$. Several examples serve to motivate the constructions.

Chapter 4. *Imbedding J_A in Hilbert Space*: The mapping $J_A \rightarrow \omega^{A'} \subset l^2(A')$ introduced in Chapter 3 is shown to be a homeomorphism. That is, J_A is imbedded into Hilbert's $l^2(A') \subset l^2(A)$ space as $\omega^{A'}$, which satisfies the no-carry property. We also review Sierpiński's original formulation of his triangle, which is the basis for Milutinović's [1992] construction: Milutinović's space M_A is a topological copy of J_A that resides in the standard simplex Δ^A of $l^2(A)$. (For Δ^A see Appendix 2.)

Chapter 5. *Infinite IFS with Attractor $\omega^{A'}$*: Neighborhoods of subsets of metric spaces are discussed and illustrated for motivation of the definition of the Hausdorff metric h on the set $\mathcal{B}_X$ of all non-empty, bounded, and closed subsets of X. The *when* and *why* $(\mathcal{B}_X, h)$ is complete, and, the properties of the related pseudo-metric h^* are detailed. The definitions of a "bounded (not necessarily finite) IFS" and the Hutchinson operator for such IFSs are discussed. We introduce the J_A IFS and then show that its attractor is $\omega^{A'}$.

Chapter 6. *Dimension Zero*: Each of the subspaces of rationals and irrationals of J_A are shown to be zero-dimensional and dense. The $n = 0$ case of the J_A^{n+1} Imbedding Theorem is established. We also show that the subspace $J_A^{\ell}(n)$ of ℓ-tuples in J_A^{ℓ} with at most n rational coordinates is n-dimensional. Each of these "general" J_A results is applied for $|A| = 2$, which yields corresponding statements about the unit interval.

Chapter 7. *Decompositions*: We present a careful development of the decompositions that are key to the proof of the J_A^{n+1} Imbedding Theorem. Given an arbitrary n-dimensional metric space X, we systematically decompose X so that the decompositions have enough properties to distinguish individual points and allow an imbedding of X into $J_A^{n+1}(n)$. Extensive graphics, none of which have previously appeared in the literature, serve to motivate (a) the idea of the dimension function "diml"; (b) "nodes" of a cover and the "nodal properties"; and (c) the constructions used in the proofs of the lemmas of the Decomposition Theorem 39.1. New and additional proofs of the lemmas and theorems are provided. For example, the proof of the Decomposition Lemma 38.9, which has been extensively applied by Ivanšić and Milutinović in their J_A-related research, contains new details. The unproven but implied claims in Lipscomb [1975] whose proofs have not previously appeared in the literature are provided in this chapter. The proof of the Decomposition Theorem 39.1 is illustrated by a new sequence of graphics that decompose the unit interval step-by-step according to the constructions used in the proof.

Chapter 8. *The J_A^{n+1} Imbedding Theorem*: We prove the J_A^{n+1} Imbedding Theorem. The presentation is a greatly extended version of the one that appears in Lipscomb [1975]. Nagata's [1960] and [1963] General Imbedding Theorems are discussed in the comment section (§46).

Chapter 9. *Minimal-Exponent Question*: The question of whether the exponent "$n+1$" used in the J_A^{n+1} Imbedding Theorem is minimal is discussed.

A very brief review of Vietoris homology and a short recap of the homology sequence of n-spheres S^n is presented. We then follow Borsuk and prove his theorem *the 2-sphere S^2 is not topologically contained in the Cartesian product of two one-dimensional spaces*. Details underlying Borsuk's proof are added, and the obvious application is that the index "$n + 1$" in the J_A^{n+1} Imbedding Theorem cannot be reduced.

Chapter 10. *The J_A^∞ Imbedding Theorem*: The proof of the J_A^∞ Imbedding Theorem is the focus. However, in §51 we compare the J_A^{n+1} and J_A^∞ imbedding theorems with two pairs of their predecessors — the classical (separable metric) pair of Urysohn [1925a] and Nöbeling [1931], and, the general (not necessarily separable) metric pair of Kowalsky [1957] and Nagata [1963].

Chapter 11. *1992–2007 J_A-Related Research*: The chronological and historical context appears in the graphic labeled Figure 55.1. The graphic spans 1875 to 2007 and provides the backdrop for the literature that relates (some more than others) in some form to J_A. The narrative part of the chapter provides a unifying survey of the J_A-related research that has heretofore only appeared in research articles. Milutinović's work, and, Ivanšić and Milutinović's joint work are featured. An example of a *Klavžar-Milutinović* graph (i.e., a graph whose structure is based on the adjacent-endpoint relation applied to finite product sets) is illustrated in §61.

Chapter 12. *Isotopy Moves J_5 into 3-Space*: We discuss the problem of deciding which J_{n+1} can be viewed in 3-space as attractors of finite IFSs. The J_5 case is detailed, and the only remaining open cases (i.e., the J_6, J_7, and J_8 cases) are identified.

Chapter 13. *From 2-Web IFS to 2-Simplex IFS, 2-Space and the 1-Sphere*: The inverse problem of constructing manifolds from fractals in the case of ω^2 is solved. That is, the ω^2 IFS is minimally extended to a 3-simplex IFS. The fibers of the corresponding address map for the 3-simplex are characterized, and the desired representations are obtained.

Chapter 14. *From 3-Web IFS to 3-Simplex IFS, 3-Space and the 2-Sphere*: The inverse problem of extending the ω^3 IFS to one whose attractor is a 3-simplex is solved. The fibers of the corresponding address map for the 3-simplex are characterized, and an application yields the desired representations. In §97, the open problem of extending the 4-web IFS to one whose attractor is a 4-simplex is discussed.

Finally, the book contains three appendices: Appendix 1. *Background Basics*; Appendix 2. *The Standard Simplex Δ^A in $l^2(A)$*; and Appendix 3. *Measures and Fractal Dimension*.

CHAPTER 1

Construction of $J_A = J_\alpha$

Following its emergence during the early 1900s, topological dimension theory evolved into an elegant body of mathematics within the context of separable (weight $\leq \aleph_0$) metric spaces. By the 1940s, when this now classical theory was well established, an extension to more general spaces seemed improbable. Nevertheless, by the mid-1960s a surprisingly new and natural theory for general (weight $\geq \aleph_0$) metric spaces was rapidly maturing.[1]

One of the remaining problems, however, concerned the absence of a theory of universal spaces (for n-dimensional weight $\alpha \geq \aleph_0$ metric spaces) that was analogous to the classical (weight $\alpha \leq \aleph_0$) theory.[2]

In the classical theory it is the product space I^{2n+1} of $2n + 1$ copies of the unit interval I that contains the universal space. And as it turned out, an analogous result surfaced in 1975: The product space J_α^{n+1} of $n+1$ copies of the one-dimensional J_α contains the universal (weight $\alpha \geq \aleph_0$) space.

Originally, circa 1970, J_α was introduced in the context of point-set topology. By the early 1990s, an infinite iterated function system operating on Hilbert's $\ell^2(A)$ space (cardinality $|A| = \alpha$) provided a homeomorphic copy ω^A of J_A; and by 2007, ω^A was shown to be the attractor of such a system. In this chapter, we focus on the 1970s' original development of $J_A = J_\alpha$.

§1 Baire's Zero-Dimensional Spaces

Any countable product $\times_i A_i$ of discrete spaces $A_i = A$ is a *Baire* (*zero-dimensional*) *space* $N(A) = \times_i A_i$. So the elements of $N(A)$ are simply sequences $a = a_1 a_2 \cdots$ in A; and when $a \in N(A)$ has a constant tail, i.e., $a_{t+1} = a_{t+2} = \cdots$ for some index t, we may write $a = a_1 \cdots a_t \overline{a_{t+1}}$.

For a doubleton or tripleton set A, Figure 1.1 provides "geometrical approximations": Baire's space $N(\{0, 2\})$ is viewed as a Cantor set, and "Cantor subspaces" induce a "triangularly organized approximation" to $N(\{0, 1, 2\})$.

---

[1] The extension of the classical theory was initiated by Stone [1948], who recognized a symbiosis between open coverings and metric spaces. This symbiosis was further developed (in the context of general topology) by Bing [1951], Nagata [1950], and Smirnov [1951] in their metrization theorems. And on that foundation, Katětov [1952] and Morita [1954] created a significant and elegant dimension theory for general metric spaces.

[2] For a given class $\mathcal{C}$ of topological spaces, $U \in \mathcal{C}$ is universal for $\mathcal{C}$ if each member of $\mathcal{C}$ is homeomorphic to a subspace of U. The classical theorem (concerning the universal space for n-dimensional separable metric spaces and corresponding imbeddings) is due to Nöbeling [1931]. For timely details and background on the problem of extending Nöbeling's work to general metric spaces see Nagata [1965] [1967] and Lipscomb [1973].

S.L. Lipscomb, *Fractals and Universal Spaces in Dimension Theory*,
DOI 10.1007/978-0-387-85494-6_1, © Springer Science+Business Media, LLC 2009

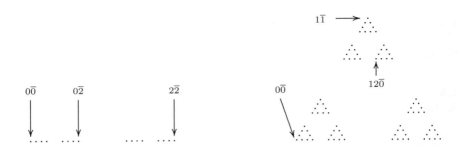

Fig. 1.1 Organized approximations to $N(\{0,2\})$ and $N(\{0,1,2\})$.

By extending the (self-similar) pattern of two groupings ($|A| = 2$) of "segments" and three groupings ($|A| = 3$) of "triangles," we may approximate $N(A)$ for $|A| = 4$ using four groupings of "tetrahedra" (Figure 1.2).

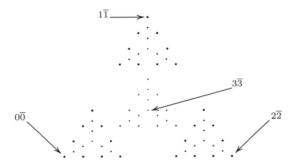

Fig. 1.2 Points on four "tetrahedra" approximate $N(\{0,1,2,3\})$.

These groupings also expose key features of a basis of $N(A)$ — the subbasis of sets $A_1 \times A_2 \times \cdots \times A_{i-1} \times \{a_i\} \times A_{i+1} \times \cdots$ yields covers $\mathcal{B}_k = \{\langle a_1, a_2, \ldots, a_k \rangle = \{a_1\} \times \cdots \times \{a_k\} \times A_{k+1} \times A_{k+2} \times \cdots\}$ of $N(A)$ (of pairwise-disjoint sets) which in turn yield a basis $\mathcal{B} = \cup_k \mathcal{B}_k$ (Figure 1.3).

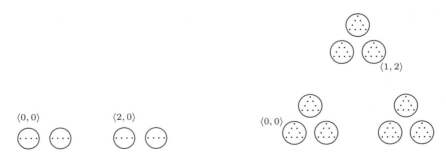

Fig. 1.3 Illustration of $\mathcal{B}_2$ relative to $N(\{0,2\})$ and $N(\{0,1,2\})$.

Obvious aspects of $\mathcal{B}$ coupled with the following theorem expose the basic properties of the Baire space $N(A)$.

1.4 Theorem (Morita [1954])(Ind $X \leq n$ and σ-locally finite bases) *Let X be a metric space. Then for $n \geq 0$, we have Ind $X \leq n$ if and only if X has a σ-locally finite basis $\mathcal{B}$ such that $G \in \mathcal{B}$ implies Ind $B(G) \leq (n-1)$.*

1.5 Theorem (basic properties of $N(A)$) *Let $|A| = \alpha \geq 1$. Then $N(A)$ and its basis $\mathcal{B}$ have the following properties:*

  (i)   *Each $\mathcal{B}_k \subset \mathcal{B}$ is an open-set partition of $N(A)$.*
  (ii) *Each member of $\mathcal{B}$ is both open and closed, and thus has empty boundary.*
  (iii) *Each $\mathcal{B}_k$ is locally finite, and $\mathcal{B}$ is σ-locally finite.*
  (iv) *When $|A| = \alpha$ is infinite, then $N(A)$ has weight α.*
  (v)   *The space $N(A)$ is metrizable and has dimension zero, i.e., $\dim N(A) = \operatorname{Ind} N(A) = 0$.*
  (vi) *The space $N(A)$ is topologically complete.*

PROOF. (i) Each $\mathcal{B}_k$ is a pairwise-disjoint open covering of $N(A)$. (ii) From (i), we have $G \in \mathcal{B}_k$ implies $G = N(A) \setminus (\cup_{B \in \mathcal{B}'_k} B)$ where $\mathcal{B}'_k = \mathcal{B}_k \setminus \{G\}$. (iii) Use (i) and $\mathcal{B} = \cup_k \mathcal{B}_k$. (iv) The weight of $N(A) \leq |\mathcal{B}| = \alpha$, and the weight of $N(A) \geq \alpha$ because $\mathcal{B}_1$ has size α and satisfies (i). (v) Since $N(A)$ is regular with a σ-locally finite basis $\mathcal{B}$, the Nagata-Smirnov Metrization Theorem applies. The zero-dimensional part follows from Theorem 1.4 with $n = 0$. (vi) Statement (vi) holds because $\times_i A_i$ is topologically complete if and only if each A_i is complete. □

The usual metric ρ for the Baire space $N(A)$ is given by

$$\rho(a,b) = \begin{cases} 1/n & \text{when } a \neq b \text{ and } n = \min\{k : a_k \neq b_k\}; \\ 0 & \text{when } a = b. \end{cases}$$

Thus, for $a = a_1 a_2 \cdots$ in $N(A)$, the set $\langle a_1, \ldots, a_k \rangle$ is the closed ball $\overline{B}_\delta(a) = \{y \in N(A) : \rho(a,y) \leq \delta\}$ centered at a with radius $\delta = 1/(k+1)$.

1.6 Theorem (Morita [1955])(dim $X \leq n$ and Baire spaces) *Let X be a metric space. Then $\dim X \leq n$ if and only if there exists a subspace S of $N(A)$ for suitable A and a closed continuous surjection $f : S \to X$ such that each fiber $f^{-1}(x)$ contains at most $n + 1$ points.*

Applying Theorem 1.6, we may bound the covering dimension of the unit interval $I = [0,1]$, i.e., $\dim I \leq 1$. In detail, let $A = \{0, 2\}$; view $S = N(A)$ as Cantor's space $\mathcal{C}$ and $f : \mathcal{C} \to \mathcal{C}/\sim (=_t I)$ as *identification of adjacent endpoints*; and note that each fiber of f contains at most two points.

§2 Adjacent-Endpoint Relation

In this section we extend the *"adjacent-endpoint relation $\sim$"* from Cantor's space $C =_t N(\{0,2\})$ to $N(A)$ where A is an arbitrary non-empty set.

First, recall that the homeomorphism $\mathcal{C} \to N(\{0,2\})$ is exposed by triadically expanding $(x \mapsto a_1 a_2 \cdots)$ each number x in C, i.e.,

$$x \in C \Longleftrightarrow x = \Sigma_{i=1}^{\infty} a_i/3^i \text{ for a unique } a_1 a_2 \cdots \in N(\{0,2\}).$$

This bijection induces a bijection of the respective topologies, matching each member of the σ-locally finite basis $\mathcal{B} = \cup_k \mathcal{B}_k$ to a closed and open member of such a basis for C. For example, $\langle 0 \rangle \leftrightarrow C \cap [0,1/3]$, $\langle 2 \rangle \leftrightarrow C \cap [2/3,1]$, $\langle 0,0 \rangle \leftrightarrow C \cap [0,1/9]$, $\langle 0,2 \rangle \leftrightarrow C \cap [2/9,1/3]$, $\langle 2,0 \rangle \leftrightarrow C \cap [2/3,7/9]$, etc.

Second, recall that "endpoints in C" correspond to eventually constant strings, e.g., $0 \leftrightarrow 00 \cdots$, $1/3 \leftrightarrow 022 \cdots$, $2/3 \leftrightarrow 200 \cdots$, $1 \leftrightarrow 22 \cdots$, etc., and that "adjacent endpoints in C" encode as "switching tails", e.g., $1/3$ and $2/3$ correspond, respectively, to $022 \cdots$ and $200 \cdots$ (See Figure 2.1).

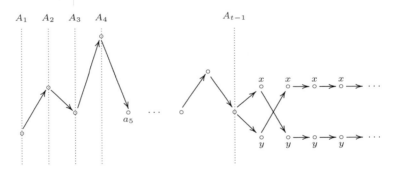

Fig. 2.1 Abstract picture of adjacent endpoints.

2.2 Definitions (Lipscomb [1973])(adjacent endpoints and the relation)
Let $N(A)$ be a Baire space. A point $a = a_1 a_2 \cdots$ in $N(A)$ is an *endpoint of $N(A)$* if there exists an index k such that $a_k = a_{k+1} = \cdots$. Distinct endpoints $a \neq b$ are *adjacent endpoints* when there exists $x \neq y$ in A such that $a = a_1 a_2 \cdots a_{t-1} x y y y \cdots$ and $b = a_1 a_2 \cdots a_{t-1} y x x x \cdots$. If a and b are adjacent endpoints, the unique index $t \geq 1$ is called the *tail index of a and b*. The relation $\sim \subset N(A) \times N(A)$ given by "$a \sim b$" when either $a = b$ or a and b are adjacent endpoints is called the *adjacent-endpoint relation*.

2.3 Theorem *Let $N(A)$ be a Baire space. Then the adjacent-endpoint relation $\sim$ is an equivalence relation on $N(A)$ with the property that each equivalence class contains at most two members.*

PROOF. The relation $\sim$ is clearly reflexive and symmetric. For transitivity, let $a \sim b \sim c$. If $a = b$ or $b = c$, then clearly $a \sim c$. Otherwise $a \neq b$ and

$b \neq c$, and both pairs have the same tail index t. So $a_1 = b_1 = c_1, \ldots,$ $a_{t-1} = b_{t-1} = c_{t-1}$, and, for index t, $a_t = b_{t+1} = c_t$, and, for indices $k \geq t$, $a_k = b_t = c_k$. Thus $a = c$, making $a \sim c$. This transitivity argument also yields $(a \neq b$ and $a \sim b)$ implies $(b = c$ or $a = c)$. $\qquad\qquad\qquad$ □

§3 J_A and the Natural Map p

We begin with the key definitions.

3.1 Definitions (Lipscomb [1973])(J_A **and its rationals and irrationals)**
Let $|A| = \alpha \geq 1$, let $N(A)$ be a Baire space, and let "$\sim$" denote the adjacent-endpoint equivalence relation on $N(A)$. Then $J_A = J_\alpha$ is the quotient space $N(A)/\sim$ and $p : N(A) \to J_A$ is *the natural mapping* given by $p(a) = [a]$ where "$[a]$" denotes the equivalence class that contains $a \in N(A)$. Moreover, $x \in J_\alpha$ is a *rational point* (or a *rational*) when $p^{-1}(x)$ is a doubleton set, and x is an *irrational point* (or an *irrational*) when $p^{-1}(x)$ is a singleton set.

So the mapping $p : N(A) \to J_A$ is surjective and $G \subset J_A$ is open in J_A if and only if $p^{-1}(G)$ is open in $N(A)$, i.e., J_A has the largest topology that makes p continuous. And since each fiber $p^{-1}(x)$ of p is either a singleton or doubleton set, we see a fortiori that each fiber of p is compact. To prove that p is also a closed mapping, we shall use the following three lemmas.

3.2 Lemma (closed mappings) *A quotient mapping $f(Y) = Z$ is closed if and only if for each fiber $f^{-1}(z)$ and each open set $G \supset f^{-1}(z)$ there is an open f-inverse set $V \subset Y$ such that $f^{-1}(z) \subset V \subset G$.*

PROOF. Suppose $f : Y \to Z$ is closed and that $G \supset f^{-1}(z)$ is open in Y. Let $F = Y \setminus G$. Then F is closed in Y, making $f(F)$ closed in Z, and, in turn, $H = Z \setminus f(F)$ open in Z. It follows that $V = f^{-1}(H)$ is the desired open f-inverse set. Conversely, suppose f is quotient and $z \in Z$ implies any open $G \supset f^{-1}(z)$ yields the specified V. Let F be any closed subset of Y such that $f(F) \neq Z$. We show that $f(F)$ is closed in Z: Consider any $z \in Z \setminus f(F)$. Then $f^{-1}(z) \cap F = \emptyset$. So $G = Y \setminus F \supset f^{-1}(z)$ being open ensures that an open f-inverse set $V = f^{-1}(H)$ exists such that $f^{-1}(z) \subset V \subset G$. But because f is a quotient map, H is open in Z. Thus, $z \in H$ and $H \cap f(F) = \emptyset$, so z is not in the closure of $f(F)$, i.e., $f(F)$ is closed. $\qquad$ □

3.3 Lemma ($p^{-1}(z)$ **is a singleton set)** *Let $p : N(A) \to J_A$ be the natural map, and let $p^{-1}(z) = \{c\} \subset G = \langle c_1, \ldots, c_k \rangle \in \mathcal{B}_k$ where $c = c_1 c_2 \cdots$. Then*

$$
V = \left\{ \begin{array}{ll} \langle c_1, \ldots, c_k \rangle \setminus \{c_1 \cdots c_k \overline{x} : x \neq c_k; x \in A\} & \text{if } c_1 = c_2 = \cdots; \\ \langle c_1, \ldots, c_k \rangle \setminus \{c_1 \cdots c_k \overline{x} : x \in A\} & \text{if } c \text{ is not an endpoint} \end{array} \right.
$$

is an open p-inverse set such that $p^{-1}(z) \subset V \subset G$.

PROOF. Clearly, $p^{-1}(z) \subset V \subset G$; and V is open because $\{c_1 \cdots c_k \overline{x} : x \in A\}$ is closed — each member of the locally finite open partition $\mathcal{B}_{k+1}$ contains at most one of the closed singleton sets $\{c_1 \cdots c_k \overline{x}\}$. Moreover, V is a p-inverse set: Suppose $a \in V$, $a \sim b$, and $a \neq b$. Then the definition of V ensures that the tail index t of a and b must satisfy $t \geq k + 1$. So $b \in V$.   □

3.4 Lemma ($p^{-1}(z)$ is a doubleton set) *Let* $p : N(A) \rightarrow J_A$ *be the natural map, and let* $p^{-1}(z) = \{c, d\} \subset G = \langle c_1, \ldots, c_k \rangle \cup \langle d_1, \ldots, d_k \rangle$ *where* k *is greater than the tail index* t *of* $c = c_1 c_2 \cdots$ *and* $d = d_1 d_2 \cdots$. *Then*

$$V = \langle c_1, \ldots, c_k \rangle \cup \langle d_1, \ldots, d_k \rangle \setminus (\cup_{\substack{x,y \in A \\ x \neq c_k, y \neq d_k}} \{c_1 \cdots c_k \overline{x}, d_1 \cdots d_k \overline{y}\})$$

is an open p-inverse set such that $p^{-1}(z) \subset V \subset G$.

PROOF. Clearly $p^{-1}(z) \subset V \subset G$; and as in the previous proof, V is open. So suppose that $a \in V$ and $a \sim b$ with tail index t'. Then $t' \geq t$: If $t' = t$, then since $a \in V$ and $k > t$, $a = c$ or $a = d$, and we are finished. If $t' > t$, then $t' \geq k + 1$ (if $t' = k$, then we would contradict $a \in V$). Thus, $b \in V$.   □

Recall the theorem "$p : X \rightarrow Y$ perfect and X metrizable implies Y is metrizable." (For more information on perfect mappings see Appendix 1.)

3.5 Theorem (p is perfect) *The natural mapping* $p : N(A) \rightarrow J_A$ *is a perfect mapping.*

PROOF. Since p is a surjective quotient mapping, an application of the previous three lemmas shows that p is also closed. Since p is a closed surjective mapping with compact fibers, p is a perfect mapping.   □

3.6 Theorem (J_A is metrizable and one-dimensional) *Let* $p : N(A) \rightarrow J_A$ *be the natural mapping. Then* J_A *is a one-dimensional metrizable space.*

PROOF. Since p is perfect and $N(A)$ is metrizable, J_A is metrizable. And since $p : N(A) \rightarrow J_A$ is at most two-to-one, Theorem 1.6 shows that $\dim J_A \leq 1$. But since $|A| = \alpha \geq 2$, Cantor's space is a topological subspace of $N(A)$. Thus the unit interval $I = [0, 1]$ is a topological subspace of J_α. So by the subspace theorem, $\dim J_A \geq 1$.   □

§4 Comments

4.1 BAIRE'S SPACES $N(A)$. These spaces are fundamental in modern dimension theory (Engelking [1978] and Nagata [1965] [1983]). For example, Theorem 1.6 (Morita [1955]) was the key to showing that $\dim J_A = 1$; $N(A)$ is universal for the class of all zero-dimensional weight $|A| \geq \aleph_0$ metric spaces

(Engelking [1978, Theorem 4.1.24]); and for $|A| = 2$, Baire's space $N(A)$ is a topological copy of the Cantor set.[3]

The Cantor set $\mathcal{C}$ is a paradigm of a fractal that is a universal space in dimension theory: On the one hand, $\mathcal{C}$ is universal for the class of zero-dimensional separable metrizable spaces (Kuratowski [1966, page 285] and Urysohn [1925b, page 77]). On the other hand, it is a classical fractal.

By the 1980s, $N(A)$ for *finite* A became popularized as code space in the context of *finite* iterated function systems and fractal geometry.

4.2 CLASSICAL ADJACENT-ENDPOINT IDENTIFICATION. It is most likely that it was Cantor who introduced classical adjacent-endpoint identification.

Consider the English translation of Cantor [1884] that appears in Edgar [1993]. On pages 15 and 16 of the translation, Cantor constructs the "Devil's Staircase." (The removal of the open horizontal line segments from the graph of the Devil's Staircase exposes the graph of $\mathcal{C} \rightarrow I = [0, 1]$.)

To set the stage for the Devil's Staircase, Cantor constructs his set $\mathcal{C}$ as the residual set of points obtained by removing a countable number of (pairwise disjoint) open intervals "$(a_\nu, \ldots, b_\nu)$." Then beginning at the bottom of page 15, Cantor states:

> A special case of this type of function was already included in an example that I mentioned in Acta Mathematica 2, page 407.

The 'Acta Mathematica 2' reference is Cantor [1883b]. Then Cantor continues as follows:

> By putting
> $$z = \frac{c_1}{3} + \frac{c_2}{3^2} + \cdots \frac{c_\rho}{3^\rho} + \cdots, \qquad (6)$$
> where the coefficients c_ρ can take any of the values 0 or 2 and where the series can have a finite or infinite number of terms ...

Cantor then represents the right-endpoints b_ν of the general open intervals by stating:

> ... all the b_ν are included in the formula
> $$b_\nu = \frac{c_1}{3} + \frac{c_2}{3^2} + \cdots \frac{c_{\mu-1}}{3^{\mu-1}} + \frac{2}{3^\mu}. \qquad (7)$$

---

[3]Initially, the Cantor set was evidently introduced by H. J. S. Smith [1875]. Hannabuss [1996] states, "... this set appeared originally in an 1875 paper by ... Henry Smith ..., some eight years before Cantor mentioned it (without giving its recursive geometrical construction) in 1883 (Cantor [1883a])...." We also have Edgar's [1993, page 11] comments, "... But Smith's sets seem to be only countable sets of endpoints, not the actual perfect sets. Of course, before 'countable' and 'uncountable' were clarified by Cantor, this distinction would not have seemed important." Today, Cantor's set is often viewed as the attractor of the iterated function system $\{w_0, w_1\}$ where each w_i is a contraction of the unit interval by $1/3$ with w_0 contracting toward "0" and w_1 contracting toward "1." In this case, $N(\{0, 1\})$ is the code space.

The points a_ν arise ... from the same formula by taking c_ρ starting with a certain ρ always equal to 2 so that, by the equation

$$1 = \frac{2}{3} + \frac{2}{3^2} + \frac{2}{3^3} + \cdots,$$

one has, by taking $c_\mu = 0$, $c_{\mu+1} = c_{\mu+2} = \cdots = 2$,

$$a_\nu = \frac{c_1}{3} + \frac{c_2}{3^2} + \cdots + \frac{c_{\mu-1}}{3^{\mu-1}} + \frac{1}{3^\mu}. \tag{8}$$

...

Cantor then specifies the classical adjacent-endpoint identification:

We now relate the variable z to another variable y defined by the formula

$$y = \frac{1}{2} \left(\frac{c_1}{2} + \frac{c_2}{2^2} + \cdots + \frac{c_\rho}{2^\rho} + \cdots \right) \tag{9}$$

in which we agree that the coefficients c_ρ have the same value as in (6).

It follows that Cantor's mapping $z \mapsto y$ is what we now call *classical* adjacent-endpoint identification. Technical details of properties (continuous closed surjection) of $z \mapsto y$ may be found in Pears [1975, page 162], who concludes his discussion with the statement:

Thus the space obtained from the Cantor set by identifying pairwise the end points of the deleted intervals is the unit interval.

Also see Pervin [1964, §8.3, Problem 3].

4.3 PREHISTORY OF ADJACENT-ENDPOINT IDENTIFICATION $N(A) \to J_A$. To understand the motivation for extending the notion of $z \mapsto y$ from $N(\{0,2\}) \to I$ to $N(A) \to J_A$ for arbitrary A, one needs some historical context of universal spaces in dimension theory prior to the 1970s.

We begin by going back to the early 1900s, when, based on extensions of the recursive scheme of cutting holes in the unit interval to create the Cantor set, other "fractals" and "universal spaces in dimension theory" emerged: Sierpiński [1916] and Menger [1926a] used recursive schemes of cutting holes in, respectively, the square I^2 and the cube I^3 to create universal spaces that are now known as classical fractals. *Sierpiński's carpet* is universal for planar compact one-dimensional metric spaces; and *Menger's sponge* (Figure 49.2) is universal for compact one-dimensional metric spaces. (For an intuitive understanding of the universality of Sierpiński's carpet see Peitgen, Jürgens, and Saupe [1992, §2.7], and for an English translation of Menger's 1926 *General Spaces and Cartesian Spaces* in the Communications to the Amsterdam Academy of Sciences see Edgar [1993].)

Menger [1926b] also stated that a compact metric space of dimension less than or equal to n could be imbedded in the Euclidean cube I^{2n+1}.

By 1931 Nöbeling [1931] had removed the compactness restriction and proved the Classical Imbedding Theorem for separable metric spaces, i.e., he specified a subspace of the Euclidean cube I^{2n+1} that is universal for n-dimensional separable metric spaces.[4]

Three decades later, following the substantial development of a dimension theory for general (not necessarily separable) metric spaces, Nagata [1960] used an *infinite-dimensional* space — Dowker's [1947] generalized Hilbert space — to construct a subspace $F_{n\alpha}$ that is universal for n-dimensional weight-$\alpha \geq \aleph_0$ metric spaces. Three years later, however, Nagata [1963] made a more transparent construction by introducing another universal space $K_{n\alpha}$. But again, $K_{n\alpha}$ emerged as a subspace of an *infinite-dimensional* space $P(A)$ — $P(A)$ is the countable product of *star spaces* (star spaces are known in the literature as *hedgehogs with $|A| = \alpha$ prickles*).

By 1966, Nagata [1967], contrasting his universal spaces (subspaces of *infinite-dimensional* spaces) with the classical universal spaces (subspaces of *finite-dimensional* Euclidean cubes), stated:

> Comparing the general imbedding theorem with the classical one for separable metric spaces we notice that $P(A)$ has infinite dimension while every n-dimensional separable metric space is imbedded in the $(2n+1)$-dimensional Euclidean cube I^{2n+1}. This leads us to the following problem, 'Improve the general imbedding theorem finding another universal n-dimensional space instead of $P(A)$.'

Nagata's statement calls attention to the fact that Nöbeling's [1931] Classical Imbedding Theorem rests on the *one-dimensional* unit interval I as the *base space* in "I^{2n+1}". It therefore seemed (to this author) that any general imbedding theorem (analogous to Nöbeling's) would require (and be built upon) a *one-dimensional weight $\alpha \geq \aleph_0$ metric space* that would serve as an analogue of the unit interval.

So prior to the 1970s, it was Nagata's research and quotation above that served as motivation for seeking analogues of the unit interval.

Also prior to the 1970s, there were *three* well-known results that indicated how to construct such an analogue: First, the unit interval I may be obtained by Cantor's identification of adjacent endpoints $z \mapsto y$; second, Cantor's set $\mathcal{C}$ is a topological copy of Baire's space $N(\{0,2\})$; and third, Morita's Theorem (Theorem 1.6), which implies that an at most 2-to-1 closed and continuous image X of any subspace of $N(A)$ has dim $X \leq 1$.

---

[4]For more detail on the Classical Imbedding Theorem see, e.g., the "Historical and bibliographic notes" section on page 128 of Engelking [1978], and the "3. Imbedding of a compact n-dimensional space in I_{2n+1}" and "4. Imbedding of an n-dimensional space in I_{2n+1}" sections on pages 56–63 in Hurewicz and Wallman [1948].

From those three results, it seemed natural to try to obtain J_A by *extending* to $N(A)$ the idea of adjacent endpoints in $N(\{0, 2\})$.[5]

4.4 J_A IN THE CONTEXT OF FRACTALS. The graphics presented in the following chapter will elucidate how the adjacent-endpoint relation induces fractal structures. It should be noted that J_A, created within the context of dimension theory prior to 1973, emerged into a "fractal void" that existed prior to Mandelbrot's [1975] introduction of *fractal*.

It was during the 1980s that fractals (Mandelbrot [1983]), finite iterated function systems (Hutchinson [1981]), fractal geometry (Falconer [1985]), code space and address maps (Barnsley [1988]) were popularized.

4.5 CLOSING COMMENTS. As stated before, the material presented in this chapter follows Lipscomb [1973]. However, the constructions in Lemmas 3.2 and 3.3 of the desired open p-inverse sets V are new and substantially more concise than their original counterparts. The new approach specifies the V externally, i.e., as the result of removing a closed set of points from a basic open set. The original proofs in Lipscomb [1973] concerned a V defined internally, i.e., as a union of an infinite number of open sets.

To close these comments, it is instructive to look back and sample the historical view during the decades of the 1930s through the 1950s expressed in the introduction of G. T. Whyburn's [1958] article *Topological Characterization of the Sierpiński Curve*:[6]

> The universal plane curve described by Sierpiński [1916] has proven highly useful in the development of various phases of topology and analysis which have gone ahead at such a rapid pace in the intervening period of over forty years. Interest in this curve and its analogue in 3-space is currently much alive and its role in mathematics is surely by no means finished. The curve is obtained very simply as the residual set remaining when one begins with a square and applies the operation of dividing it into nine equal squares and omitting the interior of the center one, then repeats this operation on each of the surviving 8 squares, ... and so on indefinitely. Sierpiński showed that this set contains a topological image of every plane continuum having no interior point and thus it has come to be known as the *Sierpiński plane universal curve*.

---

[5]Because J_A was conceived as a generalization of the unit interval I, it seemed natural to select a notation that serves as a mnemonic of the extension — select the letter that follows the letter I, namely the letter J.

[6]Whyburn's 1958 article was based on a lecture that he first presented at the Warsaw Mathematical Colloquium in the spring of 1930, where he was introduced by Sierpiński.

CHAPTER 2

Self-Similarity and J_{n+1} for Finite n

The unit interval $[0,1] =_t J_2$ is two copies $[0,1/2]$ and $[1/2,1]$ of itself, each just touching the other. In this chapter we show that J_A is $|A|$ copies of itself, each "just touching" the others. This feature appears in the graphics where side-by-side approximations of $N(A)$ and J_A elucidate adjacent-endpoint pastings. Six figures serve to illustrate J_5. For finite n, an iterated function system $\mathcal{F}_n$ is constructed whose attractor $\omega^n \subset \mathbb{R}^n$ is homeomorphic to J_{n+1}. The homeomorphism exposes the $\mathcal{F}_n$-induced address map $\phi : N(\{0,\dots,n\}) \to \omega^n$ as adjacent-endpoint identification.

§5 Self-Similarity of J_A

For any Baire space $N(A)$, the partition $\mathcal{B}_1 = \{\langle a \rangle : a \in A\}$ contains $|A|$ pairwise-disjoint homeomorphic copies

$$\langle a \rangle = \{a\} \times A \times A \times \cdots \qquad (a \in A)$$

of $N(A)$. These copies map to $|A|$ homeomorphic copies $p(\langle a \rangle)$, $a \in A$, of J_A. Furthermore, it is clear that when $a \neq b$ the rational $\{a\overline{b}, b\overline{a}\}$ is a member of both $p(\langle a \rangle)$ and $p(\langle b \rangle)$. We can say more.

5.1 Lemma (just-touching property) *Let $a, b \in A$ be distinct, and let p be the natural mapping $N(A) \to J_A$. Then $|p(\langle a \rangle) \cap p(\langle b \rangle)| = 1$ and the unique point in the intersection is the rational $r = \{a\overline{b}, b\overline{a}\}$.*

PROOF. Let $z \in p(\langle a \rangle) \cap p(\langle b \rangle)$. Then $\langle a \rangle \cap \langle b \rangle = \emptyset$ implies that there exist $c \in \langle a \rangle$ and $d \in \langle b \rangle$ such that $p(\{c,d\}) = z$. Since p is at most 2-to-1, $p^{-1}(z) = \{c,d\}$. So $c \sim d$, $c_1 = a$, and $d_1 = b$ yield $z = \{c,d\} = r$. □

So each of the $|A|$ copies of J_A just touches the others. We shall refer to this combinatorial property as *the just-touching property*.

The "self-similarity" of J_A continues at each *level*. The partition $\mathcal{B}_2 = \{\langle a, b \rangle : a, b \in A\}$ contains $|A|^2$ pairwise-disjoint copies

$$\langle a, b \rangle = \{a\} \times \{b\} \times A \times A \times \cdots \qquad (a, b \in A)$$

of $N(A)$, and again these copies map to $|A|^2$ homeomorphic copies $p(\langle a, b \rangle)$ of J_A. For a fixed $a \in A$, the set $\{p(\langle a, x \rangle) : x \in A\}$ of $|A|$ copies of J_A satisfies the just-touching property. That is, if $b \neq c$, then $p(\langle a, b \rangle)$ just touches

S.L. Lipscomb, *Fractals and Universal Spaces in Dimension Theory*,
DOI 10.1007/978-0-387-85494-6_2, © Springer Science+Business Media, LLC 2009

$p(\langle a, c \rangle)$ in a unique rational $\{ab\bar{c}, ac\bar{b}\}$. The process continues — the *level-k copies* of J_A consist of all sets $p(\langle a_1, a_2, \ldots, a_k \rangle)$ for $a_1, a_2, \cdots, a_k \in A$. And for a fixed $(k-1)$ string $a_1 a_2 \cdots a_{k-1}$, the two level-k copies $p(\langle a_1, a_2, \ldots, a_{k-1}, b \rangle)$ and $p(\langle a_1, a_2, \ldots, a_{k-1}, c \rangle)$ meet only at $r = \{a_1 a_2 \cdots a_{k-1} b\bar{c}, a_1 a_2 \cdots a_{k-1} c\bar{b}\}$.

§6 Approximations for $n + 1 = 2, 3, 4$

We begin with Figure 6.1 where $A = \{0, 2\}$. In this case $N(A)$ is homeomorphic to Cantor's set. Note that

$$1/3 = 0/3^1 + \Sigma_{i=2}^{\infty} 2/3^i \xrightarrow{p} 1/2 = 0/2^2 + \Sigma_{i=2}^{\infty} 2/2^{i+1}$$
$$2/3 = 2/3^1 + \Sigma_{i=2}^{\infty} 0/3^i \xrightarrow{p} 1/2 = 2/2^2 + \Sigma_{i=2}^{\infty} 0/2^{i+1}.$$

It is generally true that the "holes" bounded by adjacent endpoints correspond to the dyadic rationals contained in the interior of the unit interval.

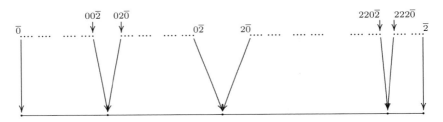

Fig. 6.1 For $|A| = 2$, the space J_2 is homeomorphic to the unit interval.

Next, we let $A = \{a, b, c\}$ be of size three (Figure 6.2). In this case we see that J_3 is a copy of the classical fractal known as Sierpiński's triangle.

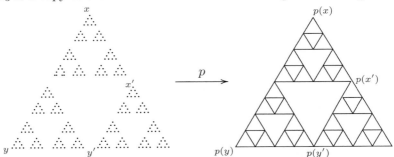

Fig. 6.2 The space J_3 is homeomorphic to Sierpiński's triangle.

In Figure 6.2 note that the "edge" $N(\{a, b\}) = [x = \bar{a}, y = \bar{b}]$, a Cantor subspace of $N(A)$, maps onto the edge $[p(x), p(y)]$, a unit interval subspace of J_3. We also see another Cantor subspace $[x' = c\bar{a}, y' = c\bar{b}] = \{c\} \times \{a, b\} \times \{a, b\} \times \cdots$ of $N(A)$ mapping onto another copy $[p(x'), p(y')]$ of the unit interval.

In Figure 6.3 below an approximation to the Baire space $N(\{0, 1, 2, 3\})$ is presented on the left side (recall Figure 1.2); and on the right side we have an approximation to J_4, the p-image of $N(\{0, 1, 2, 3\})$.

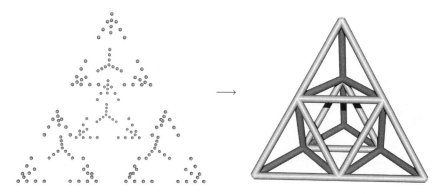

Fig. 6.3 The space J_4 is homeomorphic to the Sierpiński cheese.

So for $|A| = n + 1 = 4$, Figure 6.3 illustrates a level-1 approximation to J_4, i.e., an approximation to the classical fractal known as the Sierpiński cheese.

§7 Approximations for $n + 1 = 5$

In this section we consider a relatively recent construction that allows us to view J_5 in 3-space. We begin with a sequence of figures (Figs. 7.1 to 7.6) that illustrate the desired combinatorial structure — five congruent figures, each just touching the other four:

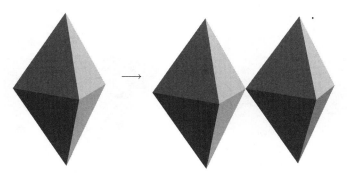

Fig. 7.1 One hexahedron, then two, each just touching the other.

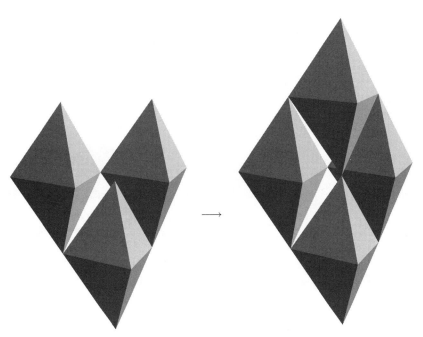

$\longrightarrow$

Fig. 7.2 Three, then four hexahedra, each just touching the others.

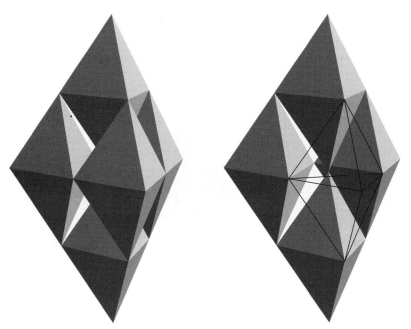

Fig. 7.3 Five hexahedra, each just touching the others.

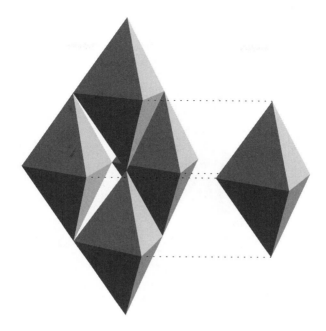

Fig. 7.4 Fitting the fifth hexahedron.

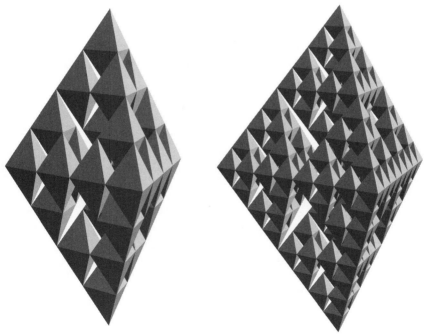

Fig. 7.5 Approximations to J_5 at levels 2 and 3.

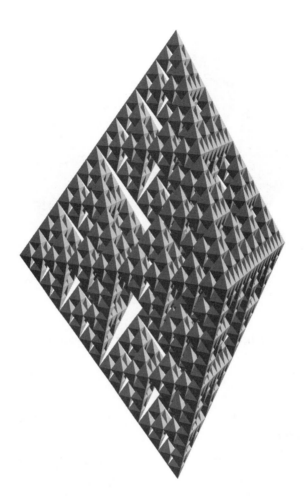

Fig. 7.6 Level-4 approximation to J_5.

§8 J_{n+1} as an Attractor ω^n of an IFS

For $n \geq 1$, let Δ_n denote the n-simplex in n-space $\mathbb{R}^n$ whose $n+1$ vertices consist of the origin $\mathbf{u}_0 = (0, \ldots, 0)$ and the terminal points of the n standard orthonormal basis vectors $\mathbf{u}_1 = (1, 0, \ldots, 0)$, $\mathbf{u}_2 = (0, 1, 0, \ldots, 0)$, $\ldots$, $\mathbf{u}_n = (0, 0, \ldots, 0, 1) \in \mathbb{R}^n$. Then Δ_n is the convex hull of its vertices and

$$\Delta_n = \{\mathbf{v} = \Sigma_0^n \lambda_i \mathbf{u}_i : \Sigma_0^n \lambda_i = 1; \ 0 \leq \text{each } \lambda_i \leq 1\}$$

where the λ_i are called the barycentric coordinates of $\mathbf{v}$.

Using $\Delta_n \subset \mathbb{R}^n$, we may generate $n+1$ scalings of Δ_n that satisfy the just-touching property.

To be sure, consider the finite iterated function system (IFS) $\mathcal{F}_n = \{w_0, w_1, \ldots, w_n\}$ where

$$w_i(\mathbf{x}) = \mathbf{u}_i + (1/2)(\mathbf{x} - \mathbf{u}_i) = (1/2)(\mathbf{x} + \mathbf{u}_i) \qquad (\mathbf{x} \in \Delta_n)$$

is a scaling by $1/2$ toward $\mathbf{u}_i$. Then by characterizing each $w_i(\Delta_n)$ as

$$w_i(\Delta_n) = \{\mathbf{v} = \Sigma_0^n \lambda_j \mathbf{u}_j : \Sigma_0^n \lambda_j = 1; 0 \leq \text{each } \lambda_j \leq 1; \text{ and } \lambda_i \geq 1/2\},$$

we may show that these $n+1$ scalings of Δ_n are just touching: For distinct $i, j \in A = \{0, 1, \ldots, n\}$, we have $|w_i(\Delta_n) \cap w_j(\Delta_n)| = 1$ with $(1/2)(\mathbf{u}_i + \mathbf{u}_j)$ being the point of intersection.

In passing, note that one may easily verify that the inverse w_i^{-1} of w_i is given by the formula $w_i^{-1}(\mathbf{x}) = 2\mathbf{x} - \mathbf{u}_i$ for $\mathbf{x} \in w_i(\Delta_n)$.

In general, let $A = \{0, 1, \ldots, n\}$, and recall that the attractor of $\mathcal{F}_n$ is the unique compact set K characterized by the equation $K = \cup_{i \in A} w_i(K)$. From this equation it follows that $K \supset w_i(K) \supset w_i \circ w_j(K)$ for every i and every j. Each $\delta = \delta_1 \delta_2 \cdots$ in code space $N(A)$ thereby determines a nested list $K \supset w_{\delta_1}(K) \supset w_{\delta_1} \circ w_{\delta_2}(K) \supset \cdots$ of compact sets whose diameters go to zero. The intersection of the sets in this nested list contains exactly one point $p_\delta \in K$. This correspondence $\delta \mapsto p_\delta$ is the *address map* $\phi : N(A) \to K$.

It turns out that the address map ϕ is identification of adjacent endpoints. To motivate the theory, we demonstrate the connection with an example.

8.1 EXAMPLE. Let ϕ be the address map induced by $\mathcal{F}_1 = \{w_0, w_1\}$. Then $\Delta_1 = [0, 1]$ is the 1-simplex, which is the unit interval, and the code space is $N(\{0, 1\})$. So $\phi : N(\{0, 1\}) \to K$ where K is the attractor of $\mathcal{F}_1$. In this example, we see that

$$[0, 1] = \Delta_1 = w_0(\Delta_1) \cup w_1(\Delta_1) = [0, 1/2] \cup [1/2, 1],$$

which shows that $K = [0, 1]$ is the unit interval. So the attractor K is homeomorphic to J_2. Now considering the sequence $\delta = 0\bar{1} \in N(\{0, 1\})$ and the k-fold composition $w_1^k = w_1 \circ \cdots \circ w_1$, we have

$$w_0(w_1^k([0, 1])) = w_0([1 - (1/2^k), 1]) = [(1/2) - (1/2^{k+1}), 1/2].$$

Thus as $k \to \infty$, we find that $\phi(0\bar{1}) = \phi(\delta) = p_\delta = 1/2$. Similarly, for the adjacent endpoint $\varepsilon = 1\bar{0}$ and $w_0^k = w_0 \circ \cdots \circ w_0$, we have

$$w_1(w_0^k([0, 1])) = w_1([0, 1/2^k]) = [1/2, 1/2 + 1/2^{k+1}]$$

which shows that $\phi(1\bar{0}) = \phi(\varepsilon) = p_\varepsilon = 1/2$. Note, for this δ and this ε, that $\delta \sim \varepsilon$ implies $\phi(\delta) = \phi(\varepsilon)$. So could it be that the fibers of the natural map $p : N(\{0, 1\}) \to J_2$ are the fibers of $\phi : N(\{0, 1\}) \to K = [0, 1]$?

8.2 Lemma (the address map ϕ) *Let $\phi : N(A) \to K$ be the address map induced by $\mathcal{F}_n$, let $w_q \in \mathcal{F}_n$, and let $\delta = \delta_1 \delta_2 \cdots \in N(A)$. Then (a) $w_q(\phi(\delta)) = \phi(q\delta_1\delta_2 \cdots)$; (b) $w_{\delta_1}^{-1}(\phi(\delta_1\delta_2 \cdots)) = \phi(\delta_2\delta_3 \cdots)$; (c) $\phi(\delta) = \mathbf{u}_q$ if and only if $\delta = \overline{q}$; and (d) ϕ is a continuous closed surjection.*

PROOF. First, consider (d): We begin by showing that ϕ is surjective. Since $x \in K = w_0(K) \cup \cdots \cup w_n(K)$, then $x \in w_{\delta_1}(K)$ for some $\delta_1 \in A$. Inductively, δ_1 yields δ_2, i.e.,

$$x \in w_{\delta_1}(K) = w_{\delta_1} \circ w_0(K) \cup \cdots \cup w_{\delta_1} \circ w_n(K)$$

implies $x \in w_{\delta_1}(K) \supset w_{\delta_1} \circ w_{\delta_2}(K)$ for some $\delta_2 \in A$. An induction argument shows that $\delta \in N(A)$ exists such that $\phi(\delta) = x$. Next, we show that ϕ is continuous. Let $\delta_n \to \delta$ in $N(A)$. Then for each $k \geq 1$, we may choose N_k such that $n > N_k$ yields $\delta_{n1} = \delta_1, \ldots, \delta_{nk} = \delta_k$. These equalities imply that both $\phi(\delta_n)$, $\phi(\delta) \in W_k = w_{\delta_1} \circ \cdots \circ w_{\delta_k}(K)$ where the diameters of the W_k go to zero as $k \to \infty$. It follows that $\phi(\delta_n) \to \phi(\delta)$. Finally, note that ϕ is closed because $N(A)$ is compact and ϕ is continuous.

Second, consider (a): By definition, $\phi(\delta)$ is the lone element in $\cap_{j=1}^{\infty} W_j$ where $W_j = w_{\delta_1} \circ \cdots \circ w_{\delta_j}(K)$. So $\phi(\delta) \in W_j$ for each j, showing that $w_q(\phi(\delta)) \in w_q(W_j) = w_q \circ w_{\delta_1} \circ \cdots \circ w_{\delta_j}(K)$ for each j. It follows that $w_q(\phi(\delta))$ is the lone element in $\cap_{j=1}^{\infty} w_q(W_j)$, i.e., $w_q(\phi(\delta)) = \phi(q\delta_1\delta_2 \cdots)$.

Third, consider (b): From (a) we have $w_{\delta_1}(\phi(\delta_2\delta_3 \cdots)) = \phi(\delta)$. Apply $w_{\delta_1}^{-1}$ to both sides.

Fourth, consider (c): Recall that a contraction has only one fixed point, and each w_q is a contraction such that $w_q(\mathbf{u}_q) = \mathbf{u}_q$. In short, $w_q(\mathbf{x}) = \mathbf{x}$ if and only if $\mathbf{x} = \mathbf{u}_q$. So now suppose $\delta = \overline{q}$. Then $w_q(\phi(\delta)) = \phi(q\delta_1\delta_2 \cdots) = \phi(\delta)$, showing that $\phi(\delta) = \mathbf{u}_q$. Conversely, suppose $\phi(\delta) = \mathbf{u}_q$. Then the only index i such that $\mathbf{u}_q \in w_i(\Delta_n)$ is $i = q$. So $\delta_1 = q$. If $\delta_1 = \cdots = \delta_{k-1} = q$, then $w_{\delta_{k-1}}^{-1} \circ \cdots \circ w_{\delta_1}^{-1}(\mathbf{u}_q) = \mathbf{u}_q$. So (b) shows that

$$\mathbf{u}_q = w_{\delta_{k-1}}^{-1} \circ \cdots \circ w_{\delta_1}^{-1}(\phi(\delta)) = \phi(\delta_k\delta_{k+1} \cdots) \in w_{\delta_k}(\Delta_n).$$

Thus, $\delta_k = q$, and by induction $\delta = \overline{q}$.   □

8.3 Definition (n-web) For $n \geq 1$, the *n-web* ω^n is the attractor K_{n+1} of $\mathcal{F}_n$ whose code space is $N(\{0, 1, \ldots, n\})$. We may also consider $\omega^n \subset \Delta^n \subset \mathbb{R}^{n+1}$ where Δ^n is the *standard* simplex (see Appendix 2) and $\mathcal{F}_n$ the obvious family of contractions.

8.4 Theorem (fibers of ϕ are the fibers of p) *Let $A = \{0, 1, \ldots, n\}$, let $\phi : N(A) \to \omega^n$ be the address map induced by $\mathcal{F}_n$, and let $\sim$ be the adjacent-endpoint relation in $N(A)$. Then $\phi(\delta) = \phi(\varepsilon)$ if and only if $\delta \sim \varepsilon$.*

PROOF. First, suppose $\delta \sim \varepsilon$. The case $\delta = \varepsilon$ is trivial. If $\delta \neq \varepsilon$, then let t

be the tail index. With $t-1$ applications of Lemma 8.2 (a) we have

(1)
$$\phi(\delta) = w_{\delta_1} \circ \cdots \circ w_{\delta_{t-1}} \left(\phi(\delta_t \overline{\varepsilon_t}) \right)$$
$$\phi(\varepsilon) = w_{\delta_1} \circ \cdots \circ w_{\delta_{t-1}} \left(\phi(\varepsilon_t \overline{\delta_t}) \right).$$

Then Lemma 8.2 (c) yields $\phi(\overline{\varepsilon_t}) = \mathbf{u}_{\varepsilon_t}$, and Lemma 8.2 (a) shows that $\phi(\delta_t \overline{\varepsilon_t}) = w_{\delta_t}(\mathbf{u}_{\varepsilon_t}) = (1/2)(\mathbf{u}_{\varepsilon_t} + \mathbf{u}_{\delta_t})$. Similarly, $\phi(\varepsilon_t \overline{\delta_t}) = w_{\varepsilon_t}(\mathbf{u}_{\delta_t}) = (1/2)(\mathbf{u}_{\delta_t} + \mathbf{u}_{\varepsilon_t})$. It follows from (1) and the fact that $w_{\delta_1} \circ \cdots \circ w_{\delta_{t-1}}$ is one-to-one, that $\phi(\delta) = \phi(\varepsilon)$.

Conversely, suppose $\phi(\delta) = \phi(\varepsilon)$.

SUBCASE 1. $\delta_1 \neq \varepsilon_1$: Then $\phi(\delta) = \phi(\varepsilon) \in w_{\delta_1}(\Delta_n) \cap w_{\varepsilon_1}(\Delta_n)$ implies

(2) $$\phi(\delta) = \phi(\varepsilon) = (1/2)(\mathbf{u}_{\delta_1} + \mathbf{u}_{\varepsilon_1}) = (1/2)(\phi(\overline{\delta_1}) + \phi(\overline{\varepsilon_1})).$$

Lemma 8.2 (b), the definition of $w_{\delta_1}^{-1}$, and (2) provide

$$\begin{aligned} \phi(\delta_2 \delta_3 \cdots) = w_{\delta_1}^{-1}(\phi(\delta)) &= 2\phi(\delta) - \mathbf{u}_{\delta_1} = 2\phi(\delta) - \phi(\overline{\delta_1}) \\ &= 2\left[(1/2)(\phi(\overline{\delta_1}) + \phi(\overline{\varepsilon_1}))\right] - \phi(\overline{\delta_1}) \\ &= \phi(\overline{\varepsilon_1}) = \mathbf{u}_{\varepsilon_1}. \end{aligned}$$

So Lemma 8.2 (c) implies $\delta_2 \delta_3 \cdots = \overline{\varepsilon_1}$, i.e., $\delta_i = \varepsilon_1$ for each $i \geq 2$. Similarly, we may also deduce that $\varepsilon_i = \delta_1$ for each $i \geq 2$, i.e., $\delta \sim \varepsilon$.

SUBCASE 2. $\delta_1 = \varepsilon_1$: Then let $t \in \{2, 3, \ldots\}$ be the smallest index such that $\delta_t \neq \varepsilon_t$. In this case, Lemma 8.2 (a) yields

$$\begin{aligned} \phi(\delta) &= w_{\delta_1} \circ \cdots \circ w_{\delta_{t-1}}(\phi(\delta_t \delta_{t+1} \cdots)) \\ \phi(\varepsilon) &= w_{\delta_1} \circ \cdots \circ w_{\delta_{t-1}}(\phi(\varepsilon_t \varepsilon_{t+1} \cdots)). \end{aligned}$$

Since $w_{\delta_1} \circ \cdots \circ w_{\delta_{t-1}}$ is one-to-one, and $\phi(\delta) = \phi(\varepsilon)$, we have $\phi(\delta_t \delta_{t+1} \cdots) = \phi(\varepsilon_t \varepsilon_{t+1} \cdots)$ where $\delta_t \neq \varepsilon_t$. An argument similar to the proof of SUBCASE 1 shows that $\delta_i = \varepsilon_t$ and $\varepsilon_i = \delta_t$ for each $i \geq t+1$. So $\delta \sim \varepsilon$.   □

8.5 Theorem (J_{n+1} is homeomorphic to ω^n) *Let $A = \{0, 1, \ldots, n\}$, let ω^n be the attractor of $\mathcal{F}_n$, let $\phi : N(A) \to \omega^n$ be the induced address map, and let $p : N(A) \to J_{n+1}$ be the natural mapping. Then $f = \phi \circ p^{-1} : J_{n+1} \to \omega^n$ is a homeomorphism.*

PROOF. By Lemma 8.4, the fibers of p are identical to the fibers of ϕ. So the mapping $f = \phi \circ p^{-1}$ is well defined and injective. Moreover, f is surjective since ϕ is surjective; f is continuous since ϕ is continuous and p is closed; and f is closed since p is continuous and ϕ is closed.   □

§9 Can We "View" J_{n+1} in 3-Space?

For $n \geq 1$, J_{n+1} is homeomorphic to the attractor $K_{n+1} = \omega^n$ of $\mathcal{F}_n$. However, since ω^n lives in n-space, when $n \geq 4$ we cannot picture ω^n, at least not directly. This quandary leads naturally to the question, "How do we picture J_{n+1}?"

From the Classical Universal Imbedding Theorem, since each J_{n+1} is a separable metric space of (topological) dimension one, it may be topologically imbedded in the Euclidean cube I^3. Such an imbedding, however, may make it humanly impossible to "see" the self-similarity.

Nevertheless, on the positive side, as illustrated in §7, the self-similarity of J_5 is clearly exposed in 3-space, making the approximations as clear as those of Sierpiński's triangle J_3.

In general, when considering J_{n+1} the subscript $n + 1$ is fundamental. It tells us that J_{n+1} contains $n + 1$ copies of itself that satisfy the just touching property, and it serves to specify the numerator when calculating the fractal dimension $D(\omega^n) = \ln(n + 1)/\ln(2)$. What we desire, then, is an imbedding into 3-space that preserves the fractal dimension and exposes the self-similarity of J_{n+1}.

If preservation of fractal dimension is important, then it is instructive to calculate the fractal dimension of the first few ω^n:

$$D(\omega^1) = 1 \quad < \quad D(\omega^2) \approx 1.58 \quad < \quad D(\omega^3) = 2 \quad < \quad D(\omega^4) \approx 2.32 \quad <$$
$$< \quad D(\omega^5) \approx 2.58 \quad < \quad D(\omega^6) \approx 2.81 \quad < \quad D(\omega^7) = 3 \quad < \cdots .$$

To *view* (imbed with fractal dimension preserved) ω^n in m-space, we see from Barnsley [1988, Theorem 2, page 202] that it is necessary that $D(\omega^n) \leq \dim(\mathbb{R}^m) = m$.

As an application, $D(\omega^2) > \dim(\mathbb{R}^1) = 1$ implies that ω^2 cannot be viewed on the real line $\mathbb{R}^1$. For $n = 3$, however, $D(\omega^3) = 2 = \dim(\mathbb{R}^2)$, and so the "necessary condition $D(\omega^3) \leq \dim(\mathbb{R}^2) = 2$" sheds no light on the fact that ω^3, the Sierpiński cheese, cannot be viewed in the plane. Nevertheless, with the aid of Figure 9.1 we see that a homeomorph of one of the two Kuratowski forbidden graphs (namely, the complete bipartite graph $K_{3,3}$) is a subspace of ω^3, and thus ω^3 cannot be imbedded in the plane. (Each vertex in $\{a, b, c\}$ is "adjacent" to each vertex in $\{A, B, C\}$ via "edges" $a1A$, $a2B$, $a3C$, bA, bB, $b4C$, $c5A$, cB, and cC.)

The next application of "the necessary condition $D(\omega^n) \leq \dim(\mathbb{R}^m) = m$" occurs at the value $n = 8$: Since

$$D(\omega^8) \approx 3.1699 > \dim(\mathbb{R}^3) = 3$$

we know that the 8-web ω^8 cannot be "viewed" in 3-space $\mathbb{R}^3$. So we are left with $n = 4, 5, 6, 7$. The $n = 4$ case was illustrated in §7, but it is an open question as to whether any n-web ω^n for $n = 5, 6, 7$ can be "viewed" in $\mathbb{R}^3$.

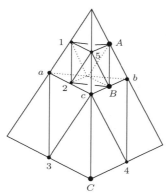

Fig. 9.1 The 3-web ω^3 cannot be imbedded in the plane.

§10 Comments

The graphics in §7 were created by Chris Dupilka, and similar graphics with relevant narratives appear in Perry and Lipscomb [2003]. As for the term *n-web*, it was introduced in Lipscomb and Perry [1992, page 1159], but it was motivated by the title *A Fractal Skewed Web* of plate 143 in Mandelbrot [1983].

As Lemma 5.1 tells us, if we "see" a copy of the 4-web ω^4 in 3-space, then we should "see" its five level-1 copies "just touching." Reasoning in reverse, Perry and Lipscomb [2003] used the self-similarity (Figure 7.3) to construct an isotopy $H : \omega^4 \times I \to \mathbb{R}^4$ rel ω^3 (homotopy with each H_t a homeomorphism that is the identity on $\omega^3 \subset \omega^4$) where each H_t is a linear transformation that preserves fractal dimension and where $H_1 : \omega^4 \to \mathbb{R}^3 \subset \mathbb{R}^4$. In other words, ω^4 may be moved into 3-space with its fractal dimension preserved (Chapter 12).

This "motion" may be intuitively explained with the aid of Figure 10.1, where cylinders represent line segments. Indeed, suppose we are observers in 3-space and "t" is a time parameter that moves from time 0 to time 1. Then at time t, we could see the part of $H_t(\omega^4) \subset \mathbb{R}^4$ that meets 3-space. In particular, when $t \approx 0$, we would see "buds" located at the points where the light-gray semitransparent cylinders meet the dark-gray opaque cylinders. As t increases, these buds begin to grow "up" (in the direction of the light-gray cylinders) toward the other ends of these cylinders. And when $t = 1$, we see the structure in Figure 10.1 that contains the light-gray semitransparent cylinders that were originally (at time $t = 0$) outside of 3-space (in $\mathbb{R}^4 \setminus \mathbb{R}^3$).

The approximation to ω^4 in Figure 10.1 should be compared with the one in Figure 7.3, and the substructure in Figure 10.1 consisting of the dark-gray cylinders should be compared with the right-side illustration in Figure 6.3.

Parts (a) and (c) of Lemma 8.2 appear in Lipscomb [2007, Lemma 9].

Finally, the phrase "just-touching" as typically used in the context of iterated function systems refers to the IFS itself, and not the attractor of the IFS (Barnsley [1988, pages 121 and 129]). In this chapter, however, we considered only one type of IFS, namely the $\mathcal{F}_n$, and we used the "just-touching" phrase to describe the attractors. It so happens that each $\mathcal{F}_n$ is also "just-touching" in the sense of Barnsley [1988, page 129].

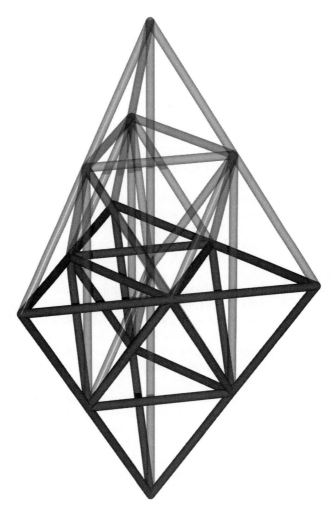

Fig. 10.1 An isotopy moves the 4-web ω^4 from 4-space into 3-space.

CHAPTER 3

No-Carry Property of $\omega^{A'}$

By the mid 1970s it was known that J_A^{n+1} for $|A| \geq \aleph_0$ contains models of all n-dimensional weight $|A|$ metric spaces. For infinite A, however, J_A did not receive a metric until 1992, when it was imbedded in Hilbert space.

In this chapter we introduce a surjection $J_A \to \omega^{A'}$ into Hilbert space. (In Chapter 4 we shall show that this surjection $J_A \to \omega^{A'}$ is an imbedding.)

The J_A-image $\omega^{A'}$ is then characterized in terms of the no-carry property, i.e., the "no-carry characterization of Sierpiński's triangle" extends to $\omega^{A'}$.

To construct the mapping, we use "star spaces," one in J_A and one in Hilbert space. To motivate the construction, we begin with three examples related to the Sierpiński triangle. Otherwise, the presentation follows Lipscomb and Perry [1992], but most of the results were also obtained independently by Milutinović [1992] (see §18).

§11 Three Examples

Imbedding J_A into Hilbert space turns out to be an extension of the "no-carry characterization of Sierpiński's triangle $T =_t J_3$." So we begin with Figure 11.1, which illustrates and specifies the no-carry property.

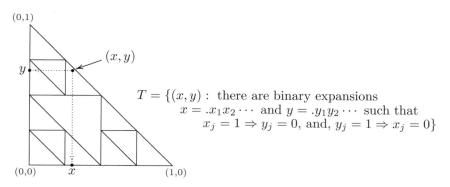

$T = \{(x, y) :$ there are binary expansions
$x = .x_1 x_2 \cdots$ and $y = .y_1 y_2 \cdots$ such that
$x_j = 1 \Rightarrow y_j = 0$, and, $y_j = 1 \Rightarrow x_j = 0\}$

Fig. 11.1 The no-carry constraint specifies a Sierpiński triangle.

Together, the two implications "$x_j = 1 \Rightarrow y_j = 0$" and "$y_j = 1 \Rightarrow x_j = 0$" are called the *no-carry conditions* — for each $j = 1, 2, \ldots$ the binary addition "$x_j + y_j$" is a "no-carry addition."

S.L. Lipscomb, *Fractals and Universal Spaces in Dimension Theory*,
DOI 10.1007/978-0-387-85494-6_3, © Springer Science+Business Media, LLC 2009

11.2 EXAMPLE. Consider the Sierpiński triangle T with vertices $(0,0)$, $(1,0)$, and $(0,1)$. Then the $x = 1/2$ and $y = 1/2$ coordinates of $(1/2, 1/2) \in T$ have binary expansions $x = .x_1 x_2 \cdots = .100 \cdots$ and $y = .y_1 y_2 \cdots = .011 \cdots$ that satisfy both "$x_j = 1$ implies $y_j = 0$" and "$y_j = 1$ implies $x_j = 0$." $\qquad\square$

For motivation of the concept "star space," let $A = \{(0,0), (1,0), (0,1)\}$ and consider $A' = A \setminus \{(0,0)\}$. For this particular A', we define

$$S(A') = \{(x,y) \in \mathbb{R}^2 : (0 \leq x \leq 1 \text{ and } y = 0) \text{ or } (0 \leq y \leq 1 \text{ and } x = 0)\}$$

as a *star space with* $|A'| = 2$ *arms*.

11.3 EXAMPLE. Using Figure 11.4, we let $A = \{z, a, b\}$ be of size three, and view the endpoints $\bar{z}$, $\bar{a}$, and $\bar{b}$ in $N(A)$ as vertices of "the $N(A)$ triangle."

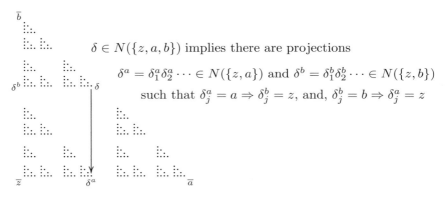

$\delta \in N(\{z, a, b\})$ implies there are projections

$\delta^a = \delta_1^a \delta_2^a \cdots \in N(\{z, a\})$ and $\delta^b = \delta_1^b \delta_2^b \cdots \in N(\{z, b\})$

such that $\delta_j^a = a \Rightarrow \delta_j^b = z$, and, $\delta_j^b = b \Rightarrow \delta_j^a = z$

Fig. 11.4 The "no-carry property" encoded in $N(\{z, a, b\})$.

Let $\delta = b\bar{a} \in N(A)$. Then viewing "$z$" as "zero" and "zeroing out all letters not equal to a" we project

$$\delta = b\bar{a} \;\mapsto\; \delta^a = z\bar{a} \in N(\{z, a\})$$

where $N(\{z, a\})$ is a copy of Cantor's set with endpoints $\bar{z}$ and $\bar{a}$. Similarly, by "zeroing out all letters not equal to b" we project

$$\delta = b\bar{a} \;\mapsto\; \delta^b = b\bar{z} \in N(\{z, b\}).$$

It follows that $\delta_j^a = a \Rightarrow \delta_j^b = z$ and $\delta_j^b = b \Rightarrow \delta_j^a = z$, i.e., these "projections" encode a "no-carry property in $N(A)$." $\qquad\square$

11.5 EXAMPLE. We illustrate how the no-carry property in $N(\{z, a, b\})$ encodes the no-carry property of the Sierpiński triangle T: First, identify the letter "a" with "1" and the letter "z" with "0", inducing a homeomorphism

$N(\{z,a\}) \leftrightarrow N(\{0,1\})$. That is, points in $N(\{z,a\}) \subset N(A)$ are identified with strings $x_1 x_2 \cdots$ of binary digits. Second, identify these binary strings $x_1 x_2 \cdots$ with points $\Sigma_1^\infty (2x_j)/3^j$ in Cantor's set $\mathcal{C}(0,1)$, obtaining a homeomorphism $N(\{0,1\}) \leftrightarrow \mathcal{C}(0,1)$. And third, use classical identification of adjacent endpoints $\Sigma_1^\infty (2x_j)/3^j \mapsto \Sigma_1^\infty x_j/2^j$ to map $\mathcal{C}(0,1) \to I = [0,1]$.

In general, $\delta^a \mapsto \sum_j \frac{x_j}{2^j} \in I$ and $\delta^b \mapsto \sum_j \frac{y_j}{2^j} \in I$ where $(\delta_j^a = a \Leftrightarrow x_j = 1)$ and $(\delta_j^b = b \Leftrightarrow y_j = 1)$:

$$\delta^a = \delta_1^a \delta_2^a \cdots \quad \leftrightarrow \quad x_1 x_2 \cdots \in N(\{0,1\}) \quad \leftrightarrow \quad \sum_{j=1}^\infty \frac{2x_j}{3^j} \to \sum_{j=1}^\infty \frac{x_j}{2^j} \in I = [0,1]$$

$$\delta^b = \delta_1^b \delta_2^b \cdots \quad \leftrightarrow \quad y_1 y_2 \cdots \in N(\{0,1\}) \quad \leftrightarrow \quad \sum_{j=1}^\infty \frac{2y_j}{3^j} \to \sum_{j=1}^\infty \frac{y_j}{2^j} \in I = [0,1].$$

And in particular, for $\delta = b\overline{a}$ we have

$$\delta^a = zaa\cdots \quad \leftrightarrow \quad 011\cdots \in N(\{0,1\}) \quad \leftrightarrow \quad \sum_{j=2}^\infty \frac{2}{3^j} = \frac{1}{3} \to \sum_{j=2}^\infty \frac{1}{2^j} = \frac{1}{2} \in I$$

$$\delta^b = bzz\cdots \quad \leftrightarrow \quad 100\cdots \in N(\{0,1\}) \quad \leftrightarrow \quad \sum_{j=1}^1 \frac{2}{3^j} = \frac{2}{3} \to \sum_{j=1}^1 \frac{1}{2^j} = \frac{1}{2} \in I,$$

i.e., $\delta \to \delta^a$ and $\delta \to \delta^b$ yield (δ^a, δ^b) which decodes as $(\frac{1}{2}, \frac{1}{2}) \in T$. $\qquad \square$

§12 Star Spaces

To illustrate the concept of "star space," let $|A| = 73$, $z \in A$ be fixed, and $A' = A \setminus \{z\}$. Then the "Cantor star" $S_\mathcal{C}(A')$, pictured on the left side of Figure 12.1, consists of the 72 Cantor spaces $N(\{z,a\})$, $a \in A'$, that meet only at the point $\overline{z}$.

$S_\mathcal{C}(A')$ $S(A')$

Fig. 12.1 Picturing $S_\mathcal{C}(A')$ and the corresponding $S(A')$ for $|A| = 73$.

Then $p : N(A) \to J_A$ maps this "Cantor star" $S_\mathcal{C}(A')$ onto a "J_A star" $S(A')$ consisting of $|A'| = 72$ unit-interval subspaces of J_A that meet only at $p(\overline{z})$.

12.2 Definition (J_A star spaces)   For any set A of size at least two, let $z \in A$, define $A' = A \setminus \{z\}$, and consider the union $\cup_{b \in A'} \mathcal{C}(z, b)$ of Cantor subspaces $\mathcal{C}(z, b) = N(\{z, b\})$ of $N(A)$. This subspace $\cup_{b \in A'} \mathcal{C}(z, b)$ of $N(A)$ is called a *Cantor star space* $S_{\mathcal{C}}(A')$ *each of whose $|A'|$ arms is a copy of Cantor's set.* Moreover, the p-image $S(A) = p(S_{\mathcal{C}}(A')) \subset J_A$ is called a J_A *star space* with $|A'|$ arms I_β ($\beta = p(\bar{b}), b \in A'$) where each arm $I_\beta = I(\zeta, \beta) = p(\mathcal{C}(z, b))$ ($\zeta = p(\bar{z})$) is homeomorphic to the unit interval.

When A is infinite, a J_A star space $S(A)$ may be viewed as *the standard star space*, which is often called *a hedgehog with $|A| = |A'|$ prickles.*[1]

§13 The Star Space in $l^2(A)$

Hilbert's $l^2(A)$ space contains a star space (centered at its zero) whose ath arm ($a \in A$) is the line segment $\{tu_a : 0 \le t \le 1\} = [\mathbf{0}, \mathbf{u_a}]$ where $\mathbf{u}_a$ is a unit vector in the standard orthonormal basis of $l^2(A)$.

A few comments are in order. Let $\mathbb{R}^A$ be the Cartesian product of $|A|$ copies of the real line $\mathbb{R}^1$. Then *Hilbert's space* $l^2(A)$ may be viewed as a metric space that has (1) *elements*: every $x = (x_a) = (x_a)_{a \in A} \in \mathbb{R}^A$ such that $x_a = 0$ for all but at most countably many $a \in A$ and $\Sigma_a x_a^2$ converges; and (2) *topology*: that induced by the metric $d(x, y) = \sqrt{\Sigma_a (x_a - y_a)^2}$.

The arms of our star subspace of $l^2(A)$ are determined by the orthonormal basis vectors $\mathbf{u}_b = (u_a^b)_{a \in A}$ where $u_a^b = 0$ when $a \ne b$ and $u_b^b = 1$. Each $\mathbf{u}_b$ provides a copy $I^b = \{t\mathbf{u}_b : 0 \le t \le 1\}$ of the unit interval — the subspace I^b of $l^2(A)$ with the induced metric makes the mapping $t \mapsto t\mathbf{u}_b$ from the unit interval I with usual metric an isometry: For each $b \in A$,

$$|t_1 - t_2| = \sqrt{(t_1 - t_2)^2} = \sqrt{\Sigma_a (t_1 u_a^b - t_2 u_a^b)^2} = d(t_1 \mathbf{u}_b, t_2 \mathbf{u}_b).$$

These isometric copies $I^a \subset l^2(A)$ of the unit interval are the "arms" of $\cup_a I^a \subset l^2(A)$. And $\cup_a I^a$ is homeomorphic to $S(A) = (\cup_a I_a, d_S)$ — the obvious mapping $\cup_a I_a \to \cup_a I^a \subset l^2(A)$ (the isometry $I_a \to I^a$ on each arm I_a of $S(A)$ with "zero" mapping to "zero") is clearly bijective. And the ε-ball centered at the *zero* in $S(A) = (\cup_a I_a, d_S)$ maps onto the ε-ball in $\cup_a I^a \subset l^2(A)$ centered at the corresponding *zero*.

---

[1] Star spaces $S(A)$ predate the introduction of J_A. Indeed, a *star space* $S(A)$ is defined as a metric space $S(A) = (\cup_a I_a, d_S)$ where the set $\cup_a I_a$ is the *star-shaped set* obtained by identifying the zeros of a disjoint union of $|A| \ge \aleph_0$ unit intervals I_a (the ath arm), and the metric d_S is given by

$$d_S(x, y) = \begin{cases} |x - y| & \text{if } x \text{ and } y \text{ belong to the same arm} \\ |x + y| & \text{if } x \text{ and } y \text{ belong to distinct arms}. \end{cases}$$

A detailed proof that a star space $S(A)$ is homeomorphic to the J_A star space for infinite $|A|$ appears in Chapter 10. Historically, star spaces appeared as the base space in product spaces $S(A)^\infty$ that were used by Kowalsky [1957] and Nagata [1963] to construct universal spaces for metric spaces and for metric spaces of finite covering dimension $\le n$, respectively. Precise statements of the Kowalsky and Nagata theorems appear in §18.

§14 Projecting $N(A)$ onto a Cantor-Star Subspace

Let $|A| \geq 2$, select any $z \in A$, and let $A' = A \setminus \{z\}$. Then for each $b \in A'$, the Baire space $N(A)$ contains the subspace $\mathcal{C}(z, b)$ of sequences whose values lie in $\{z, b\}$. With each arm $\mathcal{C}(z, b)$ of the Cantor star, we have the projection $\pi_b : N(A) \to \mathcal{C}(z, b)$ given by

$$\pi_b : \delta = \delta_1 \delta_2 \cdots \, \mapsto \, \delta^b = \delta_1^b \delta_2^b \cdots \text{ where } \delta_j^b = \begin{cases} b & \text{if } \delta_j = b, \\ z & \text{otherwise.} \end{cases}$$

Each projection π_b is open: The basic open sets $\langle a_1, \ldots, a_n \rangle$ in $N(A)$ map onto open sets in $\mathcal{C}(z, b)$, i.e.,

$$\pi_b(\langle a_1, \ldots, a_n \rangle) = \langle a_1^b, \ldots, a_n^b \rangle \subset \mathcal{C}(z, b).$$

Each π_b is continuous: Given the open-in-$\mathcal{C}(z, b)$ set $\langle x_1, \ldots, x_n \rangle$, define

$$X_i = \begin{cases} A \setminus \{b\} & \text{if } x_i = z \\ \{b\} & \text{if } x_i = b, \end{cases} \qquad (i = 1, \ldots, n)$$

and then note that $\pi_b^{-1}(\langle x_1, \ldots, x_n \rangle) = X_1 \times \cdots \times X_n \times A \times A \times \cdots$.

Whether each projection π_b is closed or not closed depends on $|A|$: When A is finite, then $N(A)$ is compact and each π_b is necessarily closed. When A is infinite, then $N(A)$ is not compact and each π_b is not a closed mapping — let $a_1, a_2, \ldots$ be a sequence in A such that $i \neq j$ implies $a_i \neq a_j$, then

$$(1) \qquad\qquad F = \{a_1 b \overline{a_1}, a_2 a_2 b \overline{a_2}, a_3 a_3 a_3 b \overline{a_3}, \ldots\}$$

is closed in $N(A)$ while $\pi_b(F)$ is not closed in $\mathcal{C}(z, b)$. We summarize these observations with the following lemma.

14.1 Lemma (properties of π_b) *The projection $\pi_b : N(A) \to \mathcal{C}(z, b)$ is a continuous open mapping. Also, π_b is closed if and only if A is finite.*

§15 Projecting J_A onto a Star Subspace

Continuing with $|A| \geq 2$, a fixed $z \in A$, and $A' = A \setminus \{z\}$, we let $p(\overline{z}) = \zeta$ be the *zero of* J_A, and, $\beta = p(\overline{b})$ for $b \in A'$, and, $p_b : \mathcal{C}(z, b) \to I(\zeta, \beta)$ the restriction of p. With these conventions the (commutative) diagram below yields the projection $\pi_\beta : J_A \to I(\zeta, \beta)$ given by $\pi_\beta = p_b \circ \pi_b \circ p^{-1}$.

$$
\begin{array}{ccc}
N(A) & \xrightarrow{\ \pi_b\ } & \mathcal{C}(z, b) \\
{\scriptstyle p}\downarrow & & \downarrow{\scriptstyle p_b} \\
J_A & \dashrightarrow[\pi_\beta] & I(\zeta, \beta)
\end{array}
$$

To be sure, π_β is well defined because $\delta, \varepsilon \in N(A)$ and $\delta \sim \varepsilon$ yield $\pi_b(\delta) \sim \pi_b(\varepsilon)$, i.e., π_b respects the adjacent-endpoint relation, making $p_b \circ \pi_b$

constant on each fiber of p. Moreover, since p is a closed map and $p_b \circ \pi_b$ is continuous, π_β is continuous.

15.1 Lemma (properties of π_β) *Let $|A| \geq 2$, fix $z \in A$, let $b \in A' = A \setminus \{z\}$, and define $\beta = p(\bar{b})$. The projection $\pi_\beta : J_A \to I(\zeta, \beta)$ into the βth-arm of the star in J_A is a well-defined and continuous mapping. Also, π_β is closed if and only if A is finite; and, π_β is open if and only if $|A| = 2$.*

PROOF. By the arguments preceding the lemma, we only need to prove the "if and only if" claims. Consider the "closed" characterization: On the one hand, suppose A is infinite. Let F be the closed subset of $N(A)$ defined by equation (1). Then $p(F)$ is a closed subset of J_A and

$$ F' = p^{-1}p(F) = F \cup \{a_1 a_1 \bar{b}, a_2 a_2 a_2 \bar{b}, a_3 a_3 a_3 a_3 \bar{b}, \ldots\} $$

is closed in $N(A)$. But $p_b(\pi_b(F')) = \pi_\beta(p(F))$ is not closed in $I(\zeta, \beta)$. On the other hand, suppose A is finite. Then $N(A)$ is compact, making π_b, and hence π_β, closed. Now consider the "open characterization.": Let $|A| = 2$. Then $A = \{z, b\}$, $N(A) = \mathcal{C}(z, b)$, π_b is the identity map $1_{\mathcal{C}(z,b)}$, and $p = p_b$. So $\pi_\beta = (p \circ 1_{\mathcal{C}(z,b)}) \circ p^{-1} = 1_{I(\zeta,\beta)}$ is the identity on $I(\zeta, \beta)$, which is an open map. On the other hand, let π_β be open. Then for $B = \langle b, z \rangle \subset N(A)$ where $b \neq z$, let $E_B = \{\delta \in B : \delta_3 = \delta_4 = \cdots\}$, and define $G = B \setminus E_B \subset N(A)$. Now G is open and $p^{-1}p(G) = G$, making $p(G)$ open in J_A. We also have $\pi_b(G) = (\langle b, z \rangle \cap \mathcal{C}(z, b)) \setminus \{bz\bar{b}\}$. But when $|A| > 2$, then $bz\bar{z} \in \pi_b(G)$ and $p_b \circ \pi_b(G)$ is the non-open half-closed interval $[p_b(bz\bar{z}), p_b(bz\bar{b})) \subset I(\zeta, \beta)$. □

If $\theta \in J_A$, then θ has *a nonzero π_β-projection into $I(\zeta, \beta)$* when $\pi_\beta(\theta) \neq \zeta$.

15.2 Lemma *Each member θ of J_A has a nonzero π_β-projection into at most a countable number of the $I(\zeta, \beta)$.*

§16 Mapping J_A into $l^2(A')$

The next lemma exhibits a homeomorphism ψ_β from the "βth arm" $I_\beta = I(\zeta, \beta)$ of the star $\cup_\beta I_\beta \subset J_A$ onto the unit interval $[0, 1]$.

Our goal is to use the isometry $[0, 1] \equiv [\mathbf{0}, \mathbf{u}_b]$ and $I(\zeta, \beta) \xleftarrow{\psi_\beta} [0, 1]$ to homeomorphically connect $I(\zeta, \beta) \longleftrightarrow [0, 1] \equiv [\mathbf{0}, \mathbf{u}_b]$ the βth-arm of the star in J_A to the bth-arm of the star in $l^2(A')$.

16.1 Lemma (matching arms of $\cup_\beta I_\beta \subset J_A$ with arms of $\cup_b I^b \subset l^2(A')$) *Let $|A| \geq 2$, fix $z \in A$, select $b \in A' = A \setminus \{z\}$, and let $\beta = p(\bar{b})$. If $\psi_b : \mathcal{C}(z, b) \to \mathcal{C}(0, 1)$ is the homeomorphism induced by identifying "z" with "0" and "b" with "1", and, $p_b : \mathcal{C}(z, b) \to I(\zeta, \beta)$ and $p_1 : \mathcal{C}(0, 1) \to [0, 1]$ the appropriate adjacent-endpoint identification maps, then $\psi_\beta : I(\zeta, \beta) \to [0, 1]$, given by*

$$ \psi_\beta = p_1 \circ \psi_b \circ p_b^{-1}, $$

is a homeomorphism that makes the following diagram commutative

$$
\begin{array}{ccccc}
N(A) & \xrightarrow{\;\pi_b\;} & \mathcal{C}(z,b) & \xrightarrow{\;\psi_b\;} & \mathcal{C}(0,1) \\
{\scriptstyle p}\big\downarrow & & {\scriptstyle p_b}\big\downarrow & & {\scriptstyle p_1}\big\downarrow \\
J_A & \xrightarrow[\;\pi_\beta\;]{} & I(\zeta,\beta) & \dashrightarrow[\;\psi_\beta\;]{} & [0,1].
\end{array}
$$

PROOF. Since $p_1 \circ \psi_b$ is a surjection that is constant on each fiber of p_b, we see that ψ_β is a well-defined surjection. Since p_b is closed and $p_1 \circ \psi_b$ is continuous it follows that ψ_β is continuous; and since $p_1 \circ \psi_b$ is closed and p_b is continuous ψ_β is closed. It is also clear, since ψ_b is injective and respects "$\sim$", that ψ_β is injective. So ψ_β is a homeomorphism. Finally, the diagram commutes because the left and right "square subdiagrams" commute. $\square$

So for each $\theta \in J_A$, we may use

$$
\theta \xrightarrow{\;\pi_\beta\;} \theta^\beta \xrightarrow{\;\psi_\beta\;} \theta_b \qquad \text{where } \beta = p(\bar{b}) \text{ for } b \in A',
$$

to define a tuple $(\theta_b) = (\theta_b)_{b \in A'}$ of numbers such that $0 \le \theta_b \le 1$. Since it turns out that each $\Sigma_b[\theta_b]^2 < \infty$, we have a mapping $J_A \to l^2(A')$.

§17 No-Carry Characterization of $\omega^{A'}$

Since each θ_b satisfies $0 \le \theta_b \le 1$, and since at most a countable number of the θ_b are nonzero (Lemmas 15.2 and 16.1), if $\sum_{b \in A'}[\theta_b]^2 < \infty$, then $(\theta_b) \in l^2(A')$. We show even more in the following lemma.

17.1 Lemma (no-carry property associated with J_A) *Let $|A| \ge 2$, let $z \in A$ be fixed, and let $A' = A \setminus \{z\}$. Then for each $\theta \in J_A$, we may choose binary expansions*

$$
.x_1^b x_2^b \cdots = \theta_b = \psi_\beta \circ \pi_\beta(\theta) \qquad (b \in A'; \beta = p(\bar{b}))
$$

such that $x_i^b = 1$ implies ($x_i^c = 0$ for each $c \in A' \setminus \{b\}$). Moreover, from this no-carry property, $\Sigma_{b \in A'}\theta_b \le 1$, and consequently $(\theta_b)_{b \in A'} \in l^2(A')$.

PROOF. Let $\phi : J_A \to N(A)$ be a "choice function" such that $\phi(\theta) = a_1 a_2 \cdots \in \theta$ selects a member $\phi(\theta)$ of the equivalence class $\theta \in J_A$. Then using the commutative diagram in Lemma 16.1, we have

$$
\theta_b = \psi_\beta \circ \pi_\beta(\theta) = p_1 \circ \psi_b \circ \pi_b(a_1 a_2 \cdots) = p_1 \circ \psi_b(a_1^b a_2^b \cdots) = \sum_{i=1}^{\infty} x_i^b/2^i,
$$

where

(2) $x_i^b = 1 \Leftrightarrow a_i^b = b \Leftrightarrow a_i = b$ and $x_i^b = 0 \Leftrightarrow a_i^b \neq b \Leftrightarrow a_i \neq b.$

Now let $i \in \{1, 2, \ldots\}$ be fixed and focus on the ith term $a_i \in A$ of $a_1 a_2 \cdots$. Since $a_i \in A$, there are two cases: First, $a_i = z$. Then $x_i^b = 0$ for every $b \in A'$. Second, there is exactly one $b \in A'$ such that $a_i = b$. For this unique b, equation (2) yields $x_i^b = 1$ and $x_i^c = 0$ for every $c \in A' \setminus \{b\}$. Thus (θ_b) satisfies the no-carry property.

So now consider the sum $\Sigma_{b \in A'} \theta_b$. Then $\Sigma_{b \in A'} \theta_b \leq 1$ because there is at most one $b \in A'$ such that the binary expansion of θ_b has the binary digit 1 in its first position, thereby contributing $1/2$ to $\Sigma_{b \in A'} \theta_b$. Similarly, the no-carry condition ensures that there is at most one $b \in A'$ such that the binary expansion of θ_b has the binary digit 1 in the second position, thereby contributing $1/4$ to $\Sigma_{b \in A'} \theta_b$, and so on (i.e., since there are at most a countable number of coordinates $\theta_b > 0$, it follows that $\sum_{b \in A'} \theta_b = \sum_{\theta_b \neq 0} \theta_b \leq \sum_{i=1}^{\infty} 1/2^i = 1$). And finally, it follows from ($0 \leq \theta_b \leq 1$ implies $0 \leq [\theta_b]^2 \leq \theta_b$) that $\sum_{b \in A'} [\theta_b]^2$ converges, i.e., that $(\theta_b)_{b \in A'} \in l^2(A')$. $\square$

Lemma 17.1 allows us to define $\omega^{A'}$ as a subset of $l^2(A')$.

17.2 Definition ($\omega^{A'}$)   Let $|A| \geq 2$, let $z \in A$ be fixed, and let $A' = A \setminus \{z\}$. Then define

$$\omega^{A'} = \{(\theta_b) \in l^2(A') : \theta \in J_A\}$$

where for each $b \in A'$, the bth-coordinate $\theta_b = \psi_\beta \circ \pi_\beta(\theta)$ with $\beta = p(\bar{b})$.

17.3 Theorem (no-carry characterization of $\omega^{A'}$)   *For the subspace $\omega^{A'}$ of $l^2(A')$ defined above, we have $\omega^{A'} = \{(x_b) \in l^2(A') : \text{each } x_b \text{ has a binary expansion } x_b = .x_1^b x_2^b \cdots, \text{ and, } x_j^b = 1 \implies x_j^c = 0 \text{ when } c \neq b\}$. In other words, $(x_b) \in \omega^{A'}$ if and only if (x_b) satisfies the no-carry condition.*

PROOF.   From Lemma 17.1 and Definition 17.2, $(x_b) \in \omega^{A'}$ implies (x_b) satisfies the no-carry condition. Conversely, suppose $(x_b) \in l^2(A')$ satisfies the no-carry condition. We construct a $\theta \in J_A$ such that for each $b \in A'$,

$$\theta \overset{\psi_\beta \circ \pi_\beta}{\longmapsto} \theta_b = x_b = .x_1^b x_2^b \cdots \qquad (\beta = p(\bar{b})).$$

To begin, let $i \in \{1, 2, \ldots\}$ be fixed. Then use the following rules to define the coordinate a_i of $\delta = a_1 a_2 \ldots \in N(A)$.

(3) $(a_i = b \iff x_i^b = 1)$ and $(a_i = z \iff x_i^b = 0$ for every $b \in A')$.

These rules are well defined because the no-carry condition on (x_b) ensures that there is at most one b such that $x_i^b = 1$. We now define θ as the equivalence class in J_A such that $\delta \in \theta$. Next, let $b \in A'$ be arbitrary but fixed, and let the index i range over the values $1, 2, \ldots$. If $x_b = 0$, then for every index i, we have $x_i^b = 0$, which forces (equation (3)) every $a_i \neq b$, which, in turn, shows that $\pi_b(\delta) = \bar{z}$. So the Lemma 16.1 diagram yields

$$\theta_b = \psi_\beta \circ \pi_\beta(\theta) = \psi_\beta \circ p_b \circ \pi_b(\delta) = \psi_\beta(\zeta) = 0 = x_b.$$

In the other case, there is an $x_i^b \neq 0$, which means that $x_i^b = 1$, making $a_i = b$. Indeed, the index set K of values of i such that $x_i^b = 1$ is also the index set of values of i such that $a_i = b$. So the Lemma 16.1 diagram yields

$$\theta_b = \psi_\beta \circ \pi_\beta(\theta) = \psi_\beta \circ p_b \circ \pi_b(\delta) = \psi_\beta \circ p_b(a_1^b a_2^b \cdots) = \Sigma_{i \in K} 1/2^i = x_b.$$

It follows that $\theta_b = x_b$ for each $b \in A'$, making $(\theta_b) = (x_b)$. □

17.4 Corollary *The following restrictions on $(x_b) \in l^2(A')$ are equivalent:*

    (i) $(x_b) \in \omega^{A'}$.
    (ii)   *there is a $\theta \in J_A$ such that $(\theta_b) = (x_b)$.*
    (iii) *(x_b) satisfies the no-carry condition.*

PROOF. It follows from the definition of $\omega^{A'}$ (Definition 17.2) that "(ii) ⇔ (i)." And the last sentence in Theorem 17.3 is the statement, "(i) ⇔ (iii)."□

§18 Comments

The "no-carry property" (Figure 11.1) has roots in the work of Kummer [1852], whose number-theoretical criterion, Kummer's Criterion, exposes a Sierpiński triangle pattern of even binomial coefficients in Pascal's triangle. A discussion of how Kummer's work relates to the Sierpiński triangle (also called Sierpiński's gasket) may be found in Peitgen, Jürgens, and Saupe [1992]. Moreover, Chapter 5, Section 5.4, of Peitgen, Jürgens, and Saupe [1992] contains a derivation of the no-carry characterization of the Sierpiński triangle.

In our Chapter 4 we shall show that the mapping $J_A \to \omega^{A'}$ given by $\theta \mapsto (\theta_b)$ is a homeomorphism. Thus, the no-carry characterization of Sierpiński's triangle is a special case ($|A| = 3$) of Theorem 17.3. Moreover, Theorem 17.3 coupled with the homeomorphism $J_A \to \omega^{A'}$ gives meaning to the phrase "no-carry property of J_A."

Historically, the no-carry property in the context of J_A grew out of two articles: Lipscomb and Perry [1992] and, independently, Milutinović [1992].

Milutinović introduced a subspace M_A of the standard simplex $\Delta^A \subset l^2(A)$ that is homeomorphic to J_A. His construction generalized Sierpiński's original [1915] construction that used Δ^2. (Both schemes are detailed in Chapter 4, and, the standard simplex Δ^A is developed in Appendix 2.)

Milutinović [1992] also proved the following proposition ($\Sigma(\tau) = M_A$)

    PROPOSITION 7. *$y = (y_\lambda) \in \Sigma(\tau) \iff$ there is a sequence (μ_n), such that $\forall \lambda$, $y_\lambda = 0, \delta_{\lambda,\mu_1} \cdots \delta_{\lambda,\mu_n} \cdots$*

where his binary expansion "$0, \delta_{\lambda,\mu_1} \cdots \delta_{\lambda,\mu_n} \cdots$" of "$y_\lambda$" is given meaning by requiring that $\delta_{\lambda,\mu_j} = 1$ when $\lambda = \mu_j$, and, that $\delta_{\lambda,\mu_j} = 0$ otherwise.

The no-carry property of M_A, and hence J_A, is an obvious corollary to Milutinović's Proposition 7, but he made no mention of the fact that $\delta_{\lambda,\mu_i} = 1$ implies $\delta_{\lambda',\mu_i} = 0$ for all $\lambda' \neq \lambda$. That is, he made no explicit mention of the no-carry property.

Likewise, Lipscomb and Perry [1992] did not explicitly state the no-carry *characterization* of $\omega^{A'}$. They did state (in their Lemma 7), "... for each subscript $i \in \{1, 2, \ldots\}$ there is at most one $b \in A$ with $x_i^b = 1$." In other words, $(x_b) \in \omega^{A'}$ implies (x_b) satisfies the no-carry condition. But they did not prove the converse. So Theorem 17.3 seems to be the first explicit statement that a homeomorph of J_A satisfies the no-carry characterization.

In this chapter we unified the finite and infinite A no-carry characterizations of J_A. (If $|A| = 1$, then ω^0 is a point and the no-carry property of J_A is trivial.) In addition, since $|A'| < |A|$ when A is finite, we used the A' notation so that the dimension of the imbedding superspace was kept to a minimum. If A is infinite, however, an application of Proposition A9.6 (A infinite) yields $\omega^{A'} =_t \omega^A$, and then an application of Milutinović's Proposition 7 yields $M_A = \omega^A$.

Historically, star spaces also played a fundamental role in constructing universal spaces for certain classes of metric spaces.

18.1 Theorem (Kowalsky [1957]) *A topological space R is metrizable if and only if it can be imbedded in a countable product of star spaces.*

And the corresponding theorem for finite n-dimensional metric spaces dovetails nicely with Kowalsky's Theorem.

18.2 Theorem (Nagata [1963]) *A metric space R has (covering) dimension $\leq n$ if and only if it can be imbedded in the subset K_n of a countable product P of star spaces, where we denote by K_n the set of points in P at most n of whose nonvanishing coordinates are rational.*

CHAPTER 4

Imbedding J_A in Hilbert Space

In this chapter we focus on the surjection $J_A \to \omega^{A'} \subset l^2(A')$ that was introduced in §16, proving that $\theta \mapsto (\theta_b) \in \omega^{A'}$ is an imbedding. We begin with a characterization of the adjacent-endpoint relation, which is used to prove that $\theta \mapsto (\theta_b)$ is also injective. We then devote §20 to the proof that this bijection is also a homeomorphism.

In §21 we review Sierpiński's original formulation of his triangle, and then in §22 we provide the parallel formulation of Milutinović's $M_A \subset l^2(A)$, which is another homeomorph of J_A.

For the proof that $J_A \to \omega^{A'}$ is an imbedding, we continue to follow Lipscomb and Perry [1992]; for Sierpiński's formulation of his triangle, we follow Sierpiński [1915]; and for Milutinović's parallel formulation of M_A, we follow Milutinović [1992] [1993].

§19 Characterization of the Adjacent-Endpoint Relation

This section contains the statement and proof of Theorem 19.2. First, however, we state the following lemma, whose proof is pedestrian.

19.1 Lemma *Let $|A| \geq 2$, let $z \in A$ be fixed, let $A' = A \setminus \{z\}$, and, let $\delta = \delta_1 \delta_2 \cdots$ and $\varepsilon = \varepsilon_1 \varepsilon_2 \cdots$ be members of $N(A)$. Moreover, for each $b \in A'$, denote $\pi_b(\delta)$ and $\pi_b(\varepsilon)$ (and their expansions) as*

$$\pi_b(\delta) = \delta^b = \delta_1^b \delta_2^b \cdots \qquad and \qquad \pi_b(\varepsilon) = \varepsilon^b = \varepsilon_1^b \varepsilon_2^b \cdots .$$

Then

(1) $(\delta_i = b \Leftrightarrow \delta_i^b = b)$ *and* $(\delta_i^b = \varepsilon_i^b \text{ for each } b \in A' \Longrightarrow \delta_i = \varepsilon_i)$

where $i \in \{1, 2, \ldots\}$ is fixed.

In the proof of the following theorem, note that the "CASE 1" part shows that $\delta^b = \varepsilon^b$ for each $b \in A'$ implies $\delta = \varepsilon$. In other words, when points δ and ε in $N(A)$ project onto the same "values" on each of the $|A'|$ arms of a Cantor star $S_C(A')$, then those points are equal.

19.2 Theorem (characterization of $\delta \sim \varepsilon$ in $N(A)$) *Let $|A| \geq 2$, $z \in A$ be fixed, $A' = A \setminus \{z\}$, and $\delta, \varepsilon \in N(A)$. Then $\delta \sim \varepsilon$ if and only if $\pi_b(\delta) \sim \pi_b(\varepsilon)$ for every $b \in A'$.*

S.L. Lipscomb, *Fractals and Universal Spaces in Dimension Theory*,
DOI 10.1007/978-0-387-85494-6_4, © Springer Science+Business Media, LLC 2009

PROOF. If $A = \{z, b\}$ has size two, then the proof is trivial because there is only one projection π_b which is the identity mapping. So suppose $|A| \geq 3$. The proof that $\delta \sim \varepsilon$ implies $\pi_b(\delta) \sim \pi_b(\varepsilon)$ for every $b \in A'$ is straightforward. For the proof of the converse, we adopt the concise notation in Lemma 19.1.

CASE 1: ($\delta^b = \varepsilon^b$ for each $b \in A'$). Then $\delta = \varepsilon$ from (1), a fortiori, $\delta \sim \varepsilon$.

CASE 2: ($\delta^b \neq \varepsilon^b$ for some $b \in A'$). Then since $\delta^b \sim \varepsilon^b$ we may assume,

$$(2) \qquad \delta^b = \delta^b_1 \cdots \delta^b_{t-1} z \overline{b} \ \sim \ \varepsilon^b = \varepsilon^b_1 \cdots \varepsilon^b_{t-1} b \overline{z} \quad (\delta^b_i = \varepsilon^b_i, \ i < t).$$

So $\delta = \delta_1 \cdots \delta_{t-1} \delta_t \overline{b}$ and $\varepsilon = \varepsilon_1 \cdots \varepsilon_{t-1} b \varepsilon_{t+1} \varepsilon_{t+2} \cdots$ where each $\varepsilon_{t+k} \neq b$.

SUBCASE 2.1: ($\delta_t = z$). Then $\delta = \delta_1 \cdots \delta_{t-1} z \overline{b}$, and first, we consider the possibility that $a = \varepsilon_{t+k} \neq z$ for some $k \geq 1$. Then $a \in A' \setminus \{b\}$, and

$$\delta^a = \delta^a_1 \cdots \delta^a_{t-1} \overline{z} \quad \text{while} \quad \varepsilon^a = \varepsilon^a_1 \cdots \varepsilon^a_{t-1} z \varepsilon^a_{t+1} \cdots \varepsilon^a_{t+k-1} a \varepsilon^a_{t+k+1} \cdots$$

which contradicts $\delta^a \sim \varepsilon^a$. Thus $z = \varepsilon_{t+1} = \varepsilon_{t+2} = \cdots$, and so $\delta_t = z$ yields

$$(3) \qquad \delta = \delta_1 \cdots \delta_{t-1} z \overline{b} \quad \text{and} \quad \varepsilon = \varepsilon_1 \cdots \varepsilon_{t-1} b \overline{z}.$$

Now we use (3) to calculate that

$$(4) \quad \delta^c = \delta^c_1 \cdots \delta^c_{t-1} \overline{z} \ \sim \ \varepsilon^c = \varepsilon^c_1 \cdots \varepsilon^c_{t-1} \overline{z} \quad (\delta^c_i = \varepsilon^c_i, \ i < t; c \in A' \setminus \{b\}).$$

Thus, (4) and (2) show that $\varepsilon^d_i = \delta^d_i$ for each $i < t$ and all $d \in A'$. An application of (1) in Lemma 19.1 then yields $\varepsilon_i = \delta_i$ for all $i < t$. These $t - 1$ equalities and (3) show that $\delta \sim \varepsilon$ when $\delta_t = z$.

SUBCASE 2.2: ($\delta_t = a \neq z$). The expansion of δ that follows (2) yields $\delta = \delta_1 \cdots \delta_{t-1} a \overline{b}$, and therefore

$$(5) \qquad \delta^a = \delta^a_1 \cdots \delta^a_{t-1} a \overline{z} \ \sim \ \varepsilon^a = \varepsilon^a_1 \cdots \varepsilon^a_{t-1} z \overline{a} \quad (\delta^a_i = \varepsilon^a_i, \ i < t).$$

So $a = \varepsilon_{t+1} = \varepsilon_{t+2} \cdots$. Then the expansions of δ and ε that follow (2) yield

$$(6) \qquad \delta = \delta_1 \cdots \delta_{t-1} a \overline{b} \quad \text{and} \quad \varepsilon = \varepsilon_1 \cdots \varepsilon_{t-1} b \overline{a}.$$

So for each $c \in A' \setminus \{a, b\}$, we may use (6) to calculate

$$(7) \qquad \delta^c = \delta^c_1 \cdots \delta^c_{t-1} \overline{z} \ \sim \ \varepsilon^c = \varepsilon^c_1 \cdots \varepsilon^c_{t-1} \overline{z} \quad (\delta^c_i = \varepsilon^c_i, \ i < t).$$

We may also use (6) to calculate

$$(8) \qquad \delta^a = \delta^a_1 \cdots \delta^a_{t-1} a \overline{z} \ \sim \ \varepsilon^a = \varepsilon^a_1 \cdots \varepsilon^a_{t-1} z \overline{a} \quad (\delta^a_i = \varepsilon^a_i, \ i < t).$$

Thus, (8), (7), and (2) combine with (1) of Lemma 19.1 to show that $\delta_i = \varepsilon_i$ for $1 \leq i \leq t - 1$, which, in turn, coupled with (6), shows once again, and finally, that $\delta \sim \varepsilon$. $\qquad \square$

§20 The Mapping $f : J_A \to \omega^{A'}$

Let $|A| \geq 2$, let $z \in A$ be fixed, and let $A' = A \setminus \{z\}$. Then recall from §16 that $f : J_A \to \omega^{A'}$ is implicit in the definition of

(9) $$\omega^{A'} = \{(\theta_b) \in l^2(A') : \theta \in J_A\}.$$

Indeed, using the commutative diagram in Lemma 16.1 and "$[0,1] \equiv [\mathbf{0}, \mathbf{u}_b]$" to indicate the obvious isometry $t \mapsto t\mathbf{u}_b$ from $[0,1]$ to $[\mathbf{0}, \mathbf{u}_b]$, we have

$$
\begin{array}{ccccc}
N(A) & \xrightarrow{\pi_b} & \mathcal{C}(z,b) & \xrightarrow{\psi_b} & \mathcal{C}(0,1) \\
\downarrow p & & \downarrow p_b & & \downarrow p_1 \\
J_A & \xrightarrow[\pi_\beta]{} & I(\zeta, \beta) & \xrightarrow[\psi_\beta]{} & [0,1] \equiv [\mathbf{0}, \mathbf{u}_b] \subset l^2(A')
\end{array}
$$

where for $b \in A'$, the point $\beta = p(\bar{b})$ is the endpoint of the I_β arm of the star in J_A and the p-image of the endpoint $\bar{b}$ of the bth arm $\mathcal{C}(z,b)$ of the Cantor star in $N(A)$. With this structure in mind, then, we may view the bth coordinate θ_b of (θ_b) as the number given by $\theta_b = \psi_\beta \circ \pi_\beta(\theta)$.

So each $\theta \in J_A$ has $|A'|$ projections $\theta \mapsto \theta^\beta \in I(\zeta, \beta)$ into the star in J_A. And each θ^β bijectively corresponds to a number $\theta_b \in [0,1]$ on the bth arm $[0, \mathbf{u}_b]$ of the star in $l^2(A')$.

The numbers θ_b, $b \in A'$, are then used as coordinates of $(\theta_b) \in \omega^{A'} \subset l^2(A')$, which is the image $f(\theta) = (\theta_b)$ of θ. From this viewpoint, f is the "$l^2(A')$ synthesis" of the $|A'|$-mappings $\theta \mapsto \theta^\beta \mapsto \theta_b$.

20.1 Lemma ($f : J_A \to \omega^{A'}$ **is bijective**) *Let f be given by $\theta \mapsto f(\theta) = (\theta_b)$. Then f is a bijection.*

PROOF. By definition (9) of $\omega^{A'}$, the map f is surjective. To see that f is injective, let $f(\theta) = f(\rho)$, i.e., $\theta_b = \rho_b$ for each $b \in A'$. It suffices to show that $\delta \in p^{-1}(\theta)$ and $\varepsilon \in p^{-1}(\rho)$ imply $\delta \sim \varepsilon$: Since each $\psi_\beta : I(\zeta, \beta) \to [0,1]$ is bijective, we have

$$\theta_b = \rho_b \in [\mathbf{0}, \mathbf{u}_b] \quad \text{if and only if} \quad \theta^\beta = \rho^\beta \in I(\zeta, \beta).$$

And since the diagram above is commutative, we also have

(10) $$p_b \circ \pi_b(\delta) = \theta^\beta = \rho^\beta = p_b \circ \pi_b(\varepsilon) \quad (\text{whenever } \beta = p(\bar{b})).$$

But (10) yields

(11) $$\pi_b(\delta) \sim \pi_b(\varepsilon) \quad (\text{for every } b \in A').$$

Together, (11) and Theorem 19.2 show that $\delta \sim \varepsilon$. $\square$

To show that $f : J_A \to \omega^{A'}$ is both continuous and open, we begin with Figure 20.2 where we introduce, for each $b \in A'$, the mapping g_b.

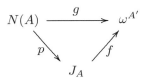

Fig. 20.2 For $b \in A'$, the mapping g_b maps $N(A)$ into $[0, 1]$.

Using the g_b mappings, we define $g : N(A) \to \omega^{A'}$ by $g = f \circ p(\delta) = (g_b(\delta))$, i.e., if $\theta = p(\delta)$, then each $g_b(\delta) = \theta_b$ and $g(\delta) = (g_b(\delta)) = (\theta_b) = f(\theta)$.

$$N(A) \xrightarrow{\quad g \quad} \omega^{A'}$$

Since f is injective and $g = f \circ p$ we see that g and p have the same fibers

$$g^{-1}\left(f(\theta)\right) = p^{-1} \circ f^{-1}\left(f(\theta)\right) = p^{-1}(\theta).$$

20.3 Proposition *The mapping $g : N(A) \to \omega^{A'}$ is continuous.*

PROOF. Let $\delta = a_1 a_2 \cdots \in N(A)$. It suffices to show that if $\delta_n \to \delta$ in $N(A)$, then $d(g(\delta_n), g(\delta)) \to 0$ where d is the $l^2(A')$ metric. To begin, let $C_\delta = \{a_1, a_2, \ldots\}$, let $\delta_n = a_{n1} a_{n2} \ldots$, and let $C_{\delta_n} = \{a_{n1}, a_{n2}, \ldots\}$. Next, let $k > 0$ be fixed, and define three subsets of A':

$$A_k = A' \cap \{a_1, \ldots, a_k\}; \ A'_k = A' \cap (C_\delta \setminus A_k); \ \text{and} \ A'_{nk} = A' \cap (C_{\delta_n} \setminus C_\delta).$$

These sets are pairwise disjoint, and, $b \in A' \setminus (A_k \cup A'_k \cup A'_{nk}) \Rightarrow g_b(\delta_n) = g_b(\delta) = 0$. Since $\delta_n \to \delta$, an N exists where $n > N$ implies $a_{ni} = a_i$ for $i \leq k$. So $n > N$, and, $g_b(\delta) = \Sigma_1^\infty x_i^b / 2^i$ where $x_i^b = 1 \Leftrightarrow a_i = b$, yield

$$\begin{aligned}
(d(g(\delta_n), g(\delta)))^2 &= \textstyle\sum_{b \in A'} |g_b(\delta_n) - g_b(\delta)|^2 \\
&\leq \textstyle\sum_{b \in A_k} |\cdots| + \sum_{b \in A'_k} |\cdots| + \sum_{b \in A'_{nk}} |\cdots| \\
&\leq \frac{k}{2^k} + \frac{1}{2^k} + \frac{1}{2^k} = \frac{k+2}{2^k},
\end{aligned}$$

which completes the proof. $\square$

20.4 Lemma *The mapping $f : J_A \to \omega^{A'}$ is continuous.*

PROOF. Let $G \subset \omega^{A'}$ be open. Then since $g : N(A) \to \omega^{A'}$ is continuous, $g^{-1}(G)$ is open in $N(A)$. But a subset of $N(A)$ is a g-inverse set if and only if it is a p-inverse set. Since p is a quotient mapping, $p(g^{-1}(G))$ is therefore open in J_A. Thus, $f^{-1}(G) = p(g^{-1}(G))$ is open in J_A. $\qquad\square$

20.5 Lemma *The mapping $f^{-1} : \omega^{A'} \to J_A$ is continuous.*

PROOF. For the sequence $\delta_1 \delta_2 \cdots$ and point δ in $N(A)$, suppose $g(\delta_n) \to g(\delta)$ in $\omega^{A'}$. The corresponding g fibers are given by

$$R = g^{-1} g(\delta) = p^{-1} p(\delta) \quad \text{and} \quad R_n = g^{-1} g(\delta_n) = p^{-1}(p(\delta_n)) \qquad (n = 1, 2 \ldots).$$

It suffices to show that "$R_n \to R$"; i.e., for every open p-inverse set $V \supset R$, there is an N such that $n > N$ implies $R_n \subset V$. (Lemmas 3.3 and 3.4 show that the open p-inverse sets form a local basis at R.)

So let the p-inverse set $V \supset R$ be fixed, and suppose "$R_n \not\to R$." Then there is an infinite subset M of N and a sequence $\{\varepsilon_m : m \in M\}$ of points each of which satisfies $\varepsilon_m \in R_m \setminus V$. There are two possibilities:

CASE I. The sequence $\{\varepsilon_m : m \in M\}$ has a convergent subsequence $\varepsilon_{m_1}, \varepsilon_{m_2} \cdots$. Since $\varepsilon_{m_k} \to \varepsilon$, we have $\varepsilon \notin V \supset R$. So $g(\varepsilon) \neq g(\delta)$, and we may select disjoint open $E, D \subset \omega^{A'}$ with $g(\varepsilon) \in E$ and $g(\delta) \in D$. Then, since $\varepsilon \in p^{-1} p(\varepsilon) \subset g^{-1}(E)$, there is an open p-inverse set W such that

$$\varepsilon \in p^{-1} p(\varepsilon) \subset W \subset g^{-1}(E).$$

Now $\varepsilon_{m_k} \to \varepsilon$ implies that the sequence $\{\varepsilon_{m_k}\}$ is eventually in W, and since W is a p-inverse set, the sequence $\{\delta_{m_k}\}$ where $\delta_{m_k} \sim \varepsilon_{m_k}$ for each k is also eventually in W. But then $\{g(\delta_{m_k})\}$ is eventually in E, which, because $g(\delta_{m_k}) \to g(\delta)$ implies $\{g(\delta_{m_k})\}$ is eventually in D, contradicts $g(\delta_n) \to g(\delta)$.

CASE II. The sequence $\{\varepsilon_m : m \in M\}$ has no convergent subsequence. For $\varepsilon_m = \varepsilon_{m(1)} \varepsilon_{m(2)} \cdots$, an $i \geq 1$ exists where $\{\varepsilon_{m(i)} : m \in M\}$ is infinite.[1]

For such an i, a subsequence $\{\varepsilon_{m_k}\}$ of $\{\varepsilon_m\}$ exists where $\{\varepsilon_{m_k}(i)\}$ is infinite and contains neither z nor any member of $\{\delta_1, \ldots, \delta_{i+1}\}$ nor (if $|R| = 2$) any member of the first $i + 1$ components of the endpoint adjacent to δ. But then for any k and any $b = \varepsilon_{m_k}(i)$,

$$1/2^{i+1} \leq |g_b(\varepsilon_{m_k}) - g_b(\delta)| \leq d(g(\varepsilon_{m_k}), g(\delta)),$$

which contradicts $g(\delta_n) \to g(\delta)$. $\qquad\square$

Applications of Lemma 20.1, Lemma 20.4, and Lemma 20.5 yield the following theorem.

---

[1] Otherwise, since $\{\varepsilon_{m(1)} : m \in M\}$ is finite and $\{\varepsilon_m\}$ is infinite, we may select an infinite $M_1 \subset M$ such that $\varepsilon_k, \varepsilon_\ell \in \{\varepsilon_m : m \in M_1\}$ implies $\varepsilon_k(1) = \varepsilon_\ell(1)$. Select $m_1 \in M_1$. Then similarly construct an infinite $M_2 \subset M_1$, and select $m_2 \in M_2$, continuing ad infinitum. The resulting subsequence $\{\varepsilon_{m_k} : k = 1, 2, \ldots\}$ converges.

20.6 Theorem (J_A **is homeomorphic to** $\omega^{A'}$) *Let* $|A| \geq 2$, *let* $z \in A$ *be fixed, and let* $A' = A \setminus \{z\}$. *Then the mapping* $f : J_A \to \omega^{A'}$ *given by* $\theta \mapsto (\theta_b)_{b \in A'}$ *is a homeomorphism.*

§21 Sierpiński's Recursive Construction

In his 1915 article *Sur une courbe dont tout point est un point de ramification*, Sierpiński introduced his now famous fractal. He conveyed his ideas and indexing in two illustrations, a partial rendition of which appears in Figure 21.1.

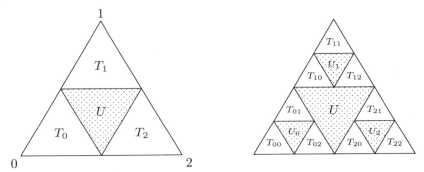

Fig. 21.1 Sierpiński's 1915 inductive construction.

With his first two (illustrated) steps we clearly see his faithful indexing: Starting with the 2-simplex Δ_2, we see the (initial) *open cuts* — first U and then, at the second step, the additional open cuts U_0, U_1, and U_2, and, we also see the (initial) *residual closed 2-simplexes* — first T_0, T_1, and T_2, and then, at the second step,

$$T_{\lambda_1,\lambda_2} \qquad (\lambda_1 = 0,1,2; \lambda_2 = 0,1,2).$$

In other words, we obtain the closed set $F_1 = \Delta_2 \setminus U = T_0 \cup T_1 \cup T_2 = \cup_{\lambda_1} T_{\lambda_1}$ and the closed set

$$F_2 = \Delta_2 \setminus (U \cup U_0 \cup U_1 \cup U_2) = \cup_{\lambda_1 \lambda_2} T_{\lambda_1 \lambda_2}.$$

And thusly Sierpiński introduced his inductive construction of nested $F_1 \supset F_2 \supset \cdots$ closed subsets of a 2-simplex. He subsequently defined his triangle as the intersection

(12) $F = F_1 \cap F_2 \cap \cdots = (\cup_{\lambda_1} T_{\lambda_1}) \cap (\cup_{\lambda_1,\lambda_2} T_{\lambda_1,\lambda_2}) \cap \cdots .^2$

---

[2]Sierpiński's indexing of the $T_{\lambda_1 \ldots \lambda_n}$ corresponds to the indexing $< \lambda_1, \ldots, \lambda_n > \in \mathcal{B}_n$ of basis elements of Baire's space $N(\{0,1,2\})$, and hence also the indexing of the corresponding 3^n copies $p(< \lambda_1, \ldots, \lambda_n >)$ of J_3.

21.2 EXAMPLE.   Let Δ^2 denote a 2-simplex with vertices 0, 1, and 2 as illustrated in Figure 21.1. Then the 2-simplex T_0 may be viewed as the image $w_0(\Delta^2)$ where $w_0 : \Delta^2 \to \Delta^2$ is a contraction (a scaling) of Δ^2 by 1/2 toward the vertex labeled 0. Similarly, each 2-simplex T_1 and T_2 may be viewed, respectively, as $w_1(\Delta^2) = T_1$ and $w_2(\Delta^2) = T_2$ where w_1 and w_2 are contractions of Δ^2 by 1/2 toward the vertices 1 and 2, respectively. In this example, the set $W = \{w_0, w_1, w_2\}$ is a *finite* iterated function system. The reason for the adjective "iterated" is justified by considering Sierpiński's indexing:

$$F_1 = \cup_{\lambda_1} T_{\lambda_1} = T_0 \cup T_1 \cup T_2 = w_0(\Delta^2) \cup w_1(\Delta^2) \cup w_2(\Delta^2)$$

and the "length-2 iterated compositions" yield

$$F_2 = \cup_{\lambda_1, \lambda_2} T_{\lambda_1 \lambda_2} = \cup_{\lambda_1 \lambda_2} \left(w_{\lambda_1} \circ w_{\lambda_2}(\Delta^2) \right).$$

For example, if $\lambda_1 = 1$ and $\lambda_2 = 0$, then $w_{\lambda_1} \circ w_{\lambda_2}(\Delta^2) = w_1(w_0(\Delta^2)) = T_{10}$, which appears in the right-side illustration of Figure 21.1. In general,

(13) $$F_n = \cup_{\lambda_1 \cdots \lambda_n} T_{\lambda_1 \cdots \lambda_n} = \cup_{\lambda_1 \cdots \lambda_n} w_{\lambda_1} \circ \cdots \circ w_{\lambda_n}(\Delta^2).$$

It follows, by substituting (13) into (12), that

$$F = F_1 \cap F_2 \cap \cdots = (\cup_{\lambda_1} w_{\lambda_1}(\Delta^2)) \cap (\cup_{\lambda_1, \lambda_2} w_{\lambda_1} \circ w_{\lambda_2}(\Delta^2)) \cap \cdots .$$

Moreover, since $F_1 \supset F_2 \supset \cdots$, we have $F = \cap_{n \geq 1} F_n = \cap_{n \geq 2} F_n = \cdots$ where $w_0(F_n) \cup w_1(F_n) \cup w_2(F_n) = F_{n+1}$. And since each $w \in W$ is one-to-one, we have $w(\cap_{n \geq 1} F_n) = \cap_{n \geq 1} w(F_n)$. These facts, combined with the fact that the intersection "$\cap_n$" distributes over "$\cup$", yield

$$w_0(F) \cup w_1(F) \cup w_2(F) = \cap_n (w_0(F_n) \cup w_1(F_n) \cup w_2(F_n)) = \cap_{n \geq 2} F_n = F.$$

In other words, Sierpiński's triangle F is the unique compact set that is the "fixed point" of the Hutchinson operator $X \to w_0(X) \cup w_1(X) \cup w_2(X)$, which is equivalent to saying that F is the attractor of the IFS W.[3]

§22 Milutinović's Subspace M_A of Hilbert Space

Example 21.2 places Sierpiński's construction within the context of a finite IFS $W = \{w_0, w_1, w_2\}$. Here, we show how Milutinović extended W to an infinite IFS $\{w_a : a \in A\}$ and thereby obtained his $M_A \subset l^2(A)$.

Let us recall (Appendix 2) that the standard orthonormal basis $\{\mathbf{u}_a : a \in A\}$ of $l^2(A)$ consists of those vectors $\mathbf{u}_b = (u_a^b) \in l^2(A)$, $b \in A$, specified by $u_a^b = 0$ when $a \neq b$ and $u_b^b = 1$; and that the *standard simplex* $\Delta^A \subset l^2(A)$ is

---

[3]Compare $F = w_0(F) \cup w_1(F) \cup w_2(F) \subset \Delta^2 \subset \mathbb{R}^3$ with the development in §8, where $\Delta_2 \subset \mathbb{R}^2$ is the superspace of the attractor $\omega^2 \subset \Delta_2 \subset \mathbb{R}^2$.

the closed convex hull of $\{\mathbf{u}_a : a \in A\}$. From A9.5, we also know that when A is infinite, then $\Delta^A = \{(x_a) \in l^2(A) : \ 0 \leq \Sigma_a x_a \leq 1; \ 0 \leq \text{each } x_a \leq 1\}$.

22.1 Definition (Milutinović's infinite IFS) Let A be infinite, let $\{\mathbf{u}_a : a \in A\}$ be the standard orthonormal basis for $l^2(A)$; and for each $a \in A$, let $w_a : l^2(A) \to l^2(A)$ be given by $w_a(\mathbf{x}) = \mathbf{u}_a + (1/2)(\mathbf{x} - \mathbf{u}_a) = (1/2)(\mathbf{x} + \mathbf{u}_a)$ for each $\mathbf{x} \in l^2(A)$. Then "W_A" denotes $\{w_a : a \in A\}$.

So for each $b \in A$, w_b is a contraction by $1/2$ toward (u_a^b). We also note that the inverse w_b^{-1} of w_b is given by $w_b^{-1}(\mathbf{x}) = 2\mathbf{x} - \mathbf{u}_b$.

22.2 Lemma (each w_a is a homeomorphism and $w_a(\Delta^A) \subset \Delta^A$) *Let W_A be as specified in Definition 22.1. Then each $w_a \in W_A$ is a homeomorphism that maps the standard simplex Δ^A into itself.*

PROOF. Since both $w_a^{-1} \circ w_a$ and $w_a \circ w_a^{-1}$ equal the identity on $l^2(A)$, it follows that w_a is bijective. To see that w_a is a homeomorphism, we show that both w_a and w_a^{-1} are continuous — for d denoting the metric on $l^2(A)$, we have $d(w_a(\mathbf{x}), w_a(\mathbf{y})) = (1/2)d(\mathbf{x}, \mathbf{y})$ and $d(w_a^{-1}(\mathbf{x}), w_a^{-1}(\mathbf{y})) = 2d(\mathbf{x}, \mathbf{y})$. To see that $w_a(\Delta^A) \subset \Delta^A$, let $(x_b)_{b \in A} \in \Delta^A$, and note that

$$0 \leq \Sigma_{b \in A} x_b \leq 1 \implies 0 \leq (1/2)\Sigma_{b \in A} x_b + (1/2) \leq 1,$$

which shows that each $w_a(\Delta^A) \subset \Delta^A$.   □

22.3 Definition (Milutinović's space M_A) Let W_A be as specified in Definition 22.1. Let $F_1 = \cup_{a \in A} w_a(\Delta^A)$; $F_2 = \cup_{(a,b) \in A^2} w_a \circ w_b(\Delta^A)$; and $F_n = \cup_{(a_1, \ldots, a_n) \in A^n} w_{a_1} \circ \cdots \circ w_{a_n}(\Delta^A)$. Then $M_A = F_1 \cap F_2 \cap \cdots = \cap_{n=1}^{\infty} F_n$.

§23 Comments

Chapters 3 and 4 document the constructions of two subspaces $\omega^{A'}$ and M_A of Hilbert space, each homeomorphic to J_A. The focus in this chapter is $\omega^{A'}$, but §18 documents a discussion of M_A and how Milutinović's Proposition 7 shows that M_A also satisfies "$(x_a) \in M_A$ if and only if $(x_a) \in \Delta^A$ has the no-carry property." Historically, both $\omega^{A'}$ and M_A were introduced (circa 1992) entirely within the context of dimension theory. The motivation was the need to increase the understanding of J_A. Except for §21, where Sierpiński's construction is cast in terms of the formula for the Hutchinson operator (from finite IFS theory), the material as presented here is basically as it appears in the literature.

CHAPTER 5

Infinite IFS with Attractor $\omega^{A'}$

In §8 we showed that $\omega^n \subset \mathbb{R}^n$ is the attractor of the finite IFS $\mathcal{F}_n$. In this chapter we show that an infinite IFS $\mathcal{F}_A$ has the attractor $\omega^{A'} \subset l^2(A')$.

In both cases, the attractors are closed and bounded, which equates to compactness in the finite case. So here the focus is the family $\mathcal{B}_X$ of non-empty closed and bounded subsets of a metric space X. The Hausdorff metric h on $\mathcal{B}_X$ is motivated and studied. The space $(\mathcal{B}_X, h)$ is complete when (X, ρ) is complete, and an infinite IFS theory evolves. The J_A system $\mathcal{F}_A$ is then defined and shown to have attractor $\omega^{A'}$. For the most part, we follow Miculescu and Mihail [2008]. For the proof that $\omega^{A'}$ is a complete and closed subspace of $l^2(A')$ we follow Perry [1996].

§24 Neighborhoods of Sets

The distance $\rho(a, b)$ between points a and b in a metric space (X, ρ) has the basic property that whenever $\varepsilon > \rho(a, b)$, then an ε-ball centered at either a or b is a neighborhood of both a and b. So, if $(\mathcal{B}_X, h)$ is a metric space, where $\mathcal{B}_X$ is any family of subsets of (X, ρ), and $\varepsilon > h(A, B)$, then an ε-ball centered at either A or B must be a neighborhood of both A and B. Figure 24.1 shows that the usual distance $\rho(A, B)$ does not produce such neighborhoods.

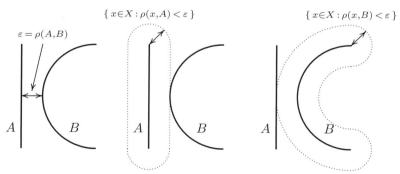

Fig. 24.1 Distance ε between A and B, and ε-neighborhoods of A and B.

24.2 EXAMPLE. (Use Figure 24.1.) Let $X = \mathbb{R} \times \mathbb{R}$ be the plane, and consider the subsets $A = \{(x, y) : x = -2 \text{ and } -1 \leq y \leq +1\}$ and $B = \{(x, y) : x^2 + y^2 = 1 \text{ and } x \leq 0\}$. Then $\rho(A, B) = \inf\{\rho(a, B) : a \in A\}$ is the "usual" distance between A and B. Thus, the "usual" distance does

S.L. Lipscomb, *Fractals and Universal Spaces in Dimension Theory*,
DOI 10.1007/978-0-387-85494-6_5, © Springer Science+Business Media, LLC 2009

not necessarily produce $\varepsilon > \rho(A, B)$ neighborhoods centered at either A or B that contain both A and B. $\qquad\square$

In the following section, we shall develop a metric h on the set of all non-empty, closed, and bounded subsets of a metric space that has the following property: If $\varepsilon > h(A, B)$, then the so-called ε-collars A_ε and B_ε contain both A and B.

§25 Hausdorff Metrics and Pseudo Metrics

We are given a metric space $X = (X, \rho)$ and we induce a metric h on $\mathcal{B}_X$ where $\mathcal{B}_X$ is the set of all non-empty, bounded, and closed subsets of X. The definition of the metric h, called the Hausdorff metric, is motivated by Figure 25.1 and Example 25.3.

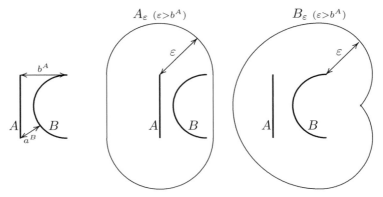

Fig. 25.1 $\quad h(A, B) = \max \{a^B, b^A\}$; and
$h(A, B) \leq \varepsilon \Leftrightarrow A \subset B_\varepsilon$ and $B \subset A_\varepsilon$.

For the following lemma, recall that the notation for the *distance between* $a \in X$ *and* $B \subset X$ is given by $\rho(a, B) = \inf_{b \in B} \rho(a, b) = \inf_b \rho(a, b)$.

25.2 Lemma ($0 \leq a^B = \sup_a \rho(a, B) < \infty$) *Let A and B be non-empty and bounded subsets of (X, ρ), and let $a^B = \sup_{a \in A} \rho(a, B) = \sup_a \rho(a, B)$. Then $0 \leq a^B < \infty$.*

PROOF. Now $0 \leq \rho(a, b)$ for each $a \in A$ and each $b \in B$ implies $0 \leq \rho(a, B)$ for each $a \in A$. So $0 \leq \sup_{a \in A} \rho(a, B) = a^B$. To see $a^B < \infty$, we show that $\{\rho(a, B) : a \in A\}$ is bounded above: Since A is bounded, its diameter $|A| = \sup_{x, y \in A} \rho(x, y)$ is finite. So we let both $a_1 \in A \neq \emptyset$ and $b_1 \in B \neq \emptyset$ be fixed. Then each $\rho(a, B) \leq \rho(a, b_1) \leq \rho(a, a_1) + \rho(a_1, b_1) \leq |A| + \rho(a_1, b_1)$, which is finite. $\qquad\square$

A similar argument shows that $0 \leq b^A = \sup_b \rho(b, A) < \infty$. To illustrate the usefulness of the values a^B and b^A, we present another example.

25.3 EXAMPLE. (Use Figure 25.1.) Again, let $X = \mathbb{R} \times \mathbb{R}$ be the plane, and consider the subsets $A = \{(x, y) : x = -2 \text{ and } -1 \leq y \leq +1\}$ and $B = \{(x, y) : x^2 + y^2 = 1 \text{ and } x \leq 0\}$. To visualize a^B, fix $a \in A$ and let ℓ_a be the line that contains a and the origin $(0, 0)$ (which is the center of the circle that contains the semicircle B). Then $\rho(a, B)$ is the length of the subsegment $[a, q]$ of ℓ_a where $\{q\} = B \cap \ell_a$. And to visualize b^A, fix $b \in B$ and let ℓ_b be the horizontal line that contains b. Then $\rho(b, A)$ is the length of the subsegment $[q, b]$ where $\{q\} = A \cap \ell_b$. So for $\varepsilon \geq h(A, B) = \max\{a^B, b^A\}$, each "$\varepsilon$-collar" contains both A and B.   □

25.4 Theorem (Hausdorff metric h) *Let (X, ρ) be a metric space, $\mathcal{B}_X$ the set of non-empty, bounded, and closed subsets of X, and $h : \mathcal{B}_X \times \mathcal{B}_X \to [0, \infty)$ a mapping given by $h(A, B) = \max\{a^B, b^A\}$ where $a^B = \sup_a \rho(a, B)$ with each $\rho(a, B) = \inf_b \rho(a, b)$. Then $(\mathcal{B}_X, h)$ is a metric space.*

PROOF. Let $A, B, C \in \mathcal{B}_X$. Then $h(A, B) \geq 0$ since $a^B \geq 0$ and $b^A \geq 0$; and $h(A, B) < \infty$ follows from Lemma 25.2. Next, observe that $h(A, B) = h(B, A)$, and that

$$h(A, B) = 0 \quad \Leftrightarrow \quad a^B = 0 = b^A \Leftrightarrow \rho(a, B) = 0 = \rho(b, A) \quad (a \in A, b \in B)$$
$$\Leftrightarrow \quad a \in \overline{B} = B \text{ and } b \in \overline{A} = A \quad (a \in A, b \in B) \Leftrightarrow A = B.$$

To prove the triangle inequality, let $a_1 \in A$ and $c_1 \in C$. Then

$$\rho(a_1, B) = \inf_b \rho(a_1, b) \leq \rho(a_1, c_1) + \inf_b \rho(c_1, b) \leq \rho(a_1, c_1) + c^B.$$

So $\rho(a_1, B) - c^B$ is a lower bound of $\rho(a_1, c_1)$ for every $c_1 \in C$, which yields $\rho(a_1, B) - c^B \leq \inf_c \rho(a_1, c) = \rho(a_1, C)$. It follows that $a \in A$ implies $\rho(a, B) \leq \rho(a, C) + c^B \leq a^C + c^B$, and, in turn, that $a^B \leq a^C + c^B$. Similarly, $b^A \leq b^C + c^A = c^A + b^C$. Taken together,

$$\begin{aligned} h(A, B) &= \max\{a^B, b^A\} \leq \max\{a^C + c^B, c^A + b^C\} \\ &\leq \max\{a^C, c^A\} + \max\{c^B, b^C\} = h(A, C) + h(C, B), \end{aligned}$$

which finishes the proof.   □

25.5 Corollary (Hausdorff pseudo metric h^*) *Let (X, ρ) be a metric space, let $\mathcal{M}_X$ be the family of all non-empty and bounded subsets of X, and let $h^* : \mathcal{M}_X \times \mathcal{M}_X \to [0, \infty)$ be a mapping given by $h^*(A, B) = \max\{a^B, b^A\}$ where $a^B = \sup_a \rho(a, B)$ with each $\rho(a, B) = \inf_b \rho(a, b)$. Then $(\mathcal{M}_X, h^*)$ is a pseudo-metric space.*

PROOF. Except for the displayed string of equivalences "$h(A, B) = 0 \Leftrightarrow \cdots \Leftrightarrow A = B$" in the proof of Theorem 25.4, the constraint that A and B be closed was not required.   □

The metric h (pseudo metric h^*) induced by (X, ρ) is called a *Hausdorff metric (pseudo metric)*. Hausdorff metrics and pseudo metrics involve the ε-collar $C_\varepsilon = \{x \in X : \rho(x, C) \leq \varepsilon\}$ of $C \in \mathcal{M}_X$ (Figure 25.1).

We note, since $C \in \mathcal{M}_X$ is both non-empty and bounded, that C_ε is also both non-empty and bounded. Moreover, since the mapping $x \to \rho(x, C)$ is continuous, a straightforward argument shows that $\overline{C_\varepsilon} \subset C_\varepsilon$. So $C \in \mathcal{M}_X$ implies $C_\varepsilon \in \mathcal{B}_X \subset \mathcal{M}_X$.

25.6 Lemma (ε-collars) *Let the metric space (X, ρ) induce $(\mathcal{M}_X, h^*)$. Then $h^*(A, B) \leq \varepsilon$ if and only if $A \subset B_\varepsilon$ and $B \subset A_\varepsilon$.*

PROOF. When $h^*(A, B) \leq \varepsilon$, both a^B and $b^A \leq \varepsilon$. So for $a_1 \in A$, we have $\rho(a_1, B) \leq \sup_a \rho(a, B) = a^B \leq \varepsilon$, which yields $a_1 \in B_\varepsilon$, i.e., $A \subset B_\varepsilon$. A similar argument shows $B \subset A_\varepsilon$. Conversely, the two inclusions $B \subset A_\varepsilon$ and $A \subset B_\varepsilon$ show, respectively, that $b^A \leq \varepsilon$ and $a^B \leq \varepsilon$, i.e., $h^*(A, B) \leq \varepsilon$. ☐

§26 Completeness of $(\mathcal{B}_X, h)$

In this section we provide a proof that whenever (X, ρ) is a complete metric space, then the induced space $(\mathcal{B}_X, h)$ is also complete.

26.1 Theorem (completeness of $(\mathcal{B}_X, h)$) *Let (X, ρ) be a complete metric space, and let $\mathcal{B}_X$ be the set of non-empty, bounded, and closed subsets of X with the induced Hausdorff metric h. Then $(\mathcal{B}_X, h)$ is complete.*

PROOF. Let $S_1, S_2, \ldots$ be a Cauchy sequence in $(\mathcal{B}_X, h)$. Then for each $\varepsilon > 0$, we have $h(S_n, S_m) < \varepsilon$ for all large n and m, say all $n, m \geq M$. So for $B = S_M$ and $k \geq M$, the ε-collar Lemma (Lemma 25.6) yields $S_k \subset B_\varepsilon$. It follows that $\cup_i S_i \subset S_1 \cup \cdots \cup S_{M-1} \cup B_\varepsilon$ is bounded. Next, let

$$A = \{ x \in X : x = \lim_{m_k \to \infty} y_{m_k} \text{ where each } m_k < m_{k+1} \text{ and each } y_{m_k} \in S_{m_k} \}.$$

First, A is bounded: This claim follows because $A \subset \overline{\cup_i S_i}$ and $\cup_i S_i$ (and hence $\overline{\cup_i S_i}$) is bounded. *Second, A is closed*: If $x \in \overline{A}$, then $x = \lim_{m \to \infty} a_m$ where each $a_m \in A$. For each $a_m \in A$, however, there exists a sequence $y^m = y_{m_1} y_{m_2} \cdots$ converging to a_m (where each $m_k < m_{k+1}$ and $y_{m_k} \in S_{m_k}$).

$$
\begin{array}{cccc}
y_{1_1} & y_{1_2} \cdots & \to & a_1 \\
y_{2_1} & y_{2_2} \cdots & \to & a_2 \\
\vdots & \vdots \quad \cdots & \to & \vdots \\
& & & x \in \overline{A}
\end{array}
$$

Without loss of generality, we may assume that each $\rho(y_{m_m}, a_m) < 1/m$ and that $m_m < (m+1)_{m+1}$ for each $m = 1, 2, \ldots$. Then, since the (diagonal) sequence $y_{1_1} y_{2_2} \cdots$ converges to x and also satisfies the specifications in the definition of A, it follows that $x \in A$. That is, $A = \overline{A}$ is closed. So we now know that $A \in \mathcal{B}_X$. Next, *we show that $A \subset B_\varepsilon$*: Recall (third sentence in this proof) that $\cup_{k \geq M} S_k \subset (S_M)_\varepsilon = B_\varepsilon$. Since $\cup_{k \geq M} S_k$ contains the tail $y_{m_M} y_{m_{M+1}} \cdots$ of any sequence $y_{m_1} y_{m_2} \cdots$ satisfying the specifications in the definition of A, and since such a tail also satisfies those same specifications, we see that

$$(1) \qquad A \subset \overline{\cup_{k \leq M} S_k} \subset \overline{B_\varepsilon} = B_\varepsilon = (S_M)_\varepsilon.$$

Now we show that $B \subset A_\varepsilon$: Select a sequence $m_0 < m_1 < m_2 \cdots$ of positive integers such that each $h(S_{m_k}, S_{m_{k+1}}) < \varepsilon/(2^k)$. Since $h(S_M, S_m) < \varepsilon = \varepsilon/(2^0)$ for each $m \geq M$, we may assume that $m_0 = M$. The list $S_{m_0}, S_{m_1}, \ldots$ yields a sequence $y_{m_0}, y_{m_1}, \ldots$ in X where each $y_{m_k} \in S_{m_k}$ and each

$$(2) \qquad \rho(y_{m_k}, y_{m_{k+1}}) < \varepsilon/(2^k).$$

Indeed, let y_{m_0} be any point of S_{m_0}, and then, with $y_{m_k} \in S_{m_k}$ defined, we may select $y_{m_{k+1}} \in S_{m_{k+1}}$ that satisfies (2) because $\inf_{y \in S_{m_{k+1}}} \rho(y_{m_k}, y) = \rho(y_{m_k}, S_{m_{k+1}})$ and

$$\rho(y_{m_k}, S_{m_{k+1}}) \leq \sup_{x \in S_{m_k}} \rho(x, S_{m_{k+1}}) \leq h(S_{m_k}, S_{m_{k+1}}) < \varepsilon/(2^k).$$

The sequence $y_{m_0}, y_{m_1}, \ldots$ is Cauchy and converges to some $a \in A$: For each $\delta > 0$, select a $k > 0$ such that $\varepsilon/(2^{k-1}) < \delta$. Then for $n \geq 0$,

$$\begin{aligned} \rho(y_{m_k}, y_{m_{k+n+1}}) &\leq \rho(y_{m_k}, y_{m_{k+1}}) + \cdots + \rho(y_{m_{k+n}}, y_{m_{k+n+1}}) \\ &< \varepsilon/(2^k) + \cdots + \varepsilon/(2^{k+n}) < \varepsilon/(2^{k-1}) < \delta. \end{aligned}$$

So the sequence $\{y_{m_k}\}$ is Cauchy and therefore converges because X is complete. Also, $y_{m_k} \to a_y \in A$ because each $y_{m_k} \in S_{m_k}$. Moreover, since each $\rho(y_{m_0}, y_{m_k}) < \varepsilon$, we have $\lim_{k \to \infty} \rho(y_{m_0}, y_{m_k}) = \rho(y_{m_0}, a_y) \leq \varepsilon$, and hence

$$\rho(y_{m_0}, A) = \inf_{a \in A} \rho(y_{m_0}, a) \leq \rho(y_{m_0}, a_y) \leq \varepsilon.$$

Since y_{m_0} may be any member of $B = S_M$, we have $\sup_{b \in B} \rho(b, A) \leq \varepsilon$. So the following inclusion holds:

$$(3) \qquad S_M = B \subset A_\varepsilon.$$

Thus, the ε-collar Lemma (Lemma 25.6) and (1) and (3) yield $h(S_M, A) \leq \varepsilon$. And, since $k \geq M$ implies $h(S_k, A) \leq h(S_k, S_M) + h(S_M, A) < 2\varepsilon$, it is clear that the sequence $S_1, S_2, \ldots$ converges to A. Thus, $(\mathcal{B}_X, h)$ is complete. $\qquad \square$

To be certain of the language used in the following lemma, recall that for a metric space (X, ρ), a *c-contraction* is a mapping $w : X \to X$ that satisfies $\rho(w(x), w(y)) \leq c\, \rho(x, y)$, $x, y \in X$, where

$$c = \sup_{x \neq y} \frac{\rho(w(x), w(y))}{\rho(x, y)} \in (0, 1).$$

Certain c-contractions are also *similitudes* — a *similitude* is a c-contraction such that $\rho(w(x), w(y)) = c\, \rho(x, y)$ for each $x, y \in X$.

For an example, recall that each member of the IFS $\mathcal{F}_n = \{w_i : i = 0, \ldots, n\}$ as defined in §8 is a similitude with $c = 1/2$.

26.2 Lemma (properties of the Hausdorff pseudo metric h^*) *Let (X, ρ) be a metric space, and $(\mathcal{M}_X, h^*)$ the induced pseudo metric space of non-empty and bounded subsets of X. Then*

   (i) $h^*(w(D), w(E)) \leq c\, h^*(D, E)$ *for each $D, E \in \mathcal{M}_X$*
        for any c-contraction $w : X \to X$; and

   (ii) $h^*(\overline{D}, \overline{E}) = h^*(D, E) \leq \sup_a h^*(D_a, E_a)$
        for $D = \cup_a D_a$, $E = \cup_a E_a$, $\{D_a\}, \{E_a\} \subset \mathcal{M}_X$, and $\overline{D}, \overline{E} \in \mathcal{M}_X$.

PROOF. First, (i): Since $\rho(w(x), w(y)) \leq c\, \rho(x, y)$, $x, y \in X$, we have

$$
\begin{aligned}
h^*(w(D), w(E)) &= \max\{w(d)^{w(E)}, w(e)^{w(D)}\} \\
&= \max\{\sup_{w(d)} \rho(w(d), w(E)), \sup_{w(e)} \rho(w(e), w(D))\} \\
&\leq \max\{c \sup_d \rho(d, E), c \sup_e \rho(e, D)\} \\
&= c \max\{d^E, e^D\} = c\, h^*(D, E).
\end{aligned}
$$

Second, (ii): Observe that $\overline{D}, \overline{E} \in \mathcal{M}_X$, $D \subset \overline{D}$, and $E \subset \overline{E}$ show not only that $D, E \in \mathcal{M}_X$ but also that $D_a, E_a \in \mathcal{M}_X$ for each a. So claim (ii) is well defined. The equality in claim (ii) follows from an application of

(4) $\qquad h^*(D, E) = \inf\{\varepsilon > 0 : D \subset E_\varepsilon \text{ and } E \subset D_\varepsilon\}.$

So first we prove (4): By the ε-collars Lemma (Lemma 25.6),

(5) $\qquad h^*(D, E) \leq \inf\{\varepsilon > 0 : D \subset E_\varepsilon \text{ and } E \subset D_\varepsilon\}.$

And the reverse inequality follows because, for any value $h^*(D, E) + \delta$ where $\delta > 0$, we may choose ε such that $h^*(D, E) \leq \varepsilon < h^*(D, E) + \delta$, yielding, by Lemma 25.6, $D \subset E_\varepsilon$ and $E \subset D_\varepsilon$, which, in turn, shows that

(6) $\qquad \inf\{\varepsilon > 0 : D \subset E_\varepsilon \text{ and } E \subset D_\varepsilon\} < h^*(D, E) + \delta.$

Thus, (5) and (6) yield (4). Now since both E_ε and D_ε are closed, we see that

(7) $\quad D \subset E_\varepsilon \Leftrightarrow \overline{D} \subset E_\varepsilon \subset (\overline{E})_\varepsilon \quad$ and $\quad E \subset D_\varepsilon \Leftrightarrow \overline{E} \subset D_\varepsilon \subset (\overline{D})_\varepsilon.$

From (7) and (4) the equality in claim (ii) holds. To see that the inequality is also true, note that $h^*(D_a, E_a) = \max\{d_a^{E_a}, e_a^{D_a}\}$ while $h^*(D, E) = \max\{d^E, e^D\}$, and then consider $d^E = \sup_{d \in D} \rho(d, E)$: For each $d \in D$ there is an a such that $d \in D_a$. So

(8) $\rho(d, E) = \inf_{e \in E} \rho(d, e) \leq \inf_{e \in E_a} \rho(d, e) = \rho(d, E_a) \leq \sup_{d \in D_a} \rho(d, E_a) = d_a^{E_a}.$

From (8), for any $d \in D$ we have $\rho(d, E) \le \sup_a \{d_a^{E_a}\}$, which shows that $d^E \le \sup_a \{d_a^{E_a}\}$. Similarly, $e^D \le \sup_a \{e_a^{D_a}\}$. It only remains to observe that $h^*(D, E) = \max\{d^E, e^D\} \le \sup_a \max\{d_a^{E_a}, e_a^{D_a}\} = \sup_a h^*(D_a, E_a)$. $\square$

§27 Hutchinson Operator for a Bounded IFS

If an IFS of c-contractions is also "bounded," then the Hutchinson operator $W : \mathcal{C}_X \to \mathcal{C}_X$ may be extended to its counterpart $\mathcal{W} : \mathcal{B}_X \to \mathcal{B}_X$.

27.1 Definition (bounded IFS) Let $(\mathcal{B}_X, h)$ be induced from the complete metric space (X, ρ). For $c \in (0, 1)$, let $\{w_a\}$ be a family of c-contractions $X \to X$ such that $Z \in \mathcal{B}_X$ implies $\cup_{a \in A} w_a(Z)$ is bounded. Then the IFS $\{w_a\}$ is called a *bounded* IFS.

27.2 Theorem (Hutchinson operator for bounded IFSs) *Let $(\mathcal{B}_X, h)$ be induced from the complete metric space (X, ρ). For $c \in (0, 1)$, let $\{w_a\}$ be a bounded IFS of c-contractions $X \to X$. Then the Hutchinson operator*

$$\mathcal{W}(Z) = \overline{\cup_{a \in A} w_a(Z)} \qquad (Z \in \mathcal{B}_X)$$

is an operator from $\mathcal{B}_X$ to $\mathcal{B}_X$ that is also a c-contraction.

PROOF. Recall that h^* is an extension of h. Then for any $D, E \in \mathcal{B}_X$

$$
\begin{aligned}
h\left(\mathcal{W}(D), \mathcal{W}(E)\right) &= h^*\left(\overline{\cup_a w_a(D)}, \overline{\cup_a w_a(E)}\right) = h^*\left(\cup_a w_a(D), \cup_a w_a(E)\right) \\
&\le \sup_a h^*(w_a(D), w_a(E)) \le \sup_a c\, h(D, E) = c\, h(D, E)
\end{aligned}
$$

where the relations among the h^* quantities follow from (ii) of Lemma 26.2, and the last inequality follows from (i) of Lemma 26.2. $\square$

§28 The Attractor of an Infinite IFS

We continue our study of the infinite bounded IFS $\{w_a\}$ as specified in Theorem 27.2. We show that each such system has an attractor.

28.1 Theorem *Let $(\mathcal{B}_X, h)$ be induced from the complete metric space (X, ρ). For $c \in (0, 1)$, let $\{w_a\}$ be a bounded IFS of c-contractions $X \to X$. Then there exists a unique $K \in \mathcal{B}_X$, called the attractor of $\{w_a\}$, characterized by the equation $K = \overline{\cup_a w_a(K)}$.*

PROOF. Since $(\mathcal{B}_X, h)$ is a complete metric space, Theorem 27.2 shows that $\mathcal{W} : \mathcal{B}_X \to \mathcal{B}_X$ given by $\mathcal{W}(Z) = \overline{\cup_a w_a(Z)}$ is a c-contraction on $\mathcal{B}_X$. Therefore, since contraction mappings on complete metric spaces have a unique fixed point, $\mathcal{W}$ has a unique fixed point $K \in \mathcal{B}_X$. That is, $\mathcal{W}(K) = K$ is characterized by the equation $K = \cup_a w_a(K)$.   □

With the mathematics developed in §27 and §28, we are in a position to formulate the infinite iterated function system that has $\omega^{A'}$ as its attractor.

§29 The J_A System

Throughout this section, the IFS $\mathcal{F}_A = \{w_a : a \in A\}$ has an index set A where $|A| \geq 2$. In addition, $z \in A$ is fixed, and $A' = A \setminus \{z\}$.

29.1 Definition (J_A system) For $l^2(A')$ with the usual metric, let $\{\mathbf{u}_a : a \in A'\}$ be the standard orthonormal basis and let $\mathbf{u}_z$ denote the zero-vector. Then for each $\mathbf{u}_a$, $a \in A$, define the $(1/2)$-similitude $w_a(\mathbf{x}) = (1/2)(\mathbf{x} + \mathbf{u}_a)$ and let $\mathcal{F}_A = \{w_a : a \in A\}$. The IFS $\mathcal{F}_A$ will be called the J_A *system*.

To see that indeed each w_a in the J_A system is a $(1/2)$-similitude, simply calculate that $\rho(w_a(\mathbf{x}), w_a(\mathbf{y})) = ||w_a(\mathbf{x}) - w_a(\mathbf{y})|| = (1/2)||\mathbf{x} - \mathbf{y}|| = (1/2)\rho(\mathbf{x}, \mathbf{y})$.

29.2 Lemma (J_A system is a bounded IFS) *For $|A| \geq 2$, let $X = l^2(A')$ and let $\mathcal{F}_A$ be the J_A system. Then for each $Z \in \mathcal{B}_X$, the set $\cup_a w_a(Z)$ is bounded, i.e., the J_A system is a bounded IFS.*

PROOF. Consider the balls B_r (with radius $r > 0$) centered at the origin of $l^2(A')$. Since Z is bounded, there is an $r > 0$ such that $Z \subset B_r$, i.e., $\mathbf{x} \in Z$ implies $||\mathbf{x}|| < r$. Thus, for any $a \in A$ and any $\mathbf{x} \in Z$,

$$||w_a(\mathbf{x})|| = ||(1/2)(\mathbf{x} + \mathbf{u}_a)|| \leq (1/2)(||\mathbf{x}|| + ||\mathbf{u}_a||) < (r+1)/2$$

shows that each $w_a(Z)$, and hence $\cup_a w_a(Z)$, is a subset of $B_{(r+1)/2}$.   □

§30 The J_A System Has Attractor $\omega^{A'}$

In this section we prove that the J_A system has attractor $\omega^{A'}$.

30.1 Lemma (convergence in $\omega^{A'}$ is ultimately convergence in $N(A)$) *Let A be infinite, and let*

$$
\begin{aligned}
\delta_1 &= \delta_{11}\delta_{12}\cdots\delta_{1j}\cdots \\
\delta_2 &= \delta_{21}\delta_{22}\cdots\delta_{2j}\cdots \\
&\vdots \quad \vdots \quad \vdots \quad \vdots \quad \vdots \quad \vdots \quad \vdots
\end{aligned}
$$

where $\{\delta_n\}$ is a sequence in $N(A)$ that has no convergent subsequence. Then there exists an index $j \in \{1, 2, \ldots\}$ such that the set $\{\delta_{1j}, \delta_{2j}, \ldots\}$ of entries in the jth column of the matrix above is an infinite subset of A.

PROOF. Suppose otherwise, and then contradict the fact that no subsequence of $\{\delta_n\}$ converges by using the argument in the footnote on page 37 with $M = \{1, 2, \ldots\}$, $\varepsilon_1 = \delta_1$, $\ldots$, and $\varepsilon_m(j) = \delta_{mj}$. $\qquad\square$

30.2 Theorem ($\omega^{A'}$ **is complete and closed in** $l^2(A')$) *Let $|A| \geq 2$, let $z \in A$ be fixed, and let $A' = A \setminus \{z\}$. Then $\omega^{A'}$ is closed and complete in $l^2(A')$.*

PROOF. If $|A| = n+1$ is finite, then from §8, $\omega^n \subset \mathbb{R}^n$ is the attractor of $\mathcal{F}_n$, which is compact, a fortiori closed and complete. Therefore, we may assume that A is infinite. Let $\{\mathbf{x}_n\}$ be a Cauchy sequence in $\omega^{A'}$, and observe that since $l^2(A')$ is complete, $\mathbf{x}_n \to \mathbf{x} \in l^2(A')$. So suppose that $\mathbf{x} \in l^2(A') \setminus \omega^{A'}$. Then no subsequence of $\{\mathbf{x}_n\}$ converges in $\omega^{A'}$.

Recall (from Chapter 4) that $g = f \circ p : N(A) \to \omega^{A'}$ where $p : N(A) \to J_A$ is perfect and $f : J_A \to \omega^{A'}$ is a homeomorphism, i.e., convergence in $\omega^{A'}$ is ultimately convergence in $N(A)$.

So select a sequence $\delta_1 \in g^{-1}(\mathbf{x_1})$, $\delta_2 \in g^{-1}(\mathbf{x_2})$, $\cdots$ in $N(A)$. Then no subsequence of $\{\delta_n\}$ converges in $N(A)$, otherwise we are finished.

From Lemma 30.1 there exists an infinite subset $\{\delta_{1j}, \delta_{2j}, \ldots\}$ of A where each δ_{kj} is the jth "coordinate" of $\delta_k \in N(A)$. Because it is only the infinite aspect of $\{\delta_{kj}\}$ that is important, we may assume that $z \notin \{\delta_{kj}\}$.

For each n, the map $g(\delta_n) = \mathbf{x}_n = (x_n^b)_{b \in A'}$ is determined by first projecting δ_n into the δ_{nj}-arm $\mathcal{C}(\overline{z}, \delta_{nj})$ of the star in $N(A)$, and then calculating, in the context of the corresponding $\mathbf{u}_{\delta_{nj}}$-arm of the star in $l^2(A')$, the corresponding coordinate $x_n^{\delta_{nj}}$ of $\mathbf{x}_n$. The calculation yields

(9) $\qquad\qquad x_n^{\delta_{nj}} \geq 1/(2^j) \qquad$ for each $n = 1, 2, \ldots$.

Another constraint follows from the fact that $\{\mathbf{x}_n\}$ is a Cauchy sequence:

(10) $\qquad ||\mathbf{x}_m - \mathbf{x}_k|| < 1/(2^{j+1}) \qquad$ for some N and all $k, m > N$.

POSSIBLE CASE I: For a fixed $k > N$ an $m > N$ exists such that the δ_{mj} coordinate of $\mathbf{x}_k$, namely, $x_k^{\delta_{mj}}$, satisfies

(11) $\qquad\qquad x_k^{\delta_{mj}} \leq 1/(2^{j+1})$.

Then (9) and (11) show that

$$x_m^{\delta_{mj}} - x_k^{\delta_{mj}} \geq 1/(2^{j+1}),$$

and so

$$||\mathbf{x}_m - \mathbf{x}_k|| \geq \left((x_m^{\delta_{mj}} - x_k^{\delta_{mj}})^2 \right)^{1/2} \geq 1/(2^{j+1}),$$

which contradicts (10).

POSSIBLE CASE II: For any $k, m > N$, the δ_{mj} coordinate of $\mathbf{x}_k$ satisfies

(12) $\qquad\qquad x_k^{\delta_{mj}} > 1/(2^{j+1})$.

Thus, for $k > N$, the inequality (12) is true for an infinite number of m, contradicting $\mathbf{x}_k \in l^2(A')$. So the sequence $\mathbf{x}_n \to \mathbf{x} \in \omega^{A'}$. □

30.3 Theorem (the J_A system has attractor $\omega^{A'}$) *Let $X = l^2(A')$, let $\mathcal{F}_A = \{w_a : a \in A\}$ be the J_A system, let $\mathcal{W} : \mathcal{B}_X \to \mathcal{B}_X$ be given by $\mathcal{W}(Z) = \cup_a w_a(Z)$, and let $\omega^{A'} \subset l^2(A')$ be the homeomorphic copy of J_A that is characterized by the no-carry property. Then $\omega^{A'} = \cup_a w_a(\omega^{A'})$, i.e., $\omega^{A'}$ is the attractor of the J_A system $\mathcal{F}_A$.*

PROOF. By Theorem 28.1, since $\omega^{A'}$ is a closed subset of $l^2(A')$, it suffices to show that $\omega^{A'} = \cup_a w_a(\omega^{A'})$: First, let $\mathbf{y} \in \omega^{A'}$. Then $\mathbf{y} = (y^b)_{b \in A'} \in l^2(A')$ and (y^b) satisfies the no-carry property, i.e., each y^b has a binary representation $.y_1^b y_2^b \cdots$ and together, for each fixed subscript i, there is at most one $b \in A'$ such that $y_i^b = 1$. So consider two cases:

CASE I: For each $b \in A'$, $y_1^b = 0$. Let $\mathbf{x} = w_z^{-1}(\mathbf{y}) = 2\mathbf{y} = (2y^b)$. Then each $x^b = 2y^b = 0.y_2^b y_3^b \cdots$, which shows that $\mathbf{x} \in \omega^{A'}$. Since $w_z(\mathbf{x}) = \mathbf{y}$, we have $\mathbf{y} \in \cup_{a \in A} w_a(\omega^{A'})$.

CASE II: There exists one and only one $d \in A'$ such that $y_1^d = 1$. Let

$$\mathbf{x} = w_d^{-1}(\mathbf{y}) = 2\mathbf{y} - \mathbf{u}_d = (2y^b) - (u_d^b) \text{ where } u_d^b = \begin{cases} 0.00\cdots & \text{if } b \neq d \\ 1.00\cdots & \text{if } b = d. \end{cases}$$

Then

$$\begin{aligned} x^d &= 2y^d - u_d^d = 1.y_2^d y_3^d \cdots - 1.000 = 0.y_2^d y_3^d \cdots & (b = d); \\ x^b &= 2y^b - u_d^b = 0.y_2^b y_3^b \cdots - 0.000 = 0.y_2^b y_3^b \cdots & (b \neq d), \end{aligned}$$

which shows that $\mathbf{x} \in \omega^{A'}$. Since $w_d(\mathbf{x}) = \mathbf{y}$, we have $\mathbf{y} \in w_d(\omega^{A'})$.

Second, let $\mathbf{y} \in \cup_{a \in A} w_a(\omega^{A'})$: Then, for some $d \in A$, we have $\mathbf{y} = w_d(\mathbf{x}) = w_d((x^b)) = (x^b/2) + (u_d^b/2)$ where, for each $b \in A'$,

$$(13) \qquad x^b/2 = .0x_1^b x_2^b \cdots \quad \text{and} \quad u_d^b/2 = \begin{cases} .000\cdots & \text{if } b \neq d \\ .100\cdots & \text{if } b = d. \end{cases}$$

From (9), $\mathbf{y} = w_d(\mathbf{x})$ satisfies the no-carry property, i.e., $\mathbf{y} \in \omega^{A'}$. □

§31 Comments

Arguably, point-set topology produced one of the greatest contributions to all of mathematics, namely the extension of the idea of convergence of numbers on a real line to abstract structures (any structure with a topology).

Such a "general convergence" is basic in Barnsley [1988], where $\mathcal{C}_X$ denotes the family of compact subsets of a complete metric space X, i.e., the "points" in $\mathcal{C}_X$ are compact subsets of X. Convergence takes place in $(\mathcal{C}_X, h)$

where h is the Hausdorff metric: A sequence $W(Z)$, $W \circ W(Z)$, ... converges to the "fractal" F, where W is a contractive (Hutchinson) operator $Z \mapsto \cup_i w_i(Z)$ on $\mathcal{C}_X$, and, $\{w_i\}$ is a finite IFS of contractive mappings $X \to X$. For *infinite* IFSs $\{w_a\}$, extensions of $\mathcal{C}_X$, W, and $\{w_i\}$ are required.

For example, let A be infinite, consider the standard orthonormal basis $B = \{\mathbf{u}_a : a \in A\} \subset l^2(A)$, and define $w_a : l^2(A) \to l^2(A)$ as the constant mapping $l^2(A) \to \{\mathbf{u}_a\}$. Then for any singleton (hence compact) set Z, we have $W(Z) = \cup_a w_a(Z) = B$ which is closed and bounded, but not compact.[1]

Historically, circa 1996, James Perry knew that J_2 was a copy of the unit interval and that J_3 was a copy of Sierpiński's triangle. And he conjectured that any J_A could be realized as an attractor of an IFS. And while he did not prove his conjecture, Perry [1996] did create an infinite IFS with an attractor.

His construction was based on the observation that ω^A, being a subset of the standard simplex Δ^A, is both a *subspace* of Δ^A and a *subset* of Tychonoff's cube I^A. (Recall that $(\theta_a) \in \omega^A$ implies that each θ_a satisfies $0 \le \theta_a \le 1$.) So he states:

> Let ω_c^A denote the space whose underlying set is that of ω^A but whose topology is induced from the Tychonoff cube I^A.

By using the compact Tychonoff cube and the $(1/2)$-contractions associated with ω^A, Perry created a hybrid of arguments that proved that ω_c^A is the attractor of an IFS $\{w_a : a \in A\}$ containing affine transformations of $\mathbb{R}^A$. Perry's IFS with attractor ω_c^A may be the first (nontrivial) example (with complete proofs) of an infinite IFS with an attractor.

At the end of the *Introduction* section in Perry [1996], he states:

> It is an open problem to construct ω^A as the attractor of an IFS containing affine transformations of $l^2(A)$.

More than a decade later, it was Miculescu and Mihail [2008] who provided a solution, the mathematics of which is the content of this chapter.

This growth, from compactness arguments within $\mathcal{C}_X$, to a hybrid of compactness arguments and arguments related to "closed and bounded" subspaces of $l^2(A)$, to "closed and bounded" arguments within $(\mathcal{B}_X, h)$, runs somewhat parallel to the growth of universal spaces in dimension theory. For example, the universal Menger sponge and Sierpiński carpet are compact as are the Euclidean cubes I^{2n+1} that are fundamental to universal spaces for n-dimensional separable metric spaces. But a J_A space is compact when A is finite and not compact when A is infinite. Nevertheless, in every case J_A is homeomorphic to $\omega^{A'}$, which is closed and bounded in $l^2(A)$. Looking back, it now seems most reasonable that it would be the mathematics of $(\mathcal{B}_{l^2(A)}, h)$ that would provide infinite IFSs with attractors homeomorphic to the noncompact J_A.

---

[1] The basis B is bounded because each $\|\mathbf{u}_a\| = 1$; it is closed in $l^2(A)$ because for distinct $a, b \in A$, each $\|\mathbf{u}_a - \mathbf{u}_b\| = \sqrt{2}$; and it is not compact because any covering of B with open balls in $l^2(A)$ of radius less than $\sqrt{2}/2$ has no finite subcover of B.

For references concerning techniques of working with compactness in the context of functional analysis (in particular $l^2(A)$), see Kolmogorov and Fomin [1957; §16–§18]. And for dimension in bicompact (compact Hausdorff) spaces, see Chapter 8 of Pears [1975], with sections on *inverse limits*, *a family of examples due to Vopěnka* (Vopěnka [1958]), and *V. V. Filippov's example* (Filippov [1970]).

An introductory model for some of the proofs within the $\mathcal{B}_X$ theory appears in Gulick [1992, Section 4.4]. For properties of the Hausdorff metric, one may review Dugunji [1966, page 205, problem 8] and Section 2.4 in Hutchinson [1981], where Federer [1969] is referenced. For the Hausdorff metric within the context of hyperspace theory and continuum theory, see, respectively, Nadler [1978] and Nadler [1992]. And for contractive mappings and similitudes in the context of fractals see Hutchinson [1981, Sections 2.2 and 2.3].

For variations on the proofs given in this chapter, see Miculescu and Mihail [2008], where Secelean [2001] is referenced. For example, the use of the no-carry characterization of $\omega^{A'}$ to prove Theorem 30.3 is new, i.e., it differs from the original proof that appears in Miculescu and Mihail [2008].

And finally, we note that Milutinović [1992, Corollary 15] provided one of the first proofs that $\omega^A = M_A \subset l^2(A)$ is closed and complete in $l^2(A)$.

CHAPTER 6

Dimension Zero

In this chapter we prove that both the J_A rationals and J_A irrationals are zero-dimensional and dense in J_A. As a corollary, using J_2, we deduce the zero-dimensionality and denseness of the rationals and irrationals in the unit interval. The $n = 0$ case of the J_A^{n+1} Imbedding Theorem is established. And for $0 \leq n \leq \ell$, we consider subspaces of $J_A^\ell(n)$ where $J_A^\ell(n)$ consists of those tuples in J_A^ℓ with at most n rational coordinates: We show that the subspace $E_A^\ell(m)$ of tuples that have exactly m rational coordinates has dimension zero. Then $J_A^\ell(n) = \cup_{m=0}^n E_A^\ell(m)$ and an application of the Decomposition Theorem within dimension theory shows that $J_A^\ell(n)$ is n-dimensional.

§32 Rationals and Irrationals

As a subspace of the unit interval, the union of the sets

$$\{\tfrac{1}{2}\}, \ \{\tfrac{1}{4}, \tfrac{3}{4}\}, \ \{\tfrac{1}{8}, \tfrac{3}{8}, \tfrac{5}{8}, \tfrac{7}{8}\}, \ \cdots, \ \{\tfrac{1}{2^n}, \tfrac{3}{2^n}, \tfrac{5}{2^n}, \tfrac{7}{2^n}, \ldots, \tfrac{2^n-3}{2^n}, \tfrac{2^n-1}{2^n}\}, \ \cdots$$

is countable and therefore zero-dimensional. Or, since each finite set in the list is closed and zero-dimensional, we may apply the Sum Theorem (A6.2). Moreover, as indicated in Figure 6.1, the unit interval I is homeomorphic to J_2 under a map that sends these dyadic rationals onto the rationals in J_2. That is, there is a homeomorphism

$$\text{rationals in } J_2 \longleftrightarrow \text{dyadic rationals in } (0,1).$$

So the J_2 rationals must also be zero-dimensional.

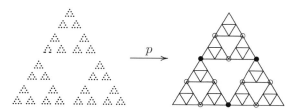

Fig. 32.1 The J_3 rationals are countable and thus zero-dimensional.

Turning to J_3, we see (Figure 32.1) a pattern for defining a list of finite sets whose union is the set of J_3 rationals. Indeed, the $3^1 = 3$ points in the first set are pictured as "black dots" and the $3^2 = 9$ points in the second set as "circles." The pattern tells us that there are 3^n points in the nth set. So the subspace of J_3 rationals, being countable, is therefore zero-dimensional.

S.L. Lipscomb, *Fractals and Universal Spaces in Dimension Theory*,
DOI 10.1007/978-0-387-85494-6_6, © Springer Science+Business Media, LLC 2009   53

32.2 Theorem (subspace of rationals in J_A is zero-dimensional) *Let $|A| \geq 2$, and let $R_A \subset J_A$ be the subspace of rationals. Then $\operatorname{Ind} R_A = 0$.*

PROOF. We shall apply the Sum Theorem. For each $t \in \{1, 2, \ldots\}$, define

$$F_t = \{z \in J_A : \ z \text{ rational}; \ t \text{ is the tail index of each member of } p^{-1}(z)\}.$$

Each F_t is closed in J_A: Since $p : N(A) \to J_A$ is a quotient map, it suffices to show that $p^{-1}(F_t)$ is closed in $N(A)$. To see that $p^{-1}(F_t)$ is closed, note that each $\delta_1 \delta_2 \cdots \in N(A) \setminus p^{-1}(F_t)$ is such that either $\delta_t = \delta_{t+1}$ or an index $k \geq t+1$ exists such that $\delta_k \neq \delta_{k+1}$. So F_t is closed in J_A. Next, we show that each F_t is zero-dimensional: Since $p^{-1}(F_t)$ is discrete ($\delta = \delta_1 \delta_2 \cdots \in p^{-1}(F_t)$ implies $\{\delta\} = \langle \delta_1, \ldots, \delta_{t+1} \rangle \cap p^{-1}(F_t)$), and since p is perfect (hence hereditarily quotient), F_t must be discrete. So each F_t is zero-dimensional. It follows from the Sum Theorem, since $R_A = \cup_t F_t$, that $\operatorname{Ind} R_A = 0$. ◻

32.3 EXAMPLE. The subspace of rationals in $\mathbb{R}$ is zero-dimensional: Let $A = \{0, 2\}$, and let R_A denote the subspace of J_A rationals. Then there exist two homeomorphisms (the first suggested by Figure 6.1):

(1) $R_A \longleftrightarrow$ dyadic rationals in $(0, 1) \longleftrightarrow$ rational reals in $\mathbb{R}$.

And since R_A is zero-dimensional (Theorem 32.2), the subspace of rational reals in $\mathbb{R}$ must also be zero-dimensional. ◻

32.4 Theorem (subspace of irrationals in J_A is zero-dimensional) *Let $|A| \geq 2$, and let $I_A \subset J_A$ be the subspace of irrationals. Then $\operatorname{Ind} I_A = 0$.*

PROOF. Since $p : N(A) \to J_A$ is perfect, p is hereditarily quotient, making the restriction p' of p to $p^{-1}(I_A)$ a quotient mapping. Since p' is also one-to-one, it is a homeomorphism. So, since p' is a continuous closed surjection from $p^{-1}(I_A) \subset N(A)$ onto I_A with singleton-set fibers, Theorem 1.6 shows that I_A is zero-dimensional. ◻

32.5 EXAMPLE. The subspace of irrationals in $\mathbb{R}$ is zero-dimensional: Let $A = \{0, 2\}$, and let I'_A denote the subspace of irrationals in J_A that contains neither $p(\overline{0})$ nor $p(\overline{2})$. Then there are two homeomorphisms (the first suggested by Figure 6.1):

(2) $I'_A \longleftrightarrow (0, 1) \setminus \{\text{dyadic rationals}\} \longleftrightarrow$ irrational reals in $\mathbb{R}$.

And since $I'_A \subset I_A$ and I_A is zero-dimensional, I'_A is zero-dimensional. Thus, since the composition of homeomorphisms in (2) is a homeomorphism, the subspace of irrational reals in $\mathbb{R}$ must be zero-dimensional. ◻

The following theorem is a result of $p : N(A) \to J_A$ being continuous.

32.6 Theorem (rationals and irrationals are dense in J_A) *Let $|A| \geq 2$. Then the rationals are dense in J_A and the irrationals are dense in J_A.*

As one might suspect, the fact that R_A and I_A are dense in J_A may imply that the rational and irrational reals are dense in $\mathbb{R}$.

32.7 EXAMPLE. The rational and the irrational reals are dense in $\mathbb{R}$: Note that the composite homeomorphism in (1) and the composite homeomorphism in (2) are restrictions of a single homeomorphism

$$J_{\{0,2\}} \setminus \{p(\overline{0}), p(\overline{2})\} \longleftrightarrow \mathbb{R}.$$

So by Theorem 32.6, the rational and irrational reals are dense in $\mathbb{R}$. $\square$

§33 J_A^{n+1} **Imbedding Theorem for** $n = 0$

For separable metric spaces, the Classical Imbedding Theorem states that the subspace of tuples in I^{2n+1} with at most n rational coordinates is universal for separable metric spaces of (covering) dimension $\leq n$. Since this theorem is well known, we shall simply state the $n = 0$ case for comparison with the same case of the J_A^{n+1} Imbedding Theorem.

33.1 Theorem ($n = 0$ case of the Classical Imbedding Theorem) *Let X be any zero-dimensional separable metric space. Then X can be imbedded in the subspace of irrationals in the unit interval I.*

Analogous to the Classical Imbedding Theorem, the J_A^{n+1} Imbedding Theorem states that the subspace of tuples in J_A^{n+1} with at most n J_A-rational coordinates is universal for weight $|A| \geq \aleph_0$ metric spaces of covering dimension $\leq n$.

33.2 Theorem ($n = 0$ case of the J_A^{n+1} Imbedding Theorem) *Let X be any zero-dimensional weight $|A| \geq \aleph_0$ metric space. Then X can be imbedded in the subspace I_A of irrationals in J_A.*

PROOF. Recall that any zero-dimensional weight $|A| \geq \aleph_0$ metric space can be imbedded in $N(A)$. So it suffices to imbed $N(A)$ into I_A. To accomplish this, let $B_1, B_2, \ldots$ be a partition of the infinite discrete space A into subspaces B_j where each $|B_j| = |A|$. Also, for each j, let $A_j = A$. These spaces induce homeomorphisms and imbeddings. In particular, homeomorphisms $q_j : A_j \to B_j$ exist because both of these discrete spaces have the same size, while the inclusion mappings $i_j : B_j \to A_j = A$ serve as imbeddings. Forming product maps, we then have the imbedding $\times_j i_j : \times_j B_j \to p^{-1}(I_A) \subset \times_j A_j = N(A)$, and, the homeomorphism $\times_j q_j : \times_j A_j \to \times_j B_j$. These mappings are illustrated in the context of a commutative diagram:

$$N(A) = \times_j A_j \xrightarrow{\times_j q_j} \times_j B_j \xrightarrow{\times_j i_j} p^{-1}(I_A)$$

$$\downarrow p'$$

$$I_A$$

where $p' = p|_{p^{-1}(I_A)} : p^{-1}(I_A) \to I_A$ is a homeomorphism. It follows that the "dashed arrow" is an imbedding of $N(A)$ into I_A.   □

§34 Subspaces of J_A^ℓ

For $0 \le n < \ell$ and $m \in \{0, 1, \ldots, n\}$, we let the subspace $J_A^\ell(n)$ consist of the ℓ-tuples in J_A^ℓ that have at most n rational coordinates, and the subspaces $E_A^\ell(m)$ consist of the ℓ-tuples with exactly m rational coordinates. In this section we calculate the dimensions of these spaces.

34.1 Lemma (Ind $E_A^\ell(m) = 0$) *Let* $|A| \ge 2$, $m \ge 0$, $\ell > m$, *and* $E_A^\ell(m) = \{z \in J_A^\ell : z$ *has exactly* m *rational coordinates}. Then* Ind $E_A^\ell(m) = 0$.

PROOF. Recall (proof of Theorem 32.2) that, for each $t \in \{1, 2, \ldots\}$,

$$F_t = \{z \in J_A : \ z \ rational;\ t \ is\ the\ tail\ index\ of\ each\ member\ of\ p^{-1}(z)\}$$

is closed in J_A and Ind $F_t = 0$. Next, we use the F_t to determine the dimension of each $E_A^\ell(m)$: Let I_A be the subspace of irrationals in J_A, and for each $S \subset \{1, 2, \ldots, \ell\}$ that contains exactly m members and each $k : S \to \{1, 2, \ldots\}$, let

$$F(S, k) = \{z \in J_A^\ell : \ z_r \in F_{k(r)} \ if \ r \in S;\ and\ z_r \in I_A \ if \ r \notin S\}.$$

Now each $F(S, k)$ is therefore the ℓ-fold product of the m zero-dimensional $F_{k(r)}$ spaces, $r \in S$, and $\ell - m$ copies of I_A. So the Product Theorem (A6.2) shows that $F(S, k)$ itself is zero-dimensional. In addition, each such $F(S, k)$ is also closed in $E_A^\ell(m)$ — if $y \in (E_A^\ell(m) - F(S, k))$, then either some index $r \in S$ exists such that y_r is irrational, or, each y_r, for $r \in S$, is rational but the tail index of at least one such y_r is not $k(r)$. So $F(S, k)$ is closed in $E_A^\ell(m)$. Finally, since the number of such pairs (S, k) is countable and

$$E_A^\ell(m) = \cup\{F(S, k) : \ |S| = m \ and \ k : S \to \{1, 2, \ldots\}\},$$

the Sum Theorem shows that $E_A^\ell(m)$ is zero-dimensional.   □

34.2 Theorem ($J_A^\ell(n)$ spaces) *Let $|A| \geq 2$, and, for integers $0 \leq n < \ell$, let the space $X = J_A^\ell(n) = \{z \in J_A^\ell : z \text{ has at most } n \text{ rational coordinates}\}$. Then $\mathrm{Ind}\, X = n$.*

PROOF. If $n = 0$, then $X = I_A^\ell$ where I_A is the subspace of irrationals in J_A. And since I_A is zero-dimensional, the Product Theorem shows that $\mathrm{Ind}\, X \leq \mathrm{Ind}\, I_A + \cdots + \mathrm{Ind}\, I_A = \ell \cdot 0 = 0$, which finishes the proof. So we may assume that $n > 0$. In this case, we begin by showing $\mathrm{Ind}\, X \geq n$: Since J_A contains a copy of the unit interval I, the product space J_A^n contains a copy of the cube I^n, which has dimension n. So the Subspace Theorem (A6.2) yields $\mathrm{Ind}\, J_A^n \geq \mathrm{Ind}\, I^n = n$. Moreover, J_A^n is homeomorphic to

$$\{q_1\} \times \cdots \times \{q_{\ell-n}\} \times J_A^n \subset J_A^\ell(n) = X$$

where $q_1, \ldots, q_{\ell-n} \in J_A$ are $\ell - n$ irrationals. Thus,

$$\mathrm{Ind}\, X \geq \mathrm{Ind}\, (\{q_1\} \times \cdots \times \{q_{\ell-n}\} \times J_A^n) = n.$$

To see that $\mathrm{Ind}\, X \leq n$, we apply the Decomposition Theorem (A6.2), i.e., $X = J_A^\ell(n) = \cup_0^n E_A^\ell(m)$ and $\mathrm{Ind}\, E_A^\ell(m) = 0$ (Lemma 34.1) for each m. So $\mathrm{Ind}\, X \leq n$ and $\mathrm{Ind}\, X \geq n$ shows that X is n-dimensional. $\qquad\square$

34.3 Corollary ($\mathrm{Ind}\, (I^{2n+1}(n)) = n$) *Let I denote the unit interval, and, for $n \geq 0$, let $I^{2n+1}(n) = \{x \in I^{2n+1} | \ x \text{ has at most } n \text{ rational coordinates}\}$. Then $\mathrm{Ind}\, (I^{2n+1}(n)) = n$.*

PROOF. Let $A = \{0, 1\}$. Then for $p : N(A) \to J_A$ and $q : N(A) \to I$ given by $(a_1, a_2, \ldots) \mapsto \Sigma_{i=1}^\infty a_i/2^i$, we have $\phi = pq^{-1} : I \to J_A$ is a homeomorphism.

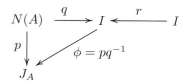

Indeed, if $\Sigma_1^\infty a_i/2^i = \Sigma_1^\infty b_i/2^i$, then let k denote the smallest index k such that $a_i \neq b_i$. We assume, without loss of generality, that $a_i = 1$. Then

$$1/2^k + \Sigma_{k+1} a_i/2^i = \Sigma_{k+1} b_i/2^i \Rightarrow \begin{cases} 1 = b_{k+1} = b_{k+2} = \cdots \text{ and} \\ 0 = a_{k+1} = a_{k+2} = \cdots. \end{cases}$$

In other words, $a_1 a_2 \cdots$ and $b_1 b_2 \cdots$ are adjacent endpoints, which shows that p and q have the same fibers (Theorem A4.3). Next, let $r : I \to I$ be a homeomorphism that preserves the natural ordering in I, and, maps the rational reals in I onto the dyadic rationals in I. Then $\psi = \phi \circ r : I \to J_A$ is a homeomorphism that maps the rational reals in I onto the rationals in J_A. So the product map $\times_i \psi_i : I^{2n+1} \to J_A^{2n+1}$ with each $\psi_i = \psi$

is a homeomorphism. Moreover, $(x_1, \ldots, x_{2n+1}) \in I^{2n+1}$ has at most n rational coordinates if and only if $(\psi(x_1), \ldots, \psi(x_{2n+1})) \in J_A^{2n+1}$ has at most n rational coordinates. Thus, since $I^{2n+1}(n)$ is homeomorphic to $J_A^{2n+1}(n)$, Theorem 34.2 shows that $\operatorname{Ind} I^{2n+1}(n) = n$. $\square$

For $\ell = n + 1$ in Theorem 34.2, we have the following corollary.

34.4 Corollary ($\operatorname{Ind} J_A^{n+1}(n) = n$) *Let* $|A| \geq 2$, *let* $n \geq 0$, *and let* $J_A^{n+1}(n) = \{z \in J_A^{n+1} : z$ *has at most* n *rational coordinates*$\}$. *Then* $\operatorname{Ind} J_A^{n+1}(n) = n$.

§35 Comments

One of the goals of this short chapter was to relate the J_A^{n+1} Imbedding Theorem to the Classical Imbedding Theorem for the case $n = 0$. The approach involved (i) proving general statements that concern J_A for any A with at least two members; and then (ii) applying these general results to J_2 to yield the corresponding results in the classical case.

For the most part, Section 33 and the proof of Theorem 33.2 follow Lipscomb [1973]. The proof of Theorem 33.2 is based on the fact that $N(A)$ is universal for the class of zero-dimensional weight $|A| \geq \aleph_0$ metric spaces. For details on the universality of $N(A)$ see Engelking [1978, Theorem 4.1.24].

CHAPTER 7

Decompositions[1]

Any n-dimensional weight $|A| \geq \aleph_0$ metric space admits an $\aleph_0 \times (n+1)$ matrix $[\mathcal{W}_{ij}]$ of decompositions that yields an imbedding into the subspace of $(n+1)$-tuples in J_A^{n+1} that have at most n rational coordinates.

We motivate and construct the decompositions $\mathcal{W}_{ij}$. The approach is a substantially expanded version of the approach in Lipscomb [1975].

§36 The Dimension Function diml

The Lebesgue or covering dimension "dim" is well known. Here, we construct another dimension function "diml" and then prove that diml $X = \dim X$ when X is a normal Hausdorff space. The "diml" concept concerns the "local order of a point," which is distinct from "order of a point." For example, consider a point p in the plane that is a point of tangency for two circles that bound two (open) 2-discs, say D_1 and D_2.

Then relative to the family $\{D_1, D_2\}$, the "order of p is zero" while the "local order of p is two."

36.1 Definition (lord$_x\,\mathcal{U}$ and lord $\mathcal{U}$)   Let X be a topological space, $\mathcal{U}$ a family of subsets of X, and $x \in X$. Then "lord$_x\,\mathcal{U}$" denotes the *local order* of $\mathcal{U}$ at x. That is,

$$lord_x\,\mathcal{U} = \min\,\{k(G_x) : x \in G_x \subset X;\ G_x \text{ open in } X\}$$

where $k(G_x)$ is the number (either a non-negative integer or ∞) of $U \in \mathcal{U}$ such that $G_x \cap U \neq \emptyset$. Moreover,

$$\mathrm{lord}\,\mathcal{U} = \sup\{\mathrm{lord}_x\,\mathcal{U} : x \in X\}$$

is the *local order of* $\mathcal{U}$.

36.2 Definition (diml X)   Let X be a topological space and n a non-negative integer. Then X has *local dimension* $\leq n$ if each locally finite open

---

[1] A locally finite pairwise-disjoint family $\mathcal{U}$ of open subsets of X is a *decomposition of* X if cl $\mathcal{U} = \{\overline{U} : U \in \mathcal{U}\}$ is a cover of X.

cover $\mathcal{V}$ of X has a local order $\leq n+1$ open refinement that covers X. If X has local dimension $\leq n$, we may write "diml $X \leq n$", and when diml $X \leq n$ and it is not true that diml $X \leq (n-1)$, then X *has local dimension* n and we may write "diml $X = n$". When no such n exists, then by definition diml $X = +\infty$. Finally, define diml $\emptyset = -1$.

From Engelking [1978, Dowker's Theorem 3.2.1], recall that for normal T_2 spaces, "dim $X \leq n$" is equivalent to "each locally finite open cover of X has a locally finite order $\leq n+1$ open refinement that covers X."

36.3 Proposition (dim X = diml X when X is normal T_2) *Let X be a normal Hausdorff space. Then* diml $X = $ dim X.

PROOF. Since a cover of local order $\leq n+1$ is necessarily a cover of order $\leq n+1$, Dowker's Theorem shows that diml $X \leq n$ implies dim $X \leq n$. For the reverse implication, suppose dim $X \leq n$; and let $\mathcal{V}$ be a locally finite open cover of X. Since dim $X \leq n$ the cover $\mathcal{V}$ has a locally finite order $\leq n+1$ open refinement $\mathcal{U}$ that covers X. From the covering characterization of normal T_2 spaces (§A2), there exists a locally finite open cover $\mathcal{U}' = \{U'_a : a \in A\}$ of X such that cl $\mathcal{U}' = \{\overline{U}'_a : a \in A\}$ precisely refines $\mathcal{U}$. If, for any $x \in X$, we define $G_x = X \setminus \cup\{\overline{U}'_a : a \in A; \ x \notin \overline{U}'_a\}$, then G_x is open and meets at most $n+1$ of the U'_a. So lord $\mathcal{U}' \leq n+1$. $\qquad\square$

§37 Nodes of a Cover

A family $\mathcal{V} = \{V_a : a \in A\}$ is *nodally indexed* if for each non-empty $V \in \mathcal{V}$ there is at most a finite number $k \geq 1$ of distinct indices $a_1, \ldots, a_k$ such that $V = V_{a_1} = \cdots = V_{a_k}$. In particular, if $\mathcal{V}$ is either *faithfully indexed* (distinct $a, b \in A$ yield distinct $V_a, V_b \in \mathcal{V}$), or, *pseudo-faithfully indexed* ($\emptyset \neq V_a = V_b \neq \emptyset$ implies $a = b$), then $\mathcal{V}$ is nodally indexed.

37.1 Definition (nodes of locally finite open covers) Let $\mathcal{V} = \{V_a : a \in A\}$ be a nodally indexed and locally finite open cover of a space X. Well order $A = (A, <)$. The *one-nodes* (1-*nodes*) *of* $\mathcal{V}$ are the members of *a nodal family* $\mathcal{O}_1 = \mathcal{O}_1(\mathcal{V}) = \{O_a : a \in A\}$ where

$$O_a = V_a \setminus (\cup\{V_b : b \neq a; \ b \in A\}) \qquad (a \in A).$$

And in general, the *k-nodes* of the cover $\mathcal{V}$ are the members of *a nodal family* $\mathcal{O}_k = \mathcal{O}_k(\mathcal{V}) = \{O_{a_1 \cdots a_k} : a_1 < \cdots < a_k; \ a_i \in A\}$ where

$$O_{a_1 \cdots a_k} = (V_{a_1} \cap \cdots \cap V_{a_k}) \setminus (\cup\{V_b : b \neq a_1, \cdots, b \neq a_k; \ b \in A\}).$$

The collection $\mathcal{O}(\mathcal{V}) = \cup_{k \geq 1} \mathcal{O}_k$ is called *the nodal family of* $\mathcal{V}$.

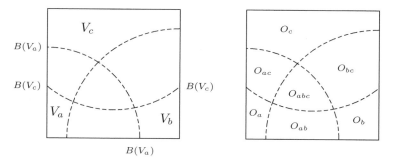

Fig. 37.2 Cover $\mathcal{V} = \{V_a, V_b, V_c\}$ and its nodes.

Nodes of the family $\mathcal{V} = \{V_a, V_b, V_c\}$ are illustrated in Figure 37.2. The three "dashed curves" specify the boundary curves of V_a, V_b, and V_c, showing that the nodes are pairwise disjoint. Collectively, however, these nodes cover X. So it remains to determine the subsets of the boundary curves included in each node.

These subsets are pictured in Figure 37.3, where the top and bottom right-side illustrations provide, respectively, examples of nodal properties (2) and (4) in Proposition 37.4.

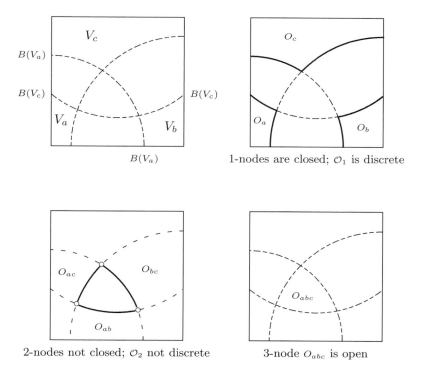

1-nodes are closed; $\mathcal{O}_1$ is discrete

2-nodes not closed; $\mathcal{O}_2$ not discrete   3-node O_{abc} is open

Fig. 37.3 Nodes of $\mathcal{V}$ partition Bdry $\mathcal{V} = B(V_a) \cup B(V_b) \cup B(V_c)$.

37.4 Proposition (properties of nodal families) *Let $\mathcal{V} = \{V_a : a \in A\}$ be a nodally indexed and locally finite open covering of a space X. Then the families of 1-nodes, 2-nodes,... satisfy the following properties:*

(1) *Any two well orderings of A yield the same nodal collections.*

(2) *$\mathcal{O}_1$ is a discrete family of closed sets with each $O_a \subset V_a$.*

(3) *$\mathcal{O}(\mathcal{V}) = \cup_{k \geq 1} \mathcal{O}_k$ is a pairwise-disjoint locally finite cover of X.*

(4) *For pseudo-faithful indexing, $\operatorname{ord} \mathcal{V} = k$ implies $O \in \mathcal{O}_k$ is open. and $x \in O \in \mathcal{O}_k$ if and only if $\operatorname{ord}_x \mathcal{V} = k$.*

(5) *If the closures of two distinct k-nodes meet, then they meet in a union of m-nodes where m ranges over indices less than k.*

(6) *The induced indexing of the members of $\mathcal{O}(\mathcal{V})$ is pseudo-faithful.*

PROOF. Standard arguments yield (1), (3), (4), and (6). For (2), show that $O_a = \overline{V_a} \setminus (\cup\{V_b : b \neq a;\ b \in A\})$, and, for the discrete part, that each x is in some V_a and V_a meets only one 1-node. For (5), let $x \in \overline{O}_{a_1 \cdots a_k} \cap \overline{O}_{b_1 \cdots b_k}$ where at least one $a_j \notin \{b_1, \ldots, b_k\}$. Then by (3), $x \in O = O_{c_1 \cdots c_m} \in \mathcal{O}(\mathcal{V})$, and by definition, $O_{c_1 \cdots c_m} \subset \cap_{i=1}^m V_{c_i}$. So x is in each V_{c_i}, which shows that each $c_i \in (\{a_j\} \cap \{b_j\})$. Since one of the $a_j \notin \{b_j\}$, we are finished. $\qquad\square$

For an example of (5), consider the bottom left-side graphic in Figure 37.3 where the point represented by the leftmost circle, say x, is contained in the closures $\overline{O}_{ab}$ and $\overline{O}_{ac}$ of two distinct 2-nodes O_{ab} and O_{ac}. Then as illustrated, these closed sets meet at x which is contained in the 1-node O_a.

Property (5) involves points common to the closures of two distinct k-nodes. But what can we say about points in the closure of a single k-node? The answer is provided by the *Nodal Closure Property*:

37.5 Corollary (Nodal Closure Property) *Let $\mathcal{V} = \{V_a : a \in A\}$ be a locally finite open cover of X, let $O = O_{a_1 \cdots a_k}$ be a k-node of $\mathcal{V}$, and let $x \in \overline{O} \setminus O$. Then $x \in O_{b_1 \cdots b_m}$ where $m < k$ and $\{b_1, \ldots, b_m\}$ is a proper subset of $\{a_1, \ldots, a_k\}$.*

PROOF. From (3), $x \in O_{b_1, \ldots, b_m}$ for some $m \geq 1$. Also, $x \notin O = O_{a_1 \cdots a_k}$ implies that either $x \notin \cap_i V_{a_i}$, or, $x \in V_b$ for some $b \notin \{a_i\}$. The latter statement, however, cannot hold because V_b open and $V_b \cap O = \emptyset$ contradicts $x \in \overline{O}$. So $\{b_1, \ldots, b_m\}$ is a proper subset of $\{a_1, \ldots, a_k\}$. $\qquad\square$

An application of the nodal closure property provides the following corollary.

37.6 Corollary (unions of nodes, closed and open) *Let $\mathcal{V} = \{V_a : a \in A\}$ be a locally finite open cover of X. Then for each $j \geq 1$, the sets $\cup\{O : O \in \overset{j}{\underset{1}{\cup}} \mathcal{O}_k\}$ and $\cup\{O : O \in \overset{\infty}{\underset{j+1}{\cup}} \mathcal{O}_k\}$ are closed and open, respectively.*

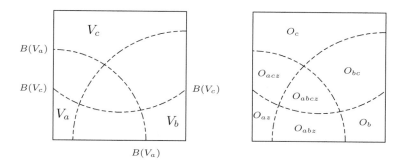

Fig. 37.7 Cover $\mathcal{V} = \{V_a, V_b, V_c, V_z\}$, $V_a = V_z$, and its nodes.

Nodes of the family $\mathcal{V} = \{V_a, V_b, V_c, V_z\}$ are illustrated in Figure 37.7. In this case $\mathcal{V}$ is nodally indexed but not faithfully indexed. Again, the nodes are pairwise disjoint; collectively cover X; and induce a partition of Bdry $\mathcal{V}$. The parts of the partition are pictured in Figure 37.8. This example provides instances of (2), (3), (5), and (6) of Proposition 37.4 for the case where the indexing is not faithful.

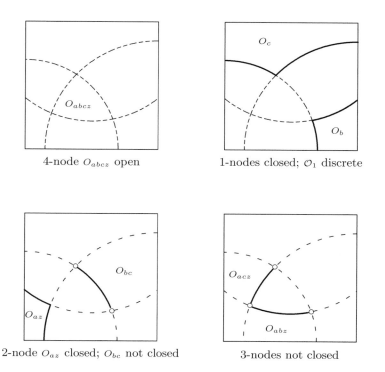

Fig. 37.8 Nodes of $\mathcal{V}$ partition Bdry $\mathcal{V} = B(V_a) \cup B(V_b) \cup B(V_c)$.

37.9 Definitions (shrinks, puffs-up, and canonical $(\mathcal{F}, \mathcal{V})$ collections) A family $\mathcal{F} = \{F_\lambda : \lambda \in \Lambda\}$ of sets *shrinks* a family $\mathcal{G} = \{G_\lambda : \lambda \in \Lambda\}$ of sets if $\mathcal{F}$ precisely refines $\mathcal{G}$, i.e., for each $\lambda \in \Lambda$, we have $F_\lambda \subset G_\lambda$. And for such a pair, the collection $\mathcal{U} = \{U_\lambda : \lambda \in \Lambda\}$ *puffs-up* $\mathcal{F}$ in $\mathcal{G}$ whenever, for each λ, we have $F_\lambda \subset U_\lambda \subset G_\lambda$. Furthermore, let $\mathcal{V} = \{V_a : a \in A\}$ be a nodally indexed locally finite open cover of X, let $\{O_\lambda : \lambda \in \Lambda\}$ be a subfamily of the nodal collection $\mathcal{O}(\mathcal{V})$, and let $\mathcal{F} = \{F_\lambda\}_\Lambda$ shrink $\{O_\lambda\}_\Lambda$, i.e., each

$$F_\lambda \subset O_\lambda = (\overset{k}{\underset{1}{\cap}} V_{a_i}) \setminus \cup \{V_b : b \neq a_1, \dots, b \neq a_k; b \in A\}$$

with $a_1 < \cdots < a_k$ determined by λ. Then the *canonical $(\mathcal{F}, \mathcal{V})$ collection* is the open family $\{G_\lambda\}_\Lambda$ where each $G_\lambda = \overset{k}{\underset{1}{\cap}} V_{a_i}$.

We close this section with the observation that if $\mathcal{F}$ shrinks a family of nodes of $\mathcal{V}$, then $\mathcal{F}$ necessarily shrinks the canonical $(\mathcal{F}, \mathcal{V})$ collection.

§38 Lemmas for the Decomposition Theorem

The proof of the main theorem of this chapter requires several lemmas. We begin with the constructions illustrated in Figure 38.1, which serve to guide the reader through the proof of Lemma 38.2.

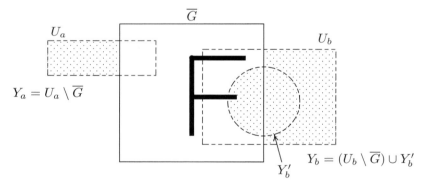

Fig. 38.1 The Y_a and Y_b constructions in the proof of Lemma 38.2.

38.2 Lemma (refining covers in a neighborhood of a dim $\leq n$ closed set) *Let $n \geq 0$; let $F \subset H$ be subspaces of X with F closed, $\dim F \leq n$, and H open; and let $\mathcal{U} = \{U_a : a \in A\}$ be a nodally indexed locally finite open cover of X. Then an open cover $\mathcal{Y}$ of X and an open subset G of X exist where $F \subset G \subset \overline{G} \subset H$; where $\mathcal{Y} = \{Y_a : a \in A\}$ precisely refines $\mathcal{U}$; and where each $Y_a = U_a \setminus \overline{G}$ or $Y_a = (U_a \setminus \overline{G}) \cup Y_a'$ with each $Y_a' \subset U_a$ defined such that $x \in \overline{G}$ implies $x \in O \in \cup_1^{n+1} \mathcal{O}_k(\mathcal{Y})$.*

PROOF. If $F = \emptyset$, then let $\mathcal{Y} = \mathcal{U}$ and $G = \emptyset$. Otherwise, let $B = \{b \in A : U_b \cap F \neq \emptyset\}$. Then $\{U_b\}_{b \in B}$ is a locally finite open cover of F. An application of the §A6.2 Theorem (refining covers of dim $\leq n$ closed subspaces) yields an open precise refinement $\mathcal{Y}' = \{Y_b' : b \in B\}$ of $\{U_b : b \in B\}$ that covers F and has the property that ord $\mathcal{Y}' \leq n+1$. Now choose G open such that $F \subset G \subset \overline{G} \subset (\cup_{b \in B} Y_b') \cap H$.

Since $\mathcal{Y}'$ is nodally indexed, for each non-empty $Y' \in \mathcal{Y}'$ there exists at most a finite number $k \geq 1$ of indices $b_1, \ldots, b_k$ such that $Y' = Y_{b_1}' = \cdots = Y_{b_k}'$. For each such Y' and all except one of $b_1, \ldots, b_k$, redefine $Y_b' = \emptyset$. With this adjustment to the indexing of $\mathcal{Y}'$, the resulting indexed family, also denoted $\mathcal{Y}'$, is pseudo-faithfully indexed, and it follows that "ord$_x \mathcal{Y}' \leq n+1$" is equivalent to "$x \in O \in \cup_1^{n+1} \mathcal{O}_k(\mathcal{Y}')$."

Next, for each $b \in B$, let $Y_b = (U_b \setminus \overline{G}) \cup Y_b'$, and, for each $a \in A \setminus B$, let $Y_a = U_a \setminus \overline{G}$. Then $\mathcal{Y} = \{Y_a : a \in A\}$ and G are as claimed. $\qquad \square$

To motivate the constructions in our next lemma (Lemma 38.5) we shall use the graphic in Figure 38.3 whose details are explained in Example 38.4.

Assumptions ($X_1 \equiv$ rationals, $X_2 \equiv$ irrationals, $X = [0,1]$, etc.)

Constructions ($\mathcal{Y}_1^1$, $\mathcal{Y}_2^1$)

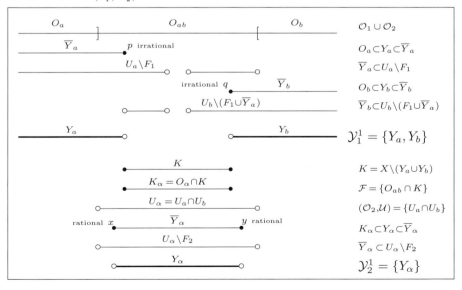

Fig. 38.3 The $\mathcal{Y}_1^1$ and $\mathcal{Y}_2^1$ constructions in the proof of Lemma 38.5.

38.4 EXAMPLE. The goal is to provide a model for the sets and constructions detailed in Lemma 38.5: Using Figure 38.3, we let $n = 1$ and $X = [0, 1]$ (X is the unit interval). Then the model for the cover $\mathcal{U}$ is $\{U_a, U_b\}$, and the models for the closed F_j are F_1 and F_2 are illustrated in the "Assumptions Box" of Figure 38.3. The "Constructions Box" provides models $\mathcal{Y}_1^1$ and $\mathcal{Y}_2^1$ of the discrete families $\mathcal{Y}_j^1$, $1 \leq j \leq n + 1$, in the conclusion of Lemma 38.5. By reading top-to-bottom in the "Constructions Box," one may track the constructions that appear in the proof of Lemma 38.5. The "vertical display format" provides a means to view the sets as subsets of X by vertically projecting onto the representation of X in the "Assumptions Box." $\qquad \square$

In Lemma 38.5 we construct the "$\mathcal{Y}_j^1$ families." The superscript "1" will currently appear superfluous. But ultimately each $\mathcal{Y}_j^1$ yields $\mathcal{U}_j^1$, which then determines $\mathcal{U}_j^2$ and $\mathcal{U}_j^3$. These families are fundamental (see Lemma 38.9).

38.5 Lemma (construction of the discrete open $\mathcal{Y}_j^1$) *For $n \geq 0$ let $\mathcal{U} = \{U_a : a \in A\}$ be a nodally indexed locally finite order $\leq n + 1$ open cover of X. For each $j \in \{1, \ldots, n + 1\}$, let $X_j \subset X$ satisfy $\dim X_j = 0$, and, let the closed $F_j \subset X$ satisfy $F_j \cap (\cup_{\mathcal{O}_j(\mathcal{U})} O) = \emptyset$. Then for each $j \in \{1, \ldots, n + 1\}$ there exists a size $\leq |A|$ discrete open family $\mathcal{Y}_j^1$ such that*

$$(7) \qquad \overline{Y} \cap F_j = \emptyset = B(Y) \cap X_j \qquad (Y \in \mathcal{Y}_j^1).$$

Moreover, $\cup_1^k \mathcal{Y}_j^1$ covers $\cup_1^k \mathcal{O}_j(\mathcal{U})$, and $\cup_1^{n+1} \mathcal{Y}_j^1$ covers X and refines $\mathcal{U}$.

PROOF. Let $j = 1$. The discrete closed family $\mathcal{O}_1(\mathcal{U}) = \mathcal{O}_1 = \{O_a : a \in A\}$ of 1-nodes shrinks the canonical $(\mathcal{O}_1, \mathcal{U})$ collection, i.e., $O_a \subset U_a$ for each $a \in A$. Moreover, since $F_1 \cap O_a = \emptyset$ when $a \in A$, each $O_a \subset U_a \setminus F_1$. Now let a_0 be the first member of (the well-ordered) $A = (A, <)$. Then for $a = a_0$, an application of the §A6.2 Theorem (open sets in the context of a 0-dimensional set) shows that an open set Y_a exists such that

$$(8) \qquad \begin{matrix} O_a \subset Y_a \subset \overline{Y}_a \subset U_a \setminus \left(F_1 \cup (\cup_{b<a} \overline{Y}_b) \right) \\ \overline{Y}_a \cap F_1 = \emptyset = B(Y_a) \cap X_1 \\ \mathcal{F}_a = \{F_1\} \cup \{\overline{Y}_b\}_{b \leq a} \cup \{O_b\}_{b>a} \text{ is closed discrete} \end{matrix}$$

where $\cup_{b<a} \overline{Y}_b = \emptyset$ when $a = a_0$.[2]

For the $j = 1$ *inductive step*, let $a_1 > a_0$ be fixed, assume that for each $a < a_1$ a Y_a that satisfies (8) has been constructed, and then consider

$$\mathcal{F}'_{a_1} = \{F_1\} \cup \{\overline{Y}_a\}_{a<a_1} \cup \{O_a\}_{a \geq a_1}.$$

Since $b < a < a_1$ implies $\overline{Y}_b$ and $\overline{Y}_a$ are disjoint, and, since each $Y_a \subset \overline{Y}_a \subset U_a \setminus \left(F_1 \cup (\cup_{b<a} \overline{Y}_b) \right)$, it follows that $\mathcal{F}'_{a_1} \setminus \{F_1\}$ is pairwise disjoint

---

[2] To see the validity of the last statement in (8), note that the clearly closed and pairwise-disjoint $\mathcal{F}_a$ is locally finite because $\mathcal{F}_a \setminus \{F_1\}$ shrinks $\mathcal{U}$.

and shrinks $\mathcal{U}$. So $\mathcal{F}'_{a_1}$ is a closed, pairwise-disjoint and locally finite family, i.e., it is closed discrete. For $a = a_1$, an open Y_a exists that satisfies (8). By transfinite construction, $\{\overline{Y}_a\}_{a \in A}$ exists with (8) satisfied by each Y_a. It follows that $\{\overline{Y}_a\}_{a \in A}$ is pairwise disjoint, closed, and locally finite, i.e., closed and discrete. Thus,

$$\mathcal{Y}_1^1 = \{Y_a : a \in A\}$$

is an open discrete collection of size $\leq |A|$ with each Y_a satisfying (8).

Next, let $k \in \{2, \ldots, n+1\}$ be fixed, and suppose, for $1 \leq j < k$, that each desired $\mathcal{Y}_j^1$ exists such that $\mathcal{Y}_{k-1} = \cup_1^{k-1} \mathcal{Y}_j^1$ is an open cover of $\cup_1^{k-1} \mathcal{O}_j(\mathcal{U})$. Then define $K = X \setminus (\cup\{Y : Y \in \mathcal{Y}_{k-1}\})$. It follows from the nodal closure property that $\mathcal{F} = \{O \cap K : O \in \mathcal{O}_k(\mathcal{U})\}$ is a discrete closed family.[3]

With the set $A_k = (A_k, <)$ of indices α of the k-nodes well ordered, the closed family $\mathcal{F}$ shrinks the open canonical $\mathcal{G} = (\mathcal{F}, \mathcal{U})$ collection, i.e., for each $\alpha \equiv a_1 \cdots a_k \in A_k$, $O_\alpha \subset U_\alpha = U_{a_1} \cap \cdots \cap U_{a_k}$. So each $O_\alpha \cap K \subset U_\alpha$. Moreover, since $F_k \cap O_\alpha = \emptyset$ for each such α, each $O_\alpha \cap K \subset U_\alpha \setminus F_k$. Now let $\alpha = \alpha_0$ be the first $\alpha \in A_k$. Then there exists an open set Y_α such that

(9)
$$O_\alpha \cap K \subset Y_\alpha \subset \overline{Y}_\alpha \subset U_\alpha \setminus \big(F_k \cup (\cup_{\beta < \alpha} \overline{Y}_\beta)\big)$$
$$\overline{Y}_\alpha \cap F_k = \emptyset = B(Y_\alpha) \cap X_k$$
$$\mathcal{F}_\alpha = \{F_k\} \cup \{\overline{Y}_\beta\}_{\beta \leq \alpha} \cup \{O_\beta \cap K\}_{\beta > \alpha} \text{ is closed discrete}$$

where $\cup_{\beta < \alpha} \overline{Y}_\beta = \emptyset$ when $\alpha = \alpha_0$. To finish, use the format of the "$j = 1$ inductive step" and the following substitutions: Substitute A_k for A; α for a; α_0 for a_0; α_1 for a_1; β for b; F_k for F_1; X_k for X_1; $O_\alpha \cap K$ for O_a; $O_\beta \cap K$ for O_b; the canonical $(\mathcal{O}_k, \mathcal{U})$ collection for $\mathcal{U}$; and (9) for (8). Thus the desired $\mathcal{Y}_k^1 = \{Y_\alpha : \alpha \in A_k\}$ exists, and by finite induction we are finished. $\square$

38.6 EXAMPLE. Consider Lemma 38.5 where $n = 0$ and where $X = X_1$ is the subspace of rationals in the unit interval. Then since $\mathcal{U}$ is a locally finite ord $\mathcal{U} = 1$ open cover of X, we have each $B(U) = \emptyset$ and $F_1 = \emptyset$ because $\mathcal{O}_1(\mathcal{U})$ covers X. In this case, we may simply define $\mathcal{Y}_1^1 = \mathcal{U}$. $\square$

38.7 Definition (property P_{n-1} and separated families)   Let $n \geq 0$. Then the closed subsets $F_1, \ldots, F_{n+1}$ of X *satisfy the property* P_{n-1} whenever $\dim (F_{j_1} \cap \cdots \cap F_{j_k}) \leq n - k$ for distinct $j_1, \ldots, j_k$. A family $\mathcal{N} = \{N_b : b \in B\}$ is a *separated family or separated collection* whenever $(\overline{N}_a \cap N_b) \cup (N_a \cap \overline{N}_b) = \emptyset$ for each distinct pair N_a and N_b of sets in $\mathcal{N}$.[4]

For an example of a separated family that is not a discrete family consider $\mathcal{N} = \{(0,1), (1,2)\}$ containing "adjacent" open intervals on the real line.

---

[3]The nodal closure property yields $\overline{O} \cap K = O \cap K$ when $O \in \mathcal{O}_k$, which makes $O \cap K$ closed because K is closed. Then (3) shows that $\mathcal{O}_k$ is locally finite, making $\mathcal{F}$ discrete.

[4]If a separated family $\mathcal{N} = \{N_b\}_B$ of subsets of a metric space (X, ρ) is also locally finite, then there exists a pairwise-disjoint open family $\mathcal{M} = \{M_b\}_B$ of sets $M_b = \{x \in X : \rho(x, N_b) < \rho(x, \cup_{a \neq b} \overline{N}_a)\}$ that puffs-up $\mathcal{N}$ and satisfies $M_b = \emptyset$ whenever $N_b = \emptyset$.

38.8 Lemma (refinements whose j-nodes avoid given F_j) *Let $n \geq 0$; let* dim $F \leq n$ *for the closed subspace F of X; let the closed $F_1, \ldots, F_{n+1} \subset F$ satisfy P_{n-1}; and let $\mathcal{U} = \{U_a : a \in A\}$ be a faithfully-indexed locally finite open cover of X. Then there is a locally finite open refinement $\mathcal{Y}$ of $\mathcal{U}$ that covers X; satisfies*

$$(10) \qquad F \subset \text{int}(\cup\{O : O \in \overset{n+1}{\underset{1}{\cup}}\mathcal{O}_k\}), \quad F_j \cap O = \emptyset \text{ when } O \in \mathcal{O}_j$$

for the nodal families $\mathcal{O}_j = \mathcal{O}_j(\mathcal{Y})$, $1 \leq j \leq n+1$; and has size $|\mathcal{Y}| \leq |A|$.

PROOF. For $n = 0$, an application of Lemma 38.2 provides the desired result: In detail, let F, X, and $\mathcal{U}$ be as specified in Lemma 38.8, and let $H = X$. Then the desired $\mathcal{Y}$ is the $\mathcal{Y}$ produced by Lemma 38.2, which is a precise locally finite open refinement of $\mathcal{U}$ that covers X and satisfies $\text{ord}_x \mathcal{Y} \leq n + 1 = 1$ for each $x \in \overline{G}$ where $G \supset F$ is open. So when $x \in G \cap Y_a$, then $x \notin Y_b$ for each $b \neq a$, i.e., x is in the 1-node $O_a \subset Y_a$. Thus, $G \cap Y_a \subset \text{int}(O_a)$, which yields the left side of (10). For the right side of (10), since $n + 1 = 1$ the lone $F_1 = F_{n+1}$ satisfies $P_{n-1} = P_{-1}$, i.e., $F_1 = \emptyset$. Thus, case $n = 0$ is valid. Now suppose that Lemma 38.8 is true for values less than $n \geq 1$, and let us focus on the nth case. Define

$$E = F_1 \cup \cdots \cup F_{n+1}, \quad E_j = \cup_k\{F_k \cap F_j : j < k \leq n+1\} \quad (1 \leq j \leq n).$$

Then dim $E \leq n - 1$ (use each dim $F_j \leq n - 1$ and the Sum Theorem) for the closed subspace E of X; the closed $E_1, \ldots, E_n \subset E$ satisfy P_{n-2}; and the given $\mathcal{U}$ covers X. Thus, with the inductive hypothesis satisfied, there is a locally finite open refinement $\mathcal{V}$ of $\mathcal{U}$ that covers X; satisfies

$$(11) \qquad E \subset \text{int}(\cup\{O : O \in \overset{n}{\underset{1}{\cup}}\mathcal{O}_k(\mathcal{V})\}), \quad E_j \cap O = \emptyset \text{ when } O \in \mathcal{O}_j(\mathcal{V})$$

for the nodal families $\mathcal{O}_j(\mathcal{V})$, $1 \leq j \leq n$; and has size $|\mathcal{V}| \leq |A|$.

Since $\mathcal{V}$ has size $\leq |A|$ we may assume that $\mathcal{V} = \{V_b : b \in B_1\}$, $|B_1| \leq |A|$, is the nodal indexing that satisfies (11). Letting $\beta \in B_j$ whenever β has length $|\beta| = j$ and $O_\beta \in \mathcal{O}_j(\mathcal{V})$, we define $B = \cup_1^{n+1}B_j$, and let

$$\mathcal{N} = \overset{n+1}{\underset{1}{\cup}}\mathcal{N}_j \text{ where } \mathcal{N}_j = \{O_\beta \cap F_j : O_\beta \in \mathcal{O}_j(\mathcal{V})\} = \{N_\beta : O_\beta \in \mathcal{O}_j\}.$$

Then $\mathcal{N} = \{N_\beta : \beta \in B\}$ is both locally finite and separated. (See 38.8.1 below for the proof.) So there exists a locally finite pairwise-disjoint open family $\mathcal{M} = \{M_\beta : \beta \in B\}$ that puffs-up $\mathcal{N}$. Let $\mathcal{G} = \{G_\beta : \beta \in B\}$ be the canonical $(\mathcal{N}, \mathcal{V})$ collection. Then $\mathcal{G}$ is locally finite and $\mathcal{N}$ shrinks $\mathcal{G}$.

Continuing the spadework, we introduce the family $\mathcal{L} = \{L_\beta : \beta \in B\}$:[5] Since each list $\beta \in B$ has a unique length $j \in \{1, \ldots, n+1\}$, the set

$$L_\beta = [M_\beta \cap G_\beta \cap \text{int}(\cup\{O : O \in \overset{n}{\underset{1}{\cup}}\mathcal{O}_k\})] \setminus [\overset{n+1}{\underset{j+1}{\cup}}F_k \cup (\cup\{O : O \in \overset{j-1}{\underset{1}{\cup}}\mathcal{O}_k\})],$$

---

[5]Intuitively, we desire a cover whose j-nodes do not meet F_j (as in (10)). Since we cannot be sure that $\mathcal{V}$ has this property, we construct $\mathcal{N}$, a family of "trouble spots" (i.e.,

where $\mathcal{O}_k = \mathcal{O}_k(\mathcal{V})$, is well defined. The family $\mathcal{L}$ is an open collection and $\mathcal{N}$ shrinks $\mathcal{L}$. (See 38.8.2 below for the proof.) So $|\mathcal{L}| \leq |A|$.

To distinguish sets in $\mathcal{L}$ from those of $\mathcal{V}$ (these families may not be disjoint), we let $B_1 \to \mathcal{V}$ and $B' \to \mathcal{L}$ be indexings where $B' = \{\beta' : \beta \in B\}$ with $\beta' = \{'\} \times \{\beta\} \neq \beta$ and $L_{\beta'} = L_\beta$. So $B_1 \cap B' = \emptyset$. Then, forming the disjoint union $B_1 \cup B'$, we define

$$(12) \qquad \mathcal{V}' = \mathcal{V} \cup \mathcal{L}, \qquad \mathcal{V}' = \{V'_\lambda : \lambda \in B_1 \cup B'\}.$$

As families of sets, since $\mathcal{L}$ shrinks $\mathcal{G}$, we see that $\mathcal{L}$ is a locally finite refinement of $\mathcal{V}$. And since $\mathcal{L}$ refines $\mathcal{V}$, and, $\mathcal{V}$ refines $\mathcal{U}$, the family $\mathcal{V}'$ refines $\mathcal{U}$. Furthermore, since both $\mathcal{V}$ and $\mathcal{L}$ are locally finite, and since $\mathcal{V}$ covers X, the union $\mathcal{V} \cup \mathcal{L} = \mathcal{V}'$ is locally finite and covers X.

The nodal families of $\mathcal{V}' = \{V'_\lambda : \lambda \in B_1 \cup B'\}$ satisfy the right side of (10). (See 38.8.3 below for the proof.) So next, we modify $\mathcal{V}'$ outside of $\cup_1^{n+1} F_j$ to obtain the desired $\mathcal{Y}$ that satisfies both statements in (10).

The modification involves yet another application of Lemma 38.2: We shall use $\mathcal{V}' = \{V'_a : a \in A\}$, $A = B_1 \cup B'$, for the $\mathcal{U}$ required by Lemma 38.2, and develop F' and H for the required F and H. First, an application of (11) and the indexing of $\mathcal{V}'$ show that

$$F_1 \cup \cdots \cup F_{n+1} = E \subset \mathrm{int}(\cup\{O : O \in \overset{n+1}{\underset{1}{\cup}} \mathcal{O}_j(\mathcal{V}')\}).$$

(See 38.8.4 for the proof.) Second, choose an open set D such that

$$E \subset D \subset \overline{D} \subset \mathrm{int}(\cup\{O : O \in \overset{n+1}{\underset{1}{\cup}} \mathcal{O}_j(\mathcal{V}')\}).$$

Third, for the "F required by Lemma 38.2" we use $F' = F \cap (X \setminus D)$, and for the "$H$ required by Lemma 38.2" we use $H = X \setminus E$. Then $F' \subset H$, F' is closed, H is open, and $\dim F' \leq n$. So for $\mathcal{V}' = \{V'_a : a \in A\}$, $A = B_1 \cup B'$, we apply Lemma 38.2, obtaining our desired $\mathcal{Y}$. (See 38.8.5 for the proof that $\mathcal{Y}$ satisfies (10).) $\qquad\square$

The triple indexing of "38.8.k" of the following five observations ($k = 1, 2, 3, 4, 5$) is used to remind the reader that each was applied in the proof of Lemma 38.8.

38.8.1 PROOF THAT $\mathcal{N}$ IS A SEPARATED FAMILY. If for each $\beta \in B$ we have $N_\beta = \emptyset$, then we are finished. Otherwise, suppose $O \cap F_j$ and $P \cap F_k$ are distinct members of $\mathcal{N}$. First, note that

$$(O \cap F_j) \cap \mathrm{cl}(P \cap F_k) \subset (O \cap F_j) \cap \overline{P} \cap F_k = (O \cap \overline{P}) \cap F_k \cap F_j.$$

---

a family of "$O_\beta \cap F_j$" where O_β is a j-node of $\mathcal{V}$). So we puff up these "spots" with the open sets in $\mathcal{L}$ and form the open cover $\mathcal{V}' = \mathcal{V} \cup \mathcal{L}$ with a nodal indexing that preserves the indexing of both $\mathcal{V}$ and $\mathcal{L}$. The hope is that since $x \in O_\beta \cap F_j$ implies $x \in O_\beta \in \mathcal{O}_j(\mathcal{V})$, then $x \in O_\gamma \in \mathcal{O}_{j+1}(\mathcal{V}')$ because x is also in L_β. As I now remember, however, the selection of $\mathcal{L}$ was nonintuitive and evolved from an iterative process of adjustments to arguments that eventually became the proof 38.8.3 given below.

If $j = k$, then since $O \cap F_j$ and $P \cap F_k$ are distinct, $O \cap P = \emptyset$. Consequently, the Nodal Closure Property yields $O \cap \overline{P} = \emptyset$. And from this equation we may show that if $j \neq k$, then

$$(O \cap \overline{P}) \cap F_k \cap F_j \subset O \cap \overline{P} \cap E_m = \emptyset \text{ where } m = \min\{j, k\},$$

because, if $j < k$, then $m = j$, $O \in \mathcal{O}_j(\mathcal{V})$, and, (11) shows $E_j \cap O = \emptyset$. On the other hand, if $j > k$, then $P \in \mathcal{O}_k(\mathcal{V})$, $O \in \mathcal{O}_j(\mathcal{V})$, and, the Nodal Closure Property yields $O \cap \overline{P} = \emptyset$. It follows that in every case the desired result holds.   □

38.8.2 PROOF THAT $\mathcal{L}$ IS AN OPEN FAMILY AND THAT $\mathcal{N}$ SHRINKS $\mathcal{L}$. Now $\cup_{j+1}^{n+1} F_k$ is closed, and Corollary 37.6 shows that $\cup\{O : O \in \cup_1^{j-1} \mathcal{O}_k\}$ is closed. So each L_β is open. For the $\mathcal{N}$ shrinks $\mathcal{L}$ proof, whenever $k \geq j + 1$ and each $O \in \mathcal{O}_j$, the right side of (11) shows that $O \cap F_j \cap F_k \subset O \cap E_j = \emptyset$. Thus, $\cup_{j+1}^{n+1} F_k$ and $N_\beta = O_\beta \cap F_j$ are disjoint. And since the nodal family $\mathcal{O}(\mathcal{V})$ is pairwise disjoint, $N_\beta \subset O_\beta \in \mathcal{O}_j$ and $\cup\{O : O \in \cup_1^{j-1} \mathcal{O}_k\}$ are also disjoint. Finally, $N_\beta \subset M_\beta \cap G_\beta$, and, $N_\beta \subset F_j \subset E$ and the left side of (11) show that $N_\beta \subset \text{int}(\cup\{O : O \in \cup_1^n \mathcal{O}_k\})$. It follows that $N_\beta \subset L_\beta$.   □

38.8.3 PROOF THAT $\mathcal{O}(\mathcal{V}')$ SATISFIES THE RIGHT-SIDE PROPERTY OF (10). Suppose otherwise. Then let $x \in F_j \cap O'$ where O' is a j-node of $\mathcal{V}'$. Since $\mathcal{L}$ is pairwise disjoint, either $x \in O_\beta \in \mathcal{O}_j(\mathcal{V})$ and x is in no member of $\mathcal{L}$, or, $x \in O_\beta \in \mathcal{O}_{j-1}(\mathcal{V})$ and x is in a member of $\mathcal{L}$. If the length $|\beta| = j$, then $x \in O_\beta \cap F_j = N_\beta \subset L_\beta = L_{\beta'}$, which contradicts "$x$ is in no member of $\mathcal{L}$." So we must have $|\beta| = j-1$ and $x \in O_\beta \cap F_j \cap L_\lambda$ for some λ. By the definition of L_λ, however, $L_\lambda \cap O_\beta = \emptyset$ unless $j - 1 \geq |\lambda|$. But then $j > |\lambda|$, and by the definition of L_λ, $L_\lambda \cap F_j = \emptyset$. So no L_λ contains x, which contradicts "x is in a member of $\mathcal{L}$." Thus, since all cases yield contradictions, we conclude that the statement "$x \in F_j \cap O'$ where O' is a j-node of $\mathcal{V}'$" is false.   □

38.8.4 PROOF THAT $\cup_1^{n+1} F_j = E \subset \text{int}\left(\cup\{O : O \in \cup_1^{n+1} \mathcal{O}_k(\mathcal{V}')\}\right)$. By (11), $x \in E$ implies $x \in G_x = \text{int}\left(\cup\{O : O \in \cup_1^n \mathcal{O}_k(\mathcal{V})\}\right)$. But $y \in G_x$ implies $y \in O \in \mathcal{O}_k(\mathcal{V})$ for some $k \leq n$. If $y \in L_\beta$ for some β, then, since $\mathcal{L}$ is pairwise disjoint, $y \in O_{k+1} \in \mathcal{O}_{k+1}(\mathcal{V}')$; or, if $y \notin L_\beta$ for every β, then $y \in O_k \in \mathcal{O}_k(\mathcal{V}')$. So $x \in G_x \subset \text{int}\left(\cup\{O : O \in \cup_1^n \mathcal{O}_k(\mathcal{V})\}\right)$.   □

38.8.5 PROOF THAT $\mathcal{Y}$ SATISFIES THE LEFT SIDE OF (10). Note that $F = [F \cap (X \setminus D)] \cup [F \cap D] = F' \cup (F \cap D)$ is a partition of F. And also recall that $F' \subset G \subset \overline{G} \subset H = X \setminus E$ and $E \subset D$. (See the graphic on the following page.) So $x \in F$ implies either $x \in G$ or $x \in D$. First, suppose $x \in G$: Then Lemma 38.2 shows that $x \in G \subset \text{int}\left(\cup\{O : O \in \cup_1^{n+1} \mathcal{O}_k(\mathcal{Y})\}\right)$. Otherwise, $x \in D$: Since $\mathcal{Y} = \{Y_a : a \in A\}$ precisely refines $\mathcal{V}' = \{V'_a : a \in A\}$, each $y \notin V'_a$ must satisfy $y \notin Y_a$, i.e., each $y \in X$ satisfies $\text{ord}_y \mathcal{Y} \leq \text{ord}_y \mathcal{V}'$. So any y satisfying $y \in O' \in \mathcal{O}_m(\mathcal{V}')$, must also satisfy $y \in O \in \mathcal{O}_k(\mathcal{Y})$ for $k \leq m$.

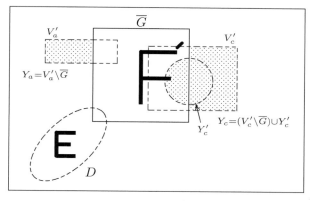

It follows, since $D \subset \cup\{O : O \in \cup_1^{n+1} \mathcal{O}_m(\mathcal{V}')\}$, that $D \subset \cup\{O : O \in \cup_1^{n+1} \mathcal{O}_k(\mathcal{Y})\}$. But D is open so $D \subset \text{int} \, (\cup\{O : O \in \cup_1^{n+1} \mathcal{O}_k(\mathcal{Y})\})$. Thus, $x \in \text{int} \, (\cup\{O : O \in \cup_1^{n+1} \mathcal{O}_k(\mathcal{Y})\})$. $\square$

38.9 Lemma ($\mathcal{U}_j$ decompositions)[6] *Let* $\dim X = n \geq 0$; $w(X) = |A| \geq \aleph_0$; *and partition* $X = \cup_1^{n+1} X_j$ *where each* $\dim X_j = 0$. *Let* $\mathcal{W}_1, \ldots, \mathcal{W}_{n+1}$ *be decompositions of* X; *each* $|\mathcal{W}_j| \leq |A|$; *each* $\text{lord } \mathcal{W}_j \leq 2$; *and each* $\text{Bdry } \mathcal{W}_j = \cup_{F \in \mathcal{F}_j} F$ *where* $|\mathcal{F}_j| \leq |A|$ *is closed discrete and shrinks an open family* $\mathcal{G}_j$. *Let* $\text{Bdry } \mathcal{W}_1, \ldots, \text{Bdry } \mathcal{W}_{n+1}$ *satisfy* P_{n-1}; *and let* $\mathcal{V}$ *be an open cover of* X. *Then pairwise-disjoint triplets* $\mathcal{U}_1^1, \mathcal{U}_1^2, \mathcal{U}_1^3; \ldots; \mathcal{U}_{n+1}^1, \mathcal{U}_{n+1}^2, \mathcal{U}_{n+1}^3$ *of discrete families exist and each decomposition* $\mathcal{U}_j = \cup_1^3 \mathcal{U}_j^k$ *of* X *satisfies:*

(13) $\text{lord } \mathcal{U}_j \leq 2$;

(14) $\text{cl}\, \mathcal{U}_j^1$ *refines* $\mathcal{V}$ *and* $\overset{n+1}{\underset{1}{\cup}} \mathcal{U}_j^1$ *covers* X;

(15) $x \in \text{Bdry } \mathcal{U}_j$ *implies distinct* $U_1, U_2 \in \mathcal{U}_j$ *where* $x \in B(U_1) \cap B(U_2)$;

(16) $\mathcal{U}_j$ *covers* X_j;

(17) $\text{Bdry } \mathcal{U}_j \cap \text{Bdry } \mathcal{W}_j = \emptyset$;

(18) $\mathcal{U}_j^1 \cup \mathcal{U}_j^3$ *refines* $\mathcal{W}_j$;

(19) $\mathcal{U}_j^1 \cup \mathcal{U}_j^2$ *is a discrete collection; and*

(20) $\mathcal{U}_j^2$ *puffs-up* $\mathcal{F}_j$ *in* $\mathcal{G}_j$ *and* $\text{cl}\, \mathcal{U}_j^2$ *shrinks* $\mathcal{G}_j$.

PROOF. Apply Lemma 38.8: Let $F_1 = \text{Bdry } \mathcal{W}_1, \ldots, F_{n+1} = \text{Bdry } \mathcal{W}_{n+1}$ where $F = X$; and let $\mathcal{U} = \{U_a : a \in A\}$ in the hypothesis of Lemma 38.8 be a locally finite open cover of X such that $\{\overline{U}_a : a \in A\}$ refines $\mathcal{V}$. Then the conclusion of Lemma 38.8 provides a locally finite open refinement $\mathcal{Y}$ of $\mathcal{U}$ that covers X and satisfies (10). And since $F = X$, $\text{ord } \mathcal{Y} \leq n + 1$. Using $\mathcal{Y}$, the X_j, and the $F_j = \text{Bdry } \mathcal{W}_j$, we apply Lemma 38.5 and thereby obtain

---

[6]Recall that "$w(X)$" denotes the weight of X; if W is open, then boundary $B(W) = \overline{W} \setminus W$; for an open family $\mathcal{W}$, $\text{Bdry } \mathcal{W} = \cup\{B(W) : W \in \mathcal{W}\}$; the definition of decomposition (footnote 1 of this chapter); and for any family $\mathcal{S}$ of subsets of X, $\text{cl}\, \mathcal{S} = \{\overline{S} : S \in \mathcal{S}\}$, $\cup \text{cl}\, \mathcal{S} = \cup\{\text{cl}(S) : S \in \mathcal{S}\} = \cup\{\overline{S} : S \in \mathcal{S}\}$.

discrete open families $\mathcal{Y}_1^1, \ldots, \mathcal{Y}_{n+1}^1$, each of size $\leq |A|$ such that

$$\overline{Y} \cap F_j = \emptyset = B(Y) \cap X_j \qquad (Y \in \mathcal{Y}_j^1), \tag{21}$$

and the union $\cup_1^{n+1} \mathcal{Y}_j^1$ covers X and refines $\mathcal{Y}$. Each of these $n+1$ collections $\mathcal{Y}_j^1$ yields a corresponding offspring, namely, the open family

$$\mathcal{U}_j^1 = \{Y \cap W : Y \in \mathcal{Y}_j^1;\ W \in \mathcal{W}_j\}. \tag{22}$$

Each $\mathcal{U}_j^1$ has size $\leq |A|$, is discrete, and satisfies $(\mathrm{Bdry}\,\mathcal{U}_j^1) \cap X_j = \emptyset$. (See 38.9.1 and 38.9.2 for the proofs.) Also, $\overline{Y \cap W} \subset \overline{Y}$ shows that $\cup \mathrm{cl}\,\mathcal{U}_j^1 \subset \cup\{\overline{Y} : Y \in \mathcal{Y}_j^1\}$. Thus, (21) yields each $(\cup \mathrm{cl}\,\mathcal{U}_j^1) \cap (\cup\{F : F \in \mathcal{F}_j\}) = \emptyset$. Now assume that $\mathcal{G}_j$ is pairwise disjoint.[7]

Then puff up $\mathcal{F}_j$ in $\mathcal{G}_j$ to a discrete open family $\mathcal{U}_j^2$, i.e., for each $F \in \mathcal{F}_j$ and corresponding $G_F \in \mathcal{G}_j$ there is an open U_F such that

$$\mathcal{U}_j^2 = \{U_F : F \in \mathcal{F}_j\} \text{ where } F \subset U_F \subset \overline{U}_F \subset G_F \cap (X \setminus \cup \mathrm{cl}\,\mathcal{U}_j^1). \tag{23}$$

Thus each $\mathcal{U}_j^2$ has size $\leq |A|$, covers Bdry $\mathcal{W}_j$, and, since $\dim X_j = 0$, we may assume that $U \in \mathcal{U}_j^2$ implies $B(U) \cap X_j = \emptyset$. From (23) we also have

$$(\cup \mathrm{cl}\,\mathcal{U}_j^1) \cap (\cup \mathrm{cl}\,\mathcal{U}_j^2) = \emptyset. \tag{24}$$

Next, for each j, $1 \leq j \leq n+1$, define

$$\mathcal{U}_j^3 = \{W \cap [X \setminus (\cup \mathrm{cl}\,\mathcal{U}_j^1 \ \cup\ \cup \mathrm{cl}\,\mathcal{U}_j^2)] : W \in \mathcal{W}_j\}. \tag{25}$$

Then each $|\mathcal{U}_j^3| \leq |A|$. With the triplets $\mathcal{U}_j^1$, $\mathcal{U}_j^2$, $\mathcal{U}_j^3$ defined, let $\mathcal{U}_j = \cup_1^3 \mathcal{U}_j^k$. The proofs that the $\mathcal{U}_j$ satisfy (13), ..., (20) appear in 38.9.5, ..., 38.9.9. □

38.9.1 Proof that each $\mathcal{U}_j^1$ is discrete. From (21) and $F_j = \mathrm{Bdry}\,\mathcal{W}_j$,

$$\cup\{\overline{Y} : Y \in \mathcal{Y}_j^1\} \cap \mathrm{Bdry}\,\mathcal{W}_j = \emptyset. \tag{26}$$

If $x \in G_x = X \setminus \cup\{\overline{Y} : Y \in \mathcal{Y}_j^1\}$, then G_x is a neighborhood of x that meets no member of $\mathcal{U}_j^1$. Otherwise, $x \in \overline{Y}_x \in \mathcal{Y}_j^1$. Then by (26), $x \in W_x \in \mathcal{W}_j$. Since $\mathcal{Y}_j^1$ is discrete there is a neighborhood G_x that meets only one $Y \in \mathcal{Y}_j^1$, namely Y_x. So $G_x \cap W_x$ meets only one member $Y_x \cap W_x$ of $\mathcal{U}_j^1$. □

38.9.2 Proof that each $(\mathrm{Bdry}\,\mathcal{U}_j^1) \cap X_j = \emptyset$. Let $x \in \mathrm{Bdry}\,\mathcal{U}_j^1$. Since $\mathcal{U}_j^1$ is discrete there is a unique $Y \in \mathcal{Y}_j^1$ and unique $W \in \mathcal{W}_j$ such that $x \in B(Y \cap W) = \mathrm{cl}\,(Y \cap W) \setminus (Y \cap W)$. So $x \in (\overline{Y} \cap \overline{W}) \setminus (Y \cap W)$ shows that $x \in \overline{Y} = B(Y) \cup Y$, which implies $x \in B(Y)$ or $x \in Y$. But $x \in Y$ and

---

[7]Since $\mathcal{F}_j$ is a separated family, we may use the "$\mathcal{M}$ construction" in footnote 4 of this chapter, and then replace each $G_F \supset F$ with $M_F \cap G_F$ where $F \subset M_F \cap G_F$ for each $F \in \mathcal{F}_j$ and corresponding $G_F \in \mathcal{G}_j$.

$x \notin Y \cap W$ imply $x \notin W$. Thus $x \in \text{Bdry } \mathcal{W}_j$, which contradicts (26). So we are left with $x \in B(Y)$. Then (21) shows $x \notin X_j$. $\qquad\square$

38.9.3 PROOF THAT EACH $\mathcal{U}_j$ IS A DECOMPOSITION. The open family $\mathcal{U}_j$ is pairwise disjoint because $\mathcal{U}_j^1$ is discrete (38.9.1); $\mathcal{U}_j^2$ is discrete (use (23) where $\mathcal{F}_j$ is discrete and $\mathcal{G}_j$ pairwise disjoint), and, $\mathcal{U}_j^1$ and $\mathcal{U}_j^2$ satisfy (25); and $\mathcal{U}_j^3$ satisfies (25) where $\mathcal{W}_j$ is pairwise disjoint. Also, cl $\mathcal{U}_j$ covers X because $x \in (\cup \text{cl} \mathcal{U}_j^1 \cup \cup \text{cl} \mathcal{U}_j^2)$, or, $x \in U \in \mathcal{U}_j^3$. $\qquad\square$

38.9.4 PROOF THAT $\text{Bdry } \mathcal{U}_j^3 \subset \text{Bdry } \mathcal{U}_j^1 \cup \text{Bdry } \mathcal{U}_j^2$. For any open sets W and R, we have $B(W \cap R) \subset (B(W) \cap \overline{R}) \cup (\overline{W} \cap B(R))$. In particular, this formula holds for $B(U) \subset \text{Bdry } \mathcal{U}_j^3$ if $U = W \cap R$ where $W \in \mathcal{W}_j$ and $R = X \setminus (\cup \text{cl} \, \mathcal{U}_j^1 \cup \cup \text{cl} \, \mathcal{U}_j^2)$. The set $(B(W) \cap \overline{R})$, however, is empty because $B(W)$ is covered by a union of members of $\mathcal{U}_j^2$. So

$$B(U) = B(W \cap R) \subset (\overline{W} \cap B(R)) \subset \text{Bdry } \mathcal{U}_j^1 \cup \text{Bdry } \mathcal{U}_j^2$$

because $B(R) \cap R = \emptyset$ and $\overline{R}$ meets no member of $\mathcal{U}_j^1 \cup \mathcal{U}_j^2$. $\qquad\square$

38.9.5 PROOF OF (13). Since $\mathcal{U}_j$ is a decomposition, $x \in U_x \in \mathcal{U}_j$ implies $\text{lord}_x \mathcal{U}_j = 1$. So let $x \in \text{Bdry } \mathcal{U}_j = \text{Bdry } \mathcal{U}_j^1 \cup \text{Bdry } \mathcal{U}_j^2$ where equality follows from 38.9.4. Since Bdry $\mathcal{W}_j$ is covered by $\mathcal{U}_j^2$, we conclude that $x \in W$ for a unique $W \in \mathcal{W}_j$. So select an open $G_x \subset W$ that contains x and meets at most one member of $\mathcal{U}_j^1 \cup \mathcal{U}_j^2$. Then $\text{lord}_x \mathcal{U}_j \leq 2$. $\qquad\square$

38.9.6 PROOF OF (14). Using "$\mathcal{A} < \mathcal{B}$" to denote "$\mathcal{A}$ refines $\mathcal{B}$," we recall

$$(27) \qquad \mathcal{U}_j^1 \overset{38.9}{<} \mathcal{Y}_j^1 < \overset{n+1}{\underset{1}{\cup}} \mathcal{Y}_j^1 \overset{38.5}{<} \mathcal{Y} \overset{38.8}{<} \mathcal{U} < \text{cl} \mathcal{U} \overset{\substack{X \text{ paracompact} \\ X \text{ normal}}}{<} \mathcal{V}$$

where the "data above $<$" indicates the assumption/lemma producing the refinement. Thus, (27) shows that cl $\mathcal{U}_j^1$ refines $\mathcal{V}$. To see that $\cup_1^{n+1} \mathcal{U}_j^1$ covers X, let $x \in X$. With Lemma 38.8, and $\mathcal{U}$ and $F = X$ as input, $\mathcal{Y}$ exists and ord $\mathcal{Y} \leq n + 1$. Using $\mathcal{Y}$ as the locally finite cover in the hypothesis of Lemma 38.5, we produce $n + 1$ families $\mathcal{Y}_j^1$ whose union covers X and refines $\mathcal{Y}$. So x is in some k-node of $\mathcal{Y}$, i.e., $x \in Y_x \in \mathcal{Y}_j^1$ for some j. Then from (7), $x \notin F_j = \text{Bdry } \mathcal{W}_j$, and so must be in some $W_x \in \mathcal{W}_j$. Thus, $x \in Y_x \cap W_x \in \mathcal{U}_j^1$. $\qquad\square$

38.9.7 PROOF OF (15). We show how to ensure that (15) is true. The idea is motivated by a simple example. Consider the decomposition $\mathcal{U} = \{(0, 1/2)(1/2, 1]\}$ of the unit interval $I = [0, 1]$. Then Bdry $\mathcal{U} = \{0, 1/2\}$, but only one of these boundary points satisfies (15). The solution? Replace $(0, 1/2)$ with $[0, 1/2)$, thereby creating $\mathcal{D} = \{[0, 1/2), (1/2, 1]\}$ that does satisfy (15). In general, for $U \in \mathcal{U}_j^k$, $k = 1, 2, 3$, define

$$D = D_U = X \setminus \cup \{\overline{V} : U \neq V \in \mathcal{U}_j\} \in \mathcal{D}_j^k, \qquad \mathcal{D}_j = \mathcal{D}_j^1 \cup \mathcal{D}_j^2 \cup \mathcal{D}_j^3.$$

Since $\mathcal{U}_j$ is a decomposition, for $D = D_U$, we have

(i) $\overline{D} = \overline{U}$,
(ii) $U \subset D$
(iii) $B(D) \subset B(U)$.

Then the $\mathcal{D}_j$ families satisfy (13), ..., (20) whenever the $\mathcal{U}_j$ satisfy (13), (14), and (16), ..., (20):

First, (i) yields that $\mathrm{cl}\,\mathcal{D}_j$ covers X whenever $\mathrm{cl}\,\mathcal{U}_j$ covers X. And the open family $\mathcal{D}_j$ is pairwise disjoint because $\mathcal{U}_j$ is pairwise disjoint. So each $\mathcal{D}_j$ is a decomposition of X. Second, let us consider properties (13), ..., (20):

For (13), $\mathrm{lord}_x\,\mathcal{U}_j \leq 2$ implies that x meets at most two members of $\mathrm{cl}\,\mathcal{U}_j$, so (i) yields $\mathrm{lord}_x\,\mathcal{D}_j \leq 2$. For (14), apply (i) and (ii).

For (15), let $x \in B(D_U) \subset \mathrm{Bdry}\,\mathcal{D}_j$. Then by (iii), $x \in B(U)$, and there must exist a another $U_1 \in \mathcal{U}_j$ such that $x \in B(U_1)$. (Otherwise, x has a neighborhood G_x that does not meet $\overline{V}$ for each $V \neq U$ in $\mathcal{U}_j$, which places $x \in \mathrm{int}(D_U)$). So $x \in B(U) \cap B(U_1)$. Moreover, $x \notin D_{U_1}$ because $U \neq U_1$, $x \in \overline{U}$, and $\overline{U} \cap D_{U_1} = \emptyset$. But $x \in \overline{D}_{U_1}$ because $x \in \overline{U}_1$ and (i) holds. Thus $x \in B(D_U) \cap B(D_{U_1})$.

For (16) and (17) use (ii) and (iii), respectively.

For (18), "$\mathrm{Bdry}\,\mathcal{U}_j \cap \mathrm{Bdry}\,\mathcal{W}_j = \emptyset$" coupled with "$\mathcal{U}_j^1 \cup \mathcal{U}_j^3$ refines $\mathcal{W}_j$" implies "$\mathrm{cl}\,(\mathcal{U}_j^1 \cup \mathcal{U}_j^3)$ refines $\mathcal{W}_j$." Then apply (i).

For (19), note that if $\mathcal{U}_j^1 \cup \mathcal{U}_j^2$ is discrete, then $\mathrm{cl}\,(\mathcal{U}_j^1 \cup \mathcal{U}_j^2)$ is discrete. Then apply (i). For (20), use (ii), and then (i) with "$\mathrm{cl}\,\mathcal{U}_j^2$ shrinks $\mathcal{G}_j$."

$\square$

38.9.8 PROOF OF (16). The comments following (22) and (23) show ($\mathrm{Bdry}\,\mathcal{U}_j^1$) $\cap X_j = \emptyset$ and, since $\mathcal{U}_j^2$ discrete, ($\mathrm{Bdry}\,\mathcal{U}_j^2$) $\cap X_j = \emptyset$. Then by 38.9.4, $\mathrm{Bdry}\,\mathcal{U}_j^3 \subset \mathrm{Bdry}\,\mathcal{U}_j^1 \cup \mathrm{Bdry}\,\mathcal{U}_j^2$. So the decomposition $\mathcal{U}_j$ covers X_j. $\square$

38.9.9 PROOFS OF (17), (18), (19), AND (20). Statement (17) follows because $\mathcal{U}_j^2$ covers $\mathrm{Bdry}\,\mathcal{W}_j$. The definitions of $\mathcal{U}_j^1$, $\mathcal{U}_j^2$, and $\mathcal{U}_j^3$ yield (18) and (20). Claims 38.9.1 and those surrounding (23) and (24) yield (19). $\square$

We close this section with an abstract illustration of a $\mathcal{U}_j$ decomposition.

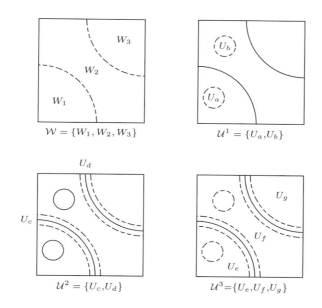

§39 The Decomposition Theorem

In this section, we state and prove the Decomposition Theorem.

39.1 Theorem (Decomposition Theorem for Finite-Dimensional Spaces)
For $n \geq 0$, *let* $X = X_1 \cup \cdots \cup X_{n+1}$ *be a partition of the n-dimensional weight* $|A| \geq \aleph_0$ *metric space X where each* $\dim X_j = 0$. *For each closed subset $F \subset X$, let* $\{G_i^F : i \geq 1\}$ *be a countable collection of open sets such that* $F = \cap_i G_i^F$. *Then there exist decompositions* $\mathcal{W}_{ij} = \{W_a : a \in A\}$, $i \geq 1$, $1 \leq j \leq n+1$, *with the following properties: For each* $i \geq 1$,

$(28)_i$ *for each j, lord $\mathcal{W}_{ij} \leq 2$;*

$(29)_i$ *for each $x \in X$, $x \in W_x \in \mathcal{W}_{ij}$ for some j where* $\mathrm{dia}(W_x) < 1/i$, *and, for each $k < i$, a $W_k \in \mathcal{W}_{kj}$ exists such that $W_x \subset W_k$;*

$(30)_i$ *if $x \in \mathrm{Bdry}\, \mathcal{W}_{ij}$, then $x \in B(W) \cap B(W')$ for $W \neq W'$ in $\mathcal{W}_{ij}$;*

$(31)_i$ *each $\mathcal{W}_{ij}$ covers X_j;*

$(32)_i$ *for $k < i$ and each j, $\mathrm{Bdry}\, \mathcal{W}_{ij} \cap \mathrm{Bdry}\, \mathcal{W}_{kj} = \emptyset$;*

$(33)_i$ *if $B(U) \cap B(V) \neq \emptyset$ for $U \neq V$ in $\mathcal{W}_{ij}$, then $W \in \{U, V\}$ exists where $W \subset W_k \in \mathcal{W}_{kj}$ for each $k < i$;*

$(34)_i$ *for $k < i$ and each j, $W \in \mathcal{W}_{ij}$ meets at most two members of $\mathcal{W}_{kj}$, and, for all but at most one of the $k < i$, $W \subset W_k \in \mathcal{W}_{kj}$;*

$(35)_i$ *for $k < i$ and each j, if $B(U) \cap B(V) \neq \emptyset$ for distinct $U, V \in \mathcal{W}_{kj}$, then $W \in \mathcal{W}_{ij}$ exists such that $B(U) \cap B(V) \subset W \subset G_i^{B(U) \cap B(V)}$;*

$(36)_i$ *for $k < i$ and each j, $W \in \mathcal{W}_{ij}$ meets $\mathrm{Bdry}\, \mathcal{W}_{kj}$ only if a unique distinct pair $U, V \in \mathcal{W}_{kj}$ exists such that $B(U) \cap B(V) \subset W$.*

PROOF. The proof is by induction on the index i. For $i = 1$ we apply Lemma 38.5: Let $\mathcal{U}$ be a size $\leq |A|$ locally finite order $\leq n + 1$ open cover of X such that mesh $\mathcal{U} < 1$. Let $F_1 = \cdots = F_{n+1} = \emptyset$, and suppose that the X_j are as specified in Theorem 39.1. Then Lemma 38.5 provides the discrete open families $\mathcal{Y}_j^1$ of size $\leq |A|$ that satisfy (7) and the covering properties in Lemma 38.5. We define

$$\mathcal{W}_{1j} = \mathcal{Y}_j^1 \cup \{X \setminus (\cup \operatorname{cl} \mathcal{Y}_j^1)\} \qquad (1 \leq j \leq n + 1).$$

Then $(32)_1, \ldots, (36)_1$ are vacuously satisfied; and $(28)_1$, $(29)_1$, and $(31)_1$ follow from those of the $\mathcal{Y}_j^1$ and the definition of $\mathcal{W}_{1j}$. Property $(30)_1$ may be obtained by modifying each $\mathcal{W}_{1j}$ as in the proof of (15) within 38.9.7.[8]

Thus, Theorem 39.1 holds for $i = 1$.

Now suppose $i > 1$ and that Theorem 39.1 is true for each $k \leq i - 1$. We shall construct the $\mathcal{W}_{ij}$ by applying Lemma 38.9: Let $\mathcal{V}$ be an open cover of X whose mesh is $\leq 1/i$. Then define

$$\mathcal{W}_j = \{W_1 \cap \cdots \cap W_{i-1} : W_k \in \mathcal{W}_{kj}; \ 1 \leq k \leq i - 1\} \qquad (1 \leq j \leq n + 1).$$

Each $\mathcal{W}_j$ is a size $\leq |A|$ decomposition of X such that

$$(37) \qquad \operatorname{Bdry} \mathcal{W}_j = \cup\{\operatorname{Bdry} \mathcal{W}_{kj} : 1 \leq k \leq i - 1\}, \qquad \operatorname{lord} \mathcal{W}_j \leq 2.[9]$$

Further, from $(31)_k$ for $1 \leq k \leq i - 1$, and the Decomposition Theorem (see §A6.2), Bdry $\mathcal{W}_1, \ldots,$ Bdry $\mathcal{W}_{n+1}$ satisfy P_{n-1}. For each j, define

$$\mathcal{F}_j = \{B(U) \cap B(V) : \text{ for some } k \leq i - 1, \ U, V \in \mathcal{W}_{kj} \text{ and } U \neq V\}.$$

Using $(30)_k$ for $1 \leq k \leq i - 1$, and applying (37), we may also show that Bdry $\mathcal{W}_j = \cup\{F : F \in \mathcal{F}_j\}$. In addition, for $1 \leq k \leq i - 1$, $(28)_k$ implies

$$\{B(U) \cap B(V) : U \neq V; U, V \in \mathcal{W}_{kj}\}$$

is a discrete collection. Then we may use $(32)_{i-1}$ to show that for each j, the closed family $\mathcal{F}_j$ is discrete. (Note that $|\mathcal{F}_j| \leq |A|$.)

Continuing, we now define the $\mathcal{G}_j$: For $k \in \{1, \ldots, i - 1\}$ fixed, let

$$F_{kj} = F_{kj}^{U,V} = B(U) \cap B(V) \in \mathcal{F}_j \qquad U, V \in \mathcal{W}_{kj}.$$

---

[8]In "38.9.7 PROOF OF (15)" replace $\mathcal{U}_j$ with $\mathcal{W}_{1j}$, and define $D_W = X \setminus \cup\{\overline{W'} : W \neq W' \in \mathcal{W}_{1j}\} \in \mathcal{D}_j$. Then for the $\mathcal{W}_{1j}$-induced decomposition $\mathcal{D}_j$, the argument that yielded (13) yields $(28)_1$; dia$(W) =$ dia$(\overline{W})$ and $\overline{W} = \overline{D}_W$ yield $(29)_1$; and $W \subset D_W$ yields $(31)_1$.

[9]The proof that Bdry $\mathcal{W}_j$ is a subset of the right-side set in (37) is straightforward. For the reverse inclusion, let $x \in$ Bdry $\mathcal{W}_{kj}$ for some $k \leq i - 1$. By $(30)_k$, $x \in B(W) \cap B(W')$ for distinct $W, W' \in \mathcal{W}_{kj}$. Then since $\mathcal{W}_{mj}$, $m \leq i - 1$, is a decomposition of X, use $(32)_\ell$ for $\ell = \max\{m, k\}$ to specify (for $m \neq k$) $W_m \in \mathcal{W}_{mj}$ such that $x \in W_m$. So for $W_k \in \{W, W'\}$, $x \in B(W_1 \cap \cdots \cap W_{i-1})$. Next, lord $\mathcal{W}_j \leq 2$: For $x \in B(W) \cap B(W') \subset$ Bdry $\mathcal{W}_{kj}$, an open G_x, $x \in G_x$, meets only the two $W, W' \in \mathcal{W}_{kj}$. With the W_m above, $x \in G = W_1 \cap \cdots \cap W_{k-1} \cap G_x \cap W_{k+1} \cap \cdots \cap W_{i-1}$, and G meets only two sets in $\mathcal{W}_j$.

If $i = 2$, let

(38) $$W_{kj} = X \supset F_{kj}.$$

If $i \geq 3$ and $k = i - 1$, use $(32)_{i-1}$ and $(33)_{i-1}$ to determine a unique[10]

(39) $$W_{kj} = \cap\{W_m : 1 \leq m < i - 1 = k; W_m \in \mathcal{W}_{mj}\} \supset F_{kj}.$$

If $i \geq 3$ and $k < i - 1$, use $(34)_{i-1}$ and $(35)_{i-1}$ to determine a unique[11]

(40) $$W_{kj} = \cap\{W_m : 1 \leq m \leq i - 1; m \neq k; W_m \in \mathcal{W}_{mj}\} \supset F_{kj}.$$

Also, since lord $\mathcal{W}_{kj} \leq 2$ an open set H exists where $F_{kj} \subset H$ and H meets at most two members of $\mathcal{W}_{kj}$. Then, for $F = F_{kj}^{U,V} = B(U) \cap B(V)$, let

$$G_F = H \cap G_i^F \cap W_{kj}$$

For each j, define $\mathcal{G}_j = \{G_F : F \in \mathcal{F}_j\}$. Then each $\mathcal{F}_j$ shrinks $\mathcal{G}_j$. Thus, we may now apply Lemma 38.9, and the $\mathcal{U}_j$ thus produced yield

$$\mathcal{W}_{ij} = \mathcal{U}_j \qquad (1 \leq j \leq n + 1).$$

(For the proof that these $\mathcal{W}_{ij}$ satisfy $(28)_i, \ldots, (36)_i$ see 39.1.1.) Thus, with the induction step complete, we are finished. ◻

39.1.1 PROOF THAT THE $\mathcal{W}_{ij}$ SATISFY $(28)_i, \ldots, (36)_i$. Note that each $\mathcal{W}_{ij} = \mathcal{U}_j = \cup_1^3 \mathcal{U}_j^k$ and $\mathcal{U}_j$ is a decomposition. Now consider the following properties:

For $(28)_i$, apply (13).

For $(29)_i$, (14) provides that $\cup_1^{n+1} \mathcal{U}_j^1$ covers X. So $x \in X$ implies $x \in W_x \in \mathcal{U}_j^1 \subset \mathcal{W}_{ij}$ for some j. And (14) also provides that each cl $\mathcal{U}_j^1$ refines $\mathcal{V}$, making $\mathrm{dia}(W_x) \leq \mathrm{mesh}\ \mathcal{V} \leq 1/i$. In addition, by (18), $\mathcal{U}_j^1$ refines $\mathcal{W}_j$, and so the definition of $\mathcal{W}_j$ shows that for each $k < i$, a $W_k \in \mathcal{W}_{kj}$ exists such that $W_x \subset W_k$.

For $(30)_i$ and $(31)_i$, apply (15) and (16), respectively.

For $(32)_i$, apply (17) and (37).

For $(33)_i$, if $B(U) \cap B(V) \neq \emptyset$ for $U \neq V$ in $\mathcal{W}_{ij}$, then from (19) we may assume that either $U \in \mathcal{U}_j^1$ and $V \in \mathcal{U}_j^3$, or, $U \in \mathcal{U}_j^2$ and $V \in \mathcal{U}_j^3$. In either case, (18) and the definition of $\mathcal{W}_j$ show that for each $k < i$, a $W_k \in \mathcal{W}_{kj}$ exists such that $V \subset W_k$.

---

[10]By $(33)_{i-1}$, $W \in \{U, V\}$ exists where $W \subset W_m \in \mathcal{W}_{mj}$ for each $m < i - 1$. Further, by $(32)_{i-1}$, $\overline{W} \subset W_m$ for each $m < i - 1$ and a fortiori, $F_{kj} \subset \overline{W}$ is a subset of each W_m.
[11]By $(35)_{i-1}$, $W \in \mathcal{W}_{(i-1)j}$ exists such that $F_{kj} \subset W$. Furthermore, $(34)_{i-1}$ shows $W \subset W_m \in \mathcal{W}_{mj}$ for each $m < i - 1$ except $m = k$.

For $(34)_i$, (18) shows that each member W of $\mathcal{U}_j^1 \cup \mathcal{U}_j^3 \subset \mathcal{W}_{ij}$ is a subset of a member of $\mathcal{W}_j$, and hence for all $k < i$, $W \subset W_k \in \mathcal{W}_{kj}$. So suppose $W \in \mathcal{U}_j^2 \subset \mathcal{W}_{ij}$. Then by (20) there is a unique $F = B(U) \cap B(V) \subset W$ where $F \in \mathcal{F}_j$ and for some $k \leq i-1$, $U, V \in \mathcal{W}_{kj}$ and $U \neq V$. Moreover, this particular $F = F_{kj} = F_{kj}^{U,V} \subset \mathcal{W}_{kj}$ where W_{kj} is given by one of (38), (39), or (40). If $W_{kj} = X$ is given by (38), then $i = 2$ and $k = 1$. By (20), $F \subset W \subset \overline{W} \subset G_F = H \cap G_1^F$ where H meets at most two members of $\mathcal{W}_{1j}$. So W meets at most two members of $\mathcal{W}_{1,j}$. Since $i = 2$, we are finished. The argument for the cases where W_{kj} is given by either (39) or (40) is similar.

For $(35)_i$, we apply (20) with $W \in \mathcal{U}_j^2 \subset \mathcal{W}_{ij}$ because Bdry $\mathcal{W}_j = \cup \{F : F \in \mathcal{F}_j\}$ and $W \subset G_F \subset G_i^F$ where $F = B(U) \cap B(V)$.

For $(36)_i$, we may apply (18) and (20) because $W \in \mathcal{W}_{ij}$ meets Bdry $\mathcal{W}_{kj}$ for $k < i$ implies $W \in \mathcal{U}_j^2$. □

39.2 EXAMPLE. Let $X = I = [0,1]$ be the unit interval, $X_1 \subset I$ the rational reals, and $X_2 = I \setminus X_1$ the irrationals. Then Figure 39.3 provides an instance of $\mathcal{W}_{11}$ and $\mathcal{W}_{12}$ as constructed in the proof of Theorem 39.1.

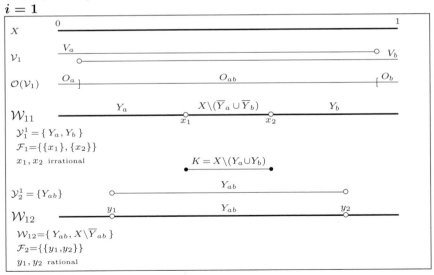

Fig. 39.3 The first step in the proof of the Decomposition Theorem.

Property $(28)_1$ is obvious; $(29)_1$ is satisfied because these decompositions refine $\mathcal{V}_1$; $(30)_1$ is satisfied at Bdry $\mathcal{W}_{11} = \{x_1, x_2\}$ and Bdry $\mathcal{W}_{12} = \{y_1, y_2\}$; and $(31)_1$ is satisfied because x_1 and x_2 are irrational, while y_1 and y_2 are rational. The other properties are vacuously satisfied. □

Turning to the inductive construction, we provide a graphic instance of the step from $i = 1$ to $i = 2$. In Figure 39.4, we focus on Bdry $\mathcal{W}_1 =$ Bdry $\mathcal{W}_{11} = \{x_1, x_2\}$, whose representation throughout the construction is indicated with dotted lines. Similarly, in Figure 39.5 we focus on Bdry $\mathcal{W}_2 =$ Bdry $\mathcal{W}_{12} = \{y_1, y_2\}$.

Assumptions $(\mathcal{W}_{11}, \mathcal{W}_{12}, \mathcal{F}_1, \mathcal{F}_2)$

$\boldsymbol{i = 2,\ j = 1}$

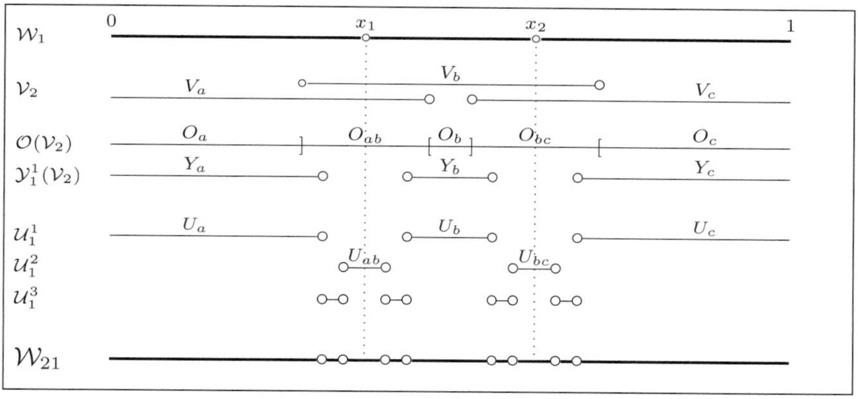

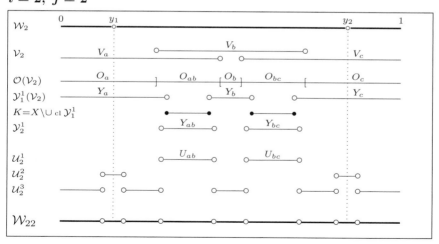

Fig. 39.4 Constructing $\mathcal{W}_{21}$ in the proof of the Decomposition Theorem.

Assumptions $(\mathcal{W}_{11}, \mathcal{W}_{12}, \mathcal{F}_1, \mathcal{F}_2)$

$\boldsymbol{i = 2,\ j = 2}$

Fig. 39.5 Constructing $\mathcal{W}_{22}$ in the proof of the Decomposition Theorem.

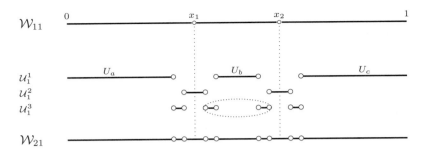

Fig. 39.6 Viewing members of $\mathcal{W}_{ij}$ for $j = 1$.

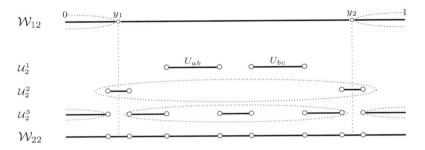

Fig. 39.7 Viewing members of $\mathcal{W}_{ij}$ for $j = 2$.

39.8 Example. The $\mathcal{W}_{ij}$ constructed in Figures 39.4 and 39.5 provide models for the properties listed in the Decomposition Theorem 39.1. In particular, within Figure 39.6 we see $\mathcal{W}_{11}$ with three members, Bdry $\mathcal{W}_{11} = \{x_1, x_2\}$, $\mathcal{W}_{21}$ with eight members (the union of the two segments in the interior of the dotted ellipse is one member of $\mathcal{U}_1^3 \subset \mathcal{W}_{21}$), and the points in Bdry $\mathcal{W}_{21}$ are represented as circles.

Similar observations hold for Figure 39.7: We note from Figure 39.3 that $\mathcal{W}_{12} = \{Y_{ab}, X \setminus Y_{ab}\}$ contains only two members, which accounts for the right and left halves of a dotted ellipse at the top of Figure 39.7. In other words, using the three segments $[0, y_1)$, (y_1, y_2), and $(y_2, 1]$ displayed at the top of Figure 39.7, we may represent the decomposition $\mathcal{W}_{12}$ as the family $\{[0, y_1) \cup (y_2, 1], (y_1, y_2)\}$. In addition, the definitions of triplets $\mathcal{U}_2^1$, $\mathcal{U}_2^2$, $\mathcal{U}_2^3$, given, respectively, by equations (22), (23), and (25), show that $\mathcal{U}_2^1$ contains two members, $\mathcal{U}_2^2$ one member, and $\mathcal{U}_2^3$ two members — the "dotted full ellipses and the dotted half ellipses" serve to indicate the unions of the intervals that form the members of these families. It follows that $\mathcal{W}_{22} = \cup_1^3 \mathcal{U}_2^k$ contains five members. □

With this graphical understanding of the four collections $\mathcal{W}_{11}$, $\mathcal{W}_{12}$, $\mathcal{W}_{21}$, and $\mathcal{W}_{22}$, we are in a position to show that these families satisfy the properties listed in the Decomposition Theorem 39.1:

For $(28)_1$ and $(28)_2$, we may simply inspect the graphics.

For $(29)_1$, observe that (trivially) the mesh $\mathcal{W}_{11} \cup \mathcal{W}_{12}$ is ≤ 1, and the statement "... for each $k < 1 \ldots W_x \subset W_k$" is vacuously satisfied.

For $(29)_2$, note that $\mathcal{U}_1^1 \cup \mathcal{U}_2^1 = \{U_a, U_b, U_c, U_{ab}, U_{bc}\} \subset \mathcal{W}_{21} \cup \mathcal{W}_{22}$ covers X, the mesh of $\mathcal{U}_1^1 \cup \mathcal{U}_2^1$ is $\leq 1/2$ (this family refines $\mathcal{V}_2$), and, $\mathcal{U}_1^1$ refines $\mathcal{W}_{11}$ and $\mathcal{U}_2^1$ refines $\mathcal{W}_{12}$.

For $(30)_i$, we may simply inspect the graphics.

For $(31)_i$, notice that the points in Bdry $\mathcal{W}_{i1}$ may be selected as irrational (X_1 is the subset of rationals) and those in Bdry $\mathcal{W}_{i2}$ as rational.

For $(32)_2$, observe that $\mathcal{U}_j^2 \subset \mathcal{W}_{2j}$ covers Bdry $\mathcal{W}_{1j}$.

For $(33)_2$, observe that $\mathcal{U}_j^1 \cup \mathcal{U}_j^3 \subset \mathcal{W}_{2j}$ and that $\mathcal{U}_j^1 \cup \mathcal{U}_j^3 \subset \mathcal{W}_{1j}$.

For $(34)_2$, note that $\mathcal{U}_j^2 \subset \mathcal{W}_{2j}$ covers Bdry $\mathcal{W}_{1j}$ and $\mathcal{U}_j^1 \cup \mathcal{U}_j^3$ refines $\mathcal{W}_{1j}$.

For $(35)_2$, note that $B(U) \cap B(V)$ equals either $\{x_1\}$ or $\{x_2\}$ for distinct $U, V \in \mathcal{W}_{11}$, and then note that $\mathcal{F}_1 = \{\{x_1\}, \{x_2\}\}$ precisely refines $\mathcal{U}_1^2 \subset \mathcal{W}_{21}$. Similarly, the lone $B(U) \cap B(V) = \{y_1, y_2\}$ because the only two distinct members $U = Y_{ab}$ and $V = X \setminus \overline{Y}_{ab}$ in $\mathcal{W}_{12}$ serve to define $\mathcal{F}_2 = \{\{y_1, y_2\}\}$, which precisely refines $\mathcal{U}_2^2$.

For $(36)_2$, the two members of $\mathcal{U}_1^2$ are the only members of $\mathcal{W}_{21}$ that meet Bdry $\mathcal{W}_{11} = \{x_1, x_2\}$, and each meets two members of $\mathcal{W}_{11}$. Similarly, the one member W of $\mathcal{U}_2^2$ is the only member of $\mathcal{W}_{22}$ that meets Bdry $\mathcal{W}_{12} = \{y_1, y_2\}$, and W meets two members of $\mathcal{W}_{12}$.

For $(32)_1$, $(33)_1$, $(34)_1$, $(35)_1$, and $(36)_1$, we note that each is vacuously satisfied.

§40 Comments

The "diml function" (§36) and the "nodes of covers" (§37) were introduced in Lipscomb [1973]. The lemmas (§38) and the Decomposition Theorem (§39) were introduced in Lipscomb [1975].

Within physics, it was the study of "coordinate systems" and "change of coordinate systems" that eventually led to the intuitive statement that *the laws of physics must be independent of coordinate systems*. Roughly, but also analogously, I view the mathematics of this chapter in a similar vein, i.e., as an example that *the laws of decompositions of metric spaces must be independent of nodal indexings*. Indeed, as illustrated in Figures 37.2 through 37.8, one cover of three sets may be nodally indexed in distinct ways, each yielding a "coordinate system" for related decompositions. A study of such indexings appears in §37 and §38. And that study yielded *one law* (the

Decomposition Theorem 39.1) concerning decompositions of metric spaces *that is independent of nodal indexings.*

So as one might suspect, the idea of nodes and nodal indexings has taken a back seat to the Decomposition Theorem. In fact, applications, extensions, modifications, and enhancements of Lemma 38.9 and the Decomposition Theorem appear in the works of Milutinović and also in the works of Ivanšić and Milutinović: For example, see Ivanšić and Milutinović [2002], [2003], [2005], and [2007]; and Milutinović [1992], [1993], and [2006].

The material in this chapter is a greatly expanded and augmented version of the original presentation, and is a direct result of comments made by the referee of Lipscomb [1975]: In June of 1974: I received a letter from Steve Armentrout, who quoted the referee:

> I believe that all of the proofs are correct, and that this paper is an outstanding contribution to dimension theory. The author's style is rather terse, and I have included a few comments for him to consider in case he wishes to do any revision of his manuscript.

The original presentation contained neither graphics, nor examples, nor the proofs that appear in the footnotes of this chapter, nor proofs 38.8.1 through 38.8.5, nor proofs 38.9.1 through 38.9.9. In addition, the precise idea of a "nodal indexing" was introduced in this chapter to add precision to the original arguments. And with the goal of making the ideas more accessible, even the original arguments were adjusted.

An understanding of the ideas presented in this chapter should serve as a gateway to all of the very substantial research referenced in this section.

It should be noted that §A6.2 Lemma (covers at points in dim $\leq n$ closed subspaces) was used to prove §A6.2 Theorem (refining covers of dim $\leq n$ subspaces), which was used to prove Lemma 38.2 (refining covers in a neighborhood of a dim $\leq n$ closed set). The idea for this sequence was initiated by the author's study of Ostrand [1971] — within §A6.2, see the paragraph in the middle of page 202.

CHAPTER 8

The J_A^{n+1} Imbedding Theorem

The Classical Imbedding Theorem states that any n-dimensional separable metric space may be imbedded in the subspace of I^{2n+1} whose tuples contain at most n rational coordinates.

In this chapter, we construct a proof of the J_A^{n+1} Imbedding Theorem that states that any n-dimensional weight $|A| \geq \aleph_0$ metric space may be imbedded in the subspace of J_A^{n+1} whose tuples contain at most n rational coordinates. The proof rests on an application of the Decomposition Theorem 39.1 that yields a commutative diagram that contains the imbedding.

§41 Mappings and the Commutative Diagram

In this section we outline the mappings and methods that produce the desired imbeddings into J_A^{n+1}. We begin with a commutative diagram.

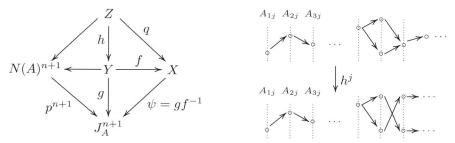

Fig. 41.1 Decomposition map q; ancestor map h; and imbedding ψ.

Within Figure 41.1, X is any weight $|A| \geq \aleph_0$ metric space of dimension $n \geq 0$. The mapping $p^{n+1} : N(A)^{n+1} \to J_A^{n+1}$ denotes the product map that has the natural map $p : N(A) \to J_A$ as each of its factors. Since p is perfect, p^{n+1} is perfect (Theorem A5.9). In addition, the two unlabeled arrows $Z \to N(A)^{n+1}$ and $Y \to N(A)^{n+1}$ denote inclusions.

The *decomposition map* $q : Z \to X$ and *ancestor map* $h : Z \to Y$ shall be defined below (both mappings are induced from an application of the Decomposition Theorem 39.1). The decomposition map q is both continuous and quotient, while the ancestor map h is a homeomorphism. The notations "q" and "h" are mnemonics, respectively, for "quotient" and "homeomorphism."

The homeomorphism h has an important property: For each $x \in X$ there is a $v_x \in J_A^{n+1}$ that has at most n rational coordinates and h maps the q-fiber $q^{-1}(x) \subset Z$ onto the p^{n+1}-fiber $(p^{n+1})^{-1}(v_x) \subset Y$. This correspondence $x \mapsto v_x$ is denoted ψ, and ψ is the desired imbedding.

S.L. Lipscomb, *Fractals and Universal Spaces in Dimension Theory*,
DOI 10.1007/978-0-387-85494-6_8, © Springer Science+Business Media, LLC 2009

As indicated in the diagram, $\psi = gf^{-1}$ where f is given by $f = q(h^{-1})$ (since h is one-to-one, f is well defined). Moreover, f is both continuous and quotient because h is a homeomorphism and q is quotient. The map g is the restriction of p^{n+1} to Y, which is closed because Y is a p^{n+1} inverse set and p^{n+1} is perfect. With these aspects of f and g exposed, an application of Theorem A4.3 shows that $\psi = gf^{-1} : X \to \psi(X)$ is a homeomorphism. And finally, it follows that $\psi = gf^{-1} = g(hq^{-1}) = p^{n+1}|_Y(hq^{-1})$.

§42 The Decomposition Map q

Since X is a weight $|A| \geq \aleph_0$ n-dimensional metric space, we may use the (dimension theory) Decomposition Theorem (§A6) to select, for each $j = 1, \ldots, n+1$, zero-dimensional subspaces $Y_j \subset X$ such that $X = \cup_1^{n+1} Y_j$. Then using the Subspace Theorem (§A6), we define $X_1 = Y_1$ and, for $j > 1$, $X_j = Y_j \setminus (\cup_{k<j} Y_k) \subset Y_j$ to obtain a partition $X = \cup_1^{n+1} X_j$ of X into zero-dimensional subspaces X_j. With these X_j, we apply the Decomposition Theorem 39.1 and thereby obtain $\mathcal{W}_{ij}$, $i \geq 1$ and $1 \leq j \leq n+1$, that satisfy $(28)_i$, $\ldots$, $(36)_i$ of the Decomposition Theorem 39.1.

Since each $\mathcal{W}_{ij}$ is of size $\leq |A| \geq \aleph_0$, we may partition A into sets A_{ij} such that $\mathcal{W}_{ij} = \{W_a : a \in A_{ij}\}$ is a faithful indexing of $\mathcal{W}_{ij}$.

Next, each $\alpha = (\alpha^1, \ldots, \alpha^{n+1}) \in N(A)^{n+1}$, where $\alpha^j = \alpha_{1j}\alpha_{2j} \cdots$ is a sequence in A, may be viewed as an $\aleph_0 \times (n+1)$ matrix $[\alpha_{ij}]$ whose entries α_{ij} are members of A. It may happen that each $\alpha_{ij} \in A_{ij}$.

In general, suppose $x \in X$. Then since each $\mathcal{W}_{ij}$ is a decomposition of X, we may select, for each i and each j, a $W_a \in \mathcal{W}_{ij}$ such that $x \in \overline{W}_a$ and then define $\alpha_{ij} = a$. The corresponding matrix $[\alpha_{ij}]$ has $n+1$ columns α^j, and each column α^j may be viewed as a sequence in A, i.e., the matrix $[\alpha_{ij}]$ corresponds to the $(n+1)$-tuple $\alpha = (\alpha^1, \ldots, \alpha^{n+1}) \in N(A)^{n+1}$ induced from the containment $x \in \cap\{\overline{W}_a : a$ is an entry of the matrix $[\alpha_{ij}]\}$.

42.1 Definition (the decomposition map $q : Z \to X$)   Let X and the corresponding $\mathcal{W}_{ij} = \{W_a : a \in A_{ij}\}$ be as constructed above. Then define $\alpha \in Z \subset N(A)^{n+1}$ and the *decomposition map* $q : Z \to X$ by

$$\alpha \in Z, \ q(\alpha) = x \quad \text{if and only if} \quad \{x\} = \cap\{\overline{W}_{\alpha_{ij}} : i \geq 1; 1 \leq j \leq n+1\}.$$

The decomposition map $q : Z \to X$ is well defined because of $(29)_i$ in Theorem 39.1.

As we shall see in the following two theorems, $(28)_i$, $\ldots$, $(36)_i$ of Theorem 39.1 expose key aspects of $q : Z \to X$.

42.2 Theorem   *Each fiber $q^{-1}(x)$ of $q : Z \to X$ is finite.*

PROOF. Either x is in no Bdry $\mathcal{W}_{ij}$, or, x is in some Bdry $\mathcal{W}_{\ell m}$. In the first case, since each $\mathcal{W}_{ij}$ is a decomposition, $q(\alpha) = q(\beta) = x$ implies the matrices $[\alpha_{ij}] \equiv \alpha$ and $[\beta_{ij}] \equiv \beta$ are equal, making $q^{-1}(x)$ a singleton set.

In the second case, $(32)_k$ and $(32)_i$ for $i > k$ show that if $x \in$ Bdry $\mathcal{W}_{km}$ then $x \notin$ Bdry $\mathcal{W}_{im}$ for all $i \neq k$. So there are only two choices for the mth column of any $[\alpha_{ij}]$ such that $q(\alpha) = x$. (Property $(30)_k$ shows that there are distinct $U, V \in \mathcal{W}_{km}$ such that $x \in B(U) \cap B(V)$; then $(28)_k$ shows that U and V are the only members of $\mathcal{W}_{km}$ whose closures contain x; and every other entry α_{im} in the mth column has x interior to $W_{\alpha_{im}} \in \mathcal{W}_{im}$.) Thus, since there are only a finite number of columns, there are only finitely many members of the fiber $q^{-1}(x)$. $\qquad\square$

42.3 Theorem *The surjection $q : Z \to X$ is both continuous and quotient.*

PROOF. The argument in the paragraph preceding Definition 42.1 shows that q is surjective. So let us show that q is continuous: For $\alpha \equiv [\alpha_{ij}]$ and $\beta \equiv [\beta_{ij}]$ in Z, we say that α is $(1/\ell)$-*close to* β whenever the first ℓ rows of $[\alpha_{ij}]$ are identical to the first ℓ rows of $[\beta_{ij}]$. Then from $(29)_\ell$, the distance between $q(\alpha)$ and $q(\beta)$ is less than $1/\ell$ whenever α is $(1/\ell)$-close to β. The continuity of q therefore follows. For the quotient aspect, suppose $q^{-1}(G)$ is open. We shall show that $G \subset X$ is open by showing that each $x \in G$ is in the interior of G. Indeed, for $\alpha \in q^{-1}(x)$, let $\langle \alpha_\ell \rangle$ be the basic open set

$$\langle \alpha_\ell \rangle = \langle \alpha_{11}, \ldots, \alpha_{\ell 1} \rangle \times \cdots \times \langle \alpha_{1(n+1)}, \ldots, \alpha_{\ell(n+1)} \rangle \subset N(A)^{n+1}.$$

Also recall that, as was shown in the proof of Theorem 42.2, if $x \in$ Bdry $\mathcal{W}_{km}$, then $x \notin$ Bdry $\mathcal{W}_{im}$ for every $i \neq k$. Thus we may choose $\ell > k$ for every k such that $x \in$ Bdry $\mathcal{W}_{kj}$ for some j. Further, since $q^{-1}(x)$ is a finite set and $q^{-1}(G)$ is open, we may assume that ℓ is large enough so that

$$(1) \qquad \cup\{\langle \alpha_\ell \rangle : \alpha \in q^{-1}(x)\} \subset q^{-1}(G).$$

Now define

$$V_x = \quad [\cap\{W : x \in W \in \mathcal{W}_{ij}; \ 1 \leq i \leq \ell \text{ and } 1 \leq j \leq n+1\}]$$
$$\cap \left[\cup\{\overline{W} : x \in B(W) \text{ and } W \in \mathcal{W}_{km} \text{ for some } k \text{ and } m\}\right].$$

Then $(30)_k$ and $(28)_k$ show that V_x is a neighborhood of x. Further, for $y \in V_x$ and $q(\beta) = y$, we have $\beta_{\ell j} = \alpha_{\ell j}$ for each j, forcing $y \in \cap_1^{n+1}\{W_a : a = \alpha_{\ell j}\}$. Then $(34)_\ell$ shows that β is a member of the left side of (1). So $\beta \in q^{-1}(G)$, which shows that $q(\beta) = y \in G$, i.e., the neighborhood V_x of x must be a subset of G. $\qquad\square$

§43 The Ancestor Map h

We now turn to the definition of $h : Z \to Y$, where it is assumed that we use the same decompositions that were used to define $q : Z \to X$. First, however, we need the concept of *ancestors*.

43.1 Definition (ancestors of $\alpha_{ij} \in A_{ij}$) Let $j \in \{1, \ldots, n+1\}$ be fixed; and let $\alpha_{kj} \in A_{kj}$ for $k = 1, \ldots, i$ be such that $\cap_{k=1}^i \overline{W}_{\alpha_{kj}} \neq \emptyset$. Then whenever

$W_{\alpha_{ij}} \cap$ Bdry $\mathcal{W}_{kj} = \emptyset$ for each $k < i$, we say that α_{ij} *has no ancestor with respect to* $\alpha_{1j}, \ldots, \alpha_{(i-1)j}$. Otherwise, as proved in the following paragraph, there is a unique $\ell < i$ and a unique $\beta_{\ell j} \in A_{\ell j}$ such that $\beta_{\ell j} \neq \alpha_{\ell j}$ and $W_{\alpha_{ij}} \cap B(W_{\alpha_{\ell j}}) \cap B(W_{\beta_{\ell j}}) \neq \emptyset$. This $\beta_{\ell j}$ is called *the 1st ancestor of* α_{ij} *with respect to* $\alpha_{1j}, \ldots, \alpha_{(i-1)j}$. The *2nd ancestor of* α_{ij} *with respect to* $\alpha_{1j}, \ldots, \alpha_{(i-1)j}$ is the 1st ancestor of $\beta_{\ell j}$ with respect to $\alpha_{1j}, \ldots, \alpha_{(\ell-1)j}$. The process of determining *ancestors* of α_{ij} terminates after at most $i - 1$ steps, and we call the final ancestor so determined *the last ancestor of* α_{ij} *with respect to* $\alpha_{1j}, \ldots, \alpha_{(i-1)j}$.

A few comments concerning *no ancestor, 1st ancestor, ...,last ancestor* are in order. That "no ancestor" is unambiguous is obvious. In contrast, the argument that "1st ancestor" is well defined is not so obvious. To address the subtlety, we prove the following: *If*

$$\text{(i)} \quad \cap_{k=1}^{i} \overline{W}_{\alpha_{kj}} \neq \emptyset \quad \text{and} \quad \text{(ii)} \ W_{\alpha_{ij}} \cap \text{Bdry } \mathcal{W}_{\ell j} \neq \emptyset \text{ for } \ell < i,$$

then ℓ *is unique, and, a unique* $\beta_{\ell j} \in A_{\ell j}$ *exists such that* $\beta_{\ell j} \neq \alpha_{\ell j}$ *and*

$$\text{(iii)} \ W_{\alpha_{ij}} \cap B(W_{\alpha_{\ell j}}) \cap B(W_{\beta_{\ell j}}) \neq \emptyset.$$

Indeed, (ii), and then $(30)_i$ of Theorem 39.1 show that $W_{\alpha_{ij}} \cap B(U) \cap B(V) \neq \emptyset$ for distinct $U, V \in \mathcal{W}_{\ell j}$; and $(34)_\ell$ of Theorem 39.1 shows that ℓ is unique. Next, suppose $W_{\alpha_{ij}} \notin \{U, V\}$. Then using W_i and W_ℓ to denote, respectively, $W_{\alpha_{ij}}$ and $W_{\alpha_{\ell j}}$, and, B_i and B_ℓ the corresponding $B(W_i)$ and $B(W_\ell)$, we consider

$$\overline{W}_i \cap \overline{W}_\ell = (B_i \cup W_i) \cap (B_\ell \cup W_\ell)$$
$$= (B_i \cap B_\ell) \cup (B_i \cap W_\ell) \cup (W_i \cap B_\ell) \cup (W_i \cap W_\ell).$$

But reading left-to-right, $(B_i \cap B_\ell) \subset$ Bdry $\mathcal{W}_{ij} \cap$ Bdry $\mathcal{W}_{\ell j} = \emptyset$ because of $(32)_i$ of Theorem 39.1; $(B_i \cap W_\ell) = \emptyset = (W_i \cap B_\ell)$ because $(34)_i$ implies $W_i \subset \overline{U} \cup \overline{V}$ and $(\overline{U} \cup \overline{V}) \cap W_\ell = \emptyset$; and $(W_i \cap W_\ell) = \emptyset$ because $(34)_i$ says that W_i meets at most two members of $\mathcal{W}_{\ell j}$. Thus, $\overline{W}_i \cap \overline{W}_\ell = \emptyset$, which contradicts (i). So $W_{\alpha_{\ell j}} \in \{U, V\}$, the unique $\beta_{\ell j} \in A_{\ell j}$ exists, $\beta_{\ell j} \neq \alpha_{\ell j}$, and (iii) is true.

To show that "2nd ancestor" is also well defined, we prove the following: *If* $\beta_{\ell j} \in A_{\ell j}$ *is the 1st ancestor of* α_{ij} *with respect to* $\alpha_{1j}, \ldots, \alpha_{(i-1)j}$, *then*

$$\text{(iv)} \quad \cap_{k=1}^{\ell-1} \overline{W}_{\alpha_{kj}} \cap \overline{W}_{\beta_{\ell j}} \neq \emptyset.$$

Indeed, from (iii) we may infer that

$(35)_i$ yields   (v) $B(W_{\alpha_{\ell j}}) \cap B(W_{\beta_{\ell j}}) \subset W_{\alpha_{ij}}$

and $(34)_i$ yields   (vi) $W_{\alpha_{ij}} \subset W_{\alpha_{kj}}$ $(1 \le k \le (\ell-1))$.

So from (vi), $\overline{W}_{\alpha_{ij}} \subset \cap_{k=1}^{\ell-1} \overline{W}_{\alpha_{kj}}$, and coupled with (v), we have

$$\emptyset \neq B(W_{\alpha_{\ell j}}) \cap B(W_{\beta_{\ell j}}) \subset \overline{W}_{\alpha_{ij}} \subset \cap_{k=1}^{\ell-1} \overline{W}_{\alpha_{kj}}, \text{which yields (iv).}$$

In short, it should now be clear that the inductive construction of ancestors is well defined, and in particular, the concept of "last ancestor" is well defined.

With the spadework of "ancestors" complete, we are ready for the definition of $h : Z \to Y$.

43.2 Definition (the ancestor map $h : Z \to Y$)   Let $\alpha \in Z$. Then $\alpha \equiv [\alpha_{ij}]$ is induced by $x = q(\alpha)$ and the $\mathcal{W}_{ij}$. For each $i \geq 1$ and each $j \leq n + 1$, define, with respect to $\alpha_{1j}, \ldots, \alpha_{(i-1)j}$,

$$\gamma_{ij} = \begin{cases} \alpha_{ij} & \text{if } \alpha_{ij} \text{ has no ancestor;} \\ \text{last ancestor of } \alpha_{ij} & \text{otherwise.} \end{cases}$$

The correspondence $\alpha_{ij} \mapsto \gamma_{ij}$ induces the mapping $h : Z \to h(Z) = Y \subset N(A)^{n+1}$ given by $\alpha \mapsto h(\alpha) = \gamma \equiv [\gamma_{ij}]$.

Since "*no ancestor* and *last ancestor* of α_{ij} with respect to $\alpha_{1j}, \ldots, \alpha_{(i-1)j}$" are well defined, $h : Z \to Y$ is well defined; and since $Y = h(Z)$ by definition, h is surjective.

To see that h is also continuous, recall (from the proof of Theorem 42.3) that β $(1/\ell)$-close to α implies $\beta_{ij} = \alpha_{ij}$ for all $i \leq \ell$ and all j, $1 \leq j \leq n+1$. Since these α_{ij} determine the corresponding γ_{ij} where $h(\alpha) = \gamma$, and these same $\alpha_{ij} = \beta_{ij}$ determine the corresponding δ_{ij} where $h(\beta) = \delta$, we have $\delta_{ij} = \gamma_{ij}$ for all $i \leq \ell$ and all j, $1 \leq j \leq n+1$, i.e., $h(\beta)$ is $(1/\ell)$-close to $h(\alpha)$. It follows that $h : Z \to Y \subset N(A)^{n+1}$ is continuous.

43.3 Lemma (an m exists where $\alpha_{im} = h(\alpha)_{im}$ for any i in an infinite R)
Let $q(\alpha) = x \in X$. Then there is an $m \in \{1, \ldots, n+1\}$ and an infinite subset $R \subset \{1, 2, \ldots\}$ such that $i \in R$ implies $h(\alpha)_{im} = \alpha_{im}$; $q(\alpha) \in W_{\alpha_{im}}$; and $\mathrm{dia}(W_{\alpha_{im}}) < 1/i$.

PROOF.   Property $(29)_i$ of Theorem 39.1 holds for every $i \in \{1, 2, \ldots\}$ and any $x \in X$. So for every $i \geq 1$ and $x = q(\alpha)$, a $j \in \{1, \ldots, n+1\}$ exists that makes $(29)_i$ true. Consequently, since j ranges over finitely many values, at least one $j \leq n+1$, say $j = m$, and an infinite $R \subset \{1, 2, \ldots\}$ exist such that $(29)_i$ is true for each $i \in R$. The inclusion condition in $(29)_i$ for $i \in R$ and the definition of *no ancestor* show that $i \in R$ implies $h(\alpha)_{im} = \alpha_{im}$. Moreover, since $i \in R$ implies that $(29)_i$ holds for $j = m$, $x = q(\alpha) \in W_{\alpha_{im}}$ where $\mathrm{dia}(W_{\alpha_{im}}) < 1/i$.   ☐

43.4 Lemma   Let $\alpha, \beta \in Z$; let $\alpha_{uv} \neq \beta_{uv}$ for some u and some v; let $i > u$; let $q(\alpha), q(\beta) \in W \in \mathcal{W}_{iv}$, and let $\emptyset \neq B(W_{\alpha_{uv}}) \cap B(W_{\beta_{uv}}) \subset W$. Then $h(\alpha)_{iv} \neq h(\beta)_{iv}$.

PROOF.   Since $q(\alpha), q(\beta) \in W \in \mathcal{W}_{iv}$, we have $\alpha_{iv} = \beta_{iv}$. Since $\emptyset \neq B(W_{\alpha_{uv}}) \cap B(W_{\beta_{uv}}) \subset W$ where $\alpha_{uv} \neq \beta_{uv}$, $(34)_i$ of Theorem 39.1 shows that β_{uv} is the 1st ancestor of α_{iv} (with respect to $\alpha_{1v}, \ldots, \alpha_{(i-1)v}$) and that

α_{uv} is the 1st ancestor of β_{iv} (with respect to $\beta_{1v}, \ldots, \beta_{(i-1)v}$). By $(33)_i$ of Theorem 39.1, either α_{uv} or β_{uv} has no ancestor, so $h(\alpha)_{iv} \neq h(\beta)_{iv}$.   □

43.5 Lemma *Let α, $\beta \in Z$ with $h(\alpha)_{im} = h(\beta)_{im} = \alpha_{im}$. Then $\beta_{im} = \alpha_{im}$.*

PROOF. By definition of the ancestor map h, with respect to $\beta_{1m}, \ldots, \beta_{(i-1)m}$

$$h(\beta)_{im} = \begin{cases} \beta_{im} & \text{if } \beta_{im} \text{ has no ancestor;} \\ \text{last ancestor of } \beta_{im} & \text{otherwise.} \end{cases}$$

Note that "$h(\beta)_{im} = \beta_{im}$" is equivalent to "$h(\beta)_{im}$ is an index in A_{im}." (A last ancestor of β_{im} would be an index in $A_{\ell m}$ where $\ell < i$ and $A_{\ell m} \cap A_{im} = \emptyset$.) It follows that $h(\beta)_{im} = \alpha_{im} \in A_{im}$ implies $\beta_{im} = \alpha_{im}$.   □

43.6 Theorem *The ancestor map $h : Z \to Y$ is a homeomorphism.*

PROOF. From the comments following Definition 43.2, h is both surjective and continuous. So it suffices to show that h is one-to-one and that h^{-1} is continuous. To see that h is one-to-one, let $\alpha, \beta \in Z$ and assume that $\alpha_{uv} \neq \beta_{uv}$ for some u and some v. If $q(\alpha) = q(\beta)$, then

$$q(\alpha) = q(\beta) \in B(W_{\alpha_{uv}}) \cap B(W_{\beta_{uv}}) \neq \emptyset$$

and we may use $(35)_{(u+1)}$ of Theorem 39.1 for $j = v$ to see that there is a $W \in \mathcal{W}_{(u+1)v}$ such that $B(W_{\alpha_{uv}}) \cap B(W_{\beta_{uv}}) \subset W$. Then Lemma 43.4 shows that $h(\alpha)_{(u+1)v} \neq h(\beta)_{(u+1)v}$. So we are left with the case where $q(\alpha) \neq q(\beta)$. In this case, Lemma 43.3 provides an $m \in \{1, \ldots, n+1\}$ and an infinite set $R \subset \{1, 2, \ldots\}$ such that $i \in R$ implies $h(\alpha)_{im} = \alpha_{im}$; $q(\alpha) \in W_{\alpha_{im}}$; and $\text{dia}(W_{\alpha_{im}}) < 1/i$. Since $q(\alpha) \neq q(\beta)$, we let $\varepsilon > 0$ denote the distance between $q(\alpha)$ and $q(\beta)$. Now choose $i \in R$ such that

$$(1/i) < (\varepsilon/2); \quad q(\alpha) \in W_{\alpha_{im}}; \quad h(\alpha)_{im} = \alpha_{im}; \quad \text{dia}(W_{\alpha_{im}}) < 1/i.$$

Then $q(\beta) \notin \overline{W}_{\alpha_{im}}$ and consequently, $h(\beta)_{im} \neq \alpha_{im} = h(\alpha)_{im}$.

To see that h^{-1} is continuous, we shall use the "$(1/i)$-*close*" terminology. (See the third sentence in the proof of Theorem 42.3.) So let $\alpha \in Z$ and $\ell \geq 1$. Then we shall show that k_ℓ exists such that $h(\beta)$ $(1/k_\ell)$-close to $h(\alpha)$ implies β $(1/\ell)$-close to α. First, however, as in the third sentence of the proof of Theorem 42.2 shows, we may assume that if $q(\alpha) \in B(W_{\alpha_{im}})$ for some i and some m, then $i < \ell$. Now consider

$$(2) \qquad G = \cap\{W_{\alpha_{ij}} : i \leq \ell; \ j \leq n+1; \text{ and } q(\alpha) \in W_{\alpha_{ij}}\}.$$

Then G is open and $q(\alpha) \in G$ implies (by Lemma 43.3 for an m and an infinite R) that there is a $k_\ell \in R$ such that $k_\ell = k > \ell$; $h(\alpha)_{km} = \alpha_{km}$; and $q(\alpha) \in W_{\alpha_{km}} \subset \overline{W}_{\alpha_{km}} \subset G$. (The inclusion $\overline{W}_{\alpha_{km}} \subset G$ follows because $q(\alpha) \in G$; $\text{dia}(W_{\alpha_{km}}) < (1/k)$; and $k \in R$ may be arbitrarily large.)

Consequently, if we suppose that $h(\beta)$ is $(1/k_\ell)$-close to $h(\alpha)$, then $h(\beta)_{ij} = h(\alpha)_{ij}$ for all $i \leq k_\ell = k$ and all $j \leq n+1$. In particular, $h(\beta)_{km} = h(\alpha)_{km} = \alpha_{km}$, which shows (by Lemma 43.5) that $\beta_{km} = \alpha_{km}$. So

$$q(\beta) \in \overline{W}_{\beta_{km}} = \overline{W}_{\alpha_{km}} \subset G.$$

It follows from the definition of G that $\beta_{ij} = \alpha_{ij}$ for all $i \leq \ell$ and all $j \leq n+1$ such that $q(\alpha) \in W_{\alpha_{ij}}$. Now consider those pairs in

$$(3) \qquad \{(u, v) : q(\alpha) \in B(W_{\alpha_{uv}}) \text{ for } u < \ell \text{ and } v \leq n+1\}$$

where the constraint $u < \ell$ may be replaced by $u \leq \ell$ because of the constraint placed on ℓ in the statement preceding (2).

We shall finish the proof that h^{-1} is continuous by showing that $\alpha_{uv} \neq \beta_{uv}$ for (u, v) in the set defined in (3) contradicts $h(\beta)_{(u+1)v} = h(\alpha)_{(u+1)v}$ for $u+1 \leq \ell < k_\ell$: Indeed, suppose $\alpha_{uv} \neq \beta_{uv}$. Now for each i, $u < i \leq \ell$, $(32)_i$ (of Theorem 39.1) for $j = v$ shows that $q(\alpha) \in W_{\alpha_{iv}} \in \mathcal{W}_{iv}$. Also, for each i, $1 \leq i < u < \ell$, $(32)_u$ yields $q(\alpha) \in W_{\alpha_{iv}}$. So $q(\beta) \in G$ implies $q(\beta) \in W_{\alpha_{iv}}$ for each i, $1 \leq i \leq \ell$ such that $i \neq u$. In other words,

$$\alpha_{1v} = \beta_{1v}, \quad \ldots, \quad \alpha_{(u-1)v} = \beta_{(u-1)v}, \quad \alpha_{(u+1)v} = \beta_{(u+1)v}, \quad \ldots, \quad \alpha_{\ell v} = \beta_{\ell v}.$$

Also, $q(\alpha) \in B(W_{\alpha_{uv}})$, $(30)_u$ for $j = v$, and $(34)_\ell$ for $j = v$ imply

$$B(W_{\alpha_{uv}}) \cap B(W_{\beta_{uv}}) \neq \emptyset \qquad \text{because } \alpha_{uv} \neq \beta_{uv}.$$

Further, $(36)_{(u+1)}$ for $j = v$ shows that

$$\emptyset \neq B(W_{\alpha_{uv}}) \cap B(W_{\beta_{uv}}) \subset W_{\alpha_{(u+1)v}} \in \mathcal{W}_{(u+1)v},$$

which yields (Lemma 43.4) $h(\alpha)_{uv} \neq h(\beta)_{uv}$. Thus, h^{-1} is continuous. $\qquad\square$

§44 Matching q-Fibers with p^{n+1}-Fibers

The adjacent-endpoint relation $\sim \,\subset N(A) \times N(A)$ was introduced within Definitions 2.2 as the $p : N(A) \to J_A$ induced equivalence relation. Now, however, we are concerned with $N(A)^{n+1}$ where n may be strictly greater than zero. So we extend the adjacent-endpoint relation as follows: For $\gamma, \delta \in N(A)^{n+1}$, we write "$\gamma \sim_\times \delta$" whenever each pair γ^j and δ^j of corresponding components are $\sim$-related. This "product relation" $\sim_\times\,\subset N(A)^{n+1} \times N(A)^{n+1}$ is the p^{n+1}-induced equivalence relation.

44.1 Lemma (h matches q-fibers with p^{n+1}-fibers) *Let $q : Z \to X$ be the decomposition map; $h : Z \to Y$ the ancestor map; and $\alpha \in q^{-1}(x)$. Then $h(\alpha) \sim_\times \gamma$ implies that there is a $\beta \in q^{-1}(x)$ such that $h(\beta) = \gamma$.*

PROOF. It suffices to prove the claim for the case where $h(\alpha)$ and γ differ at exactly one component.[1]

To begin, let $h(\alpha)$ and γ differ at only their jth components. We show below that a unique $\ell \geq 1$ and a unique $\beta_{\ell j} \in A_{\ell j}$ exist such that $\beta_{\ell j} \neq \alpha_{\ell j}$ and

$$(4) \qquad B(W_{\alpha_{\ell j}}) \cap B(W_{\beta_{\ell j}}) = \cap \{\overline{W}_{\alpha_{ij}} : i \geq 1\}.$$

Since $q(\alpha) = x$ is given by $\{x\} = \cap \{\overline{W}_{\alpha_{ij}} : i \geq 1; \; j \leq n+1\}$, equation (4) shows that $q(\alpha) \in \text{Bdry } \mathcal{W}_{\ell j}$. Next, introduce $\beta \in N(A)^{n+1}$ such that $\beta_{\ell j}$ is given in (4), but is otherwise given by $\beta_{im} = \alpha_{im}$. Then from (4), $q(\beta) = q(\alpha)$.

Using an argument similar to the proof of Lemma 43.4, we may show that the jth components of $h(\alpha)$ and $h(\beta)$ are unequal. In addition, equation (8) below and the definition of the ancestor map h show that $h(\alpha)^j \sim h(\beta)^j$. Then $h(\beta) = \gamma$ because the size of any $\sim$ equivalence class is at most two.

With these remarks, it suffices to show that (4) is true: So let $h(\alpha) \sim_{\times} \gamma$ where $h(\alpha)$ and γ differ only at their jth components. Then let t be the tail index of the sequence $h(\alpha)_{1j} h(\alpha)_{2j} \cdots$, and consider the 1st ancestor $\beta_{\ell j}$ of $\alpha_{(t+1)j}$ with respect to $\alpha_{1j}, \ldots, \alpha_{tj}$.[2]

So $\ell < t+1$ and $W_{\alpha_{(t+1)j}} \cap \text{Bdry } \mathcal{W}_{\ell j} \neq \emptyset$, i.e., for $i = t+1$,

$$W_{\alpha_{ij}} \cap B(W_{\alpha_{\ell j}}) \cap B(W_{\beta_{\ell j}}) \neq \emptyset,$$

which is statement (iii) in the first paragraph following Definition 43.1. Then with this statement (iii) again, we may use the argument in the second paragraph following Definition 43.1 to obtain (v) (in that second paragraph) for $i = t+1$. We index "(v)" as

$$(5) \qquad \emptyset \neq B(W_{\alpha_{\ell j}}) \cap B(W_{\beta_{\ell j}}) \subset W_{\alpha_{ij}}.$$

Now let $r > i = t+1$ be arbitrary but fixed. An argument that is analogous to the $i = t+1$ argument that produced (5) tells us that α_{rj} has a 1st ancestor ε_{uj} with respect to $\alpha_{1j}, \ldots, \alpha_{(r-1)j}$ such that

$$(6) \qquad \emptyset \neq B(W_{\alpha_{uj}}) \cap B(W_{\varepsilon_{uj}}) \subset W_{\alpha_{rj}}.$$

---

[1] When $h(\alpha)$ and γ differ at exactly two components, we may construct a $\delta \in N(A)^{n+1}$ such that $\delta \sim_{\times} h(\alpha)$ and $\delta \sim_{\times} \gamma$ where δ differs from each of $h(\alpha)$ and γ at exactly one component. For example, if $h(\alpha) = (\theta^1, \theta^2, \theta^3, \theta^4)$ and $\gamma = (\gamma^1, \theta^2, \theta^3, \gamma^4)$ are such points in $N(A)^4$, then consider $\delta = (\gamma^1, \theta^2, \theta^3, \theta^4)$. In such a case, the desired $\beta \in q^{-1}(x)$ may be obtained by applying the "differ by exactly one component" result twice. The general case, where $h(\alpha)$ and γ differ at exactly k components, then follows from an induction argument.

[2] To see that $\alpha_{(t+1)j}$ has an ancestor, suppose otherwise. Then $\alpha_{(t+1)j}$ has no ancestor, i.e., $h(\alpha)_{(t+1)j} = \alpha_{(t+1)j}$. Thus, $h(\alpha)_{(t+1)j} h(\alpha)_{(t+2)j} \cdots$ is a constant sequence whose first term is $\alpha_{(t+1)j}$. So for $k > 1$, the last ancestor $h(\alpha)_{(t+k)j}$ of $\alpha_{(t+k)j}$ with respect to $\alpha_{1j}, \ldots, \alpha_{(t+1)j}, \ldots, \alpha_{(t+k-1)j}$ is $\alpha_{(t+1)j}$. But this is a contradiction because no ancestor can be a member of the "with respect to list."

It turns out that $u < i = t + 1$.[3]

Then from $u < i = t + 1 < r$ and the inclusion in (6), we find that $(34)_r$ of Theorem 39.1 shows that for $i = t + 1$

$$(7) \qquad\qquad W_{\alpha_{rj}} \subset W_{\alpha_{ij}}.$$

Consequently, for $i = t + 1$, (5), (6), and (7) show that for $i = t + 1$

$$\left. \begin{array}{l} \emptyset \neq B(W_{\alpha_{\ell j}}) \cap B(W_{\beta_{\ell j}}) \\ \emptyset \neq B(W_{\alpha_{uj}}) \cap B(W_{\varepsilon_{uj}}) \end{array} \right\} \subset W_{\alpha_{ij}}.$$

Then for $i = t + 1$, $(34)_i$ of Theorem 39.1 shows that $u = \ell$. Moreover, since the corresponding arguments for each $i > t + 1$ yield (5), (6) and (7) for $i > t + 1$,

$$(8) \qquad B(W_{\alpha_{\ell j}}) \cap B(W_{\beta_{\ell j}}) \subset W_{\alpha_{ij}} \qquad \text{for all } i \geq t + 1.$$

Then (4) follows from the constraint

$$W_{\alpha_{ij}} \subset G_i(B(W_{\alpha_{\ell j}}) \cap B(W_{\beta_{\ell j}}))$$

on the $W_{\alpha_{ij}}$ in $(35)_i$ of Theorem 39.1, and the equality

$$B(W_{\alpha_{\ell j}}) \cap B(W_{\beta_{\ell j}}) = \bigcap_{t+1}^{\infty} G_i(B(W_{\alpha_{\ell j}}) \cap B(W_{\beta_{\ell j}}))$$

in the hypothesis of Theorem 39.1.   □

§45 Proof of the J_A^{n+1} Imbedding Theorem

Using the lemmas and theorems in the preceding sections of this chapter, we prove the J_A^{n+1} Imbedding Theorem.

45.1 Theorem (J_A^{n+1} Imbedding Theorem) *A metric space X of weight $|A| \geq \aleph_0$ is of dimension $\leq n$ if and only if it can be imbedded in the subset of J_A^{n+1} whose tuples have at most n rational coordinates.*

PROOF. Suppose $X \to J_A^{n+1}(n)$ is an imbedding. Then since Theorem 34.2 for $\ell = n + 1$ shows that the subspace

$$J_A^{n+1}(n) = \{z \in J_A^{n+1} : z \text{ has at most } n \text{ rational coordinates}\}$$

---

[3] If $u \geq i = t + 1$, then again α_{uj} has a 1st ancestor with respect to $\alpha_{1j}, \ldots, \alpha_{(u-1)j}$. So by $(33)_u$, ε_{uj} has no ancestor, i.e.,

$$W_{\varepsilon_{uj}} \cap \text{Bdry } W_{kj} = \emptyset \qquad (1 \leq k \leq u - 1).$$

Thus, (6) tells us that $h(\alpha)_{rj} = \varepsilon_{uj} \in A_{uj}$ where $u \geq i = t + 1$, which contradicts $h(\alpha)_{rj} = h(\alpha)_{(t+1)j}$, i.e., contradicts the fact that the last ancestor $h(\alpha)_{(t+1)j}$ of $\alpha_{(t+1)j}$ belongs to A_{kj} for some $k \leq \ell < t + 1 \leq u$.

of J_A^{n+1} has dimension n, the Subspace Theorem shows that $\dim X \leq n$. Conversely, suppose $\dim X \leq n$. Then we shall use the mappings and commutative diagram in Figure 41.1. That is, $\psi = gf^{-1}$. From the claims in §41, and their proofs in §42, §43, and §44, we may deduce that ψ is indeed an imbedding. So it only remains to show that each tuple $\psi(x) \in J_A^{n+1}$ has at most n rational coordinates. To specify at least one irrational coordinate in each $\psi(x)$ we apply Lemmas 43.3 and 44.1. In detail, let $\alpha \in q^{-1}(x)$. Then $q(\alpha) = x$, and, $\psi(x) = p^{n+1}(h(\alpha))$. (Use $f^{-1}(x) = hq^{-1}(x)$ and Lemma 44.1, which shows that h maps the q-fiber $q^{-1}(x)$ onto the p^{n+1}-fiber $(p^{n+1})^{-1}\psi(x)$.) Then an application of Lemma 43.3 for this particular α shows that there is a component, say $h(\alpha)^m$, of $h(\alpha)$ and an infinite subset $R \subset \{1, 2, \ldots\}$ such that $i \in R$ implies $h(\alpha)_{im} = \alpha_{im}$. Consequently, since $A_{im} \cap A_{km} = \emptyset$ when $i \neq k$ we see that the mth component of $h(\alpha)$ contains an infinite number α_{im} $(i \in R)$ of terms, i.e., the mth component of $h(\alpha)$ has no constant tail. Thus, the mth component of $p^{n+1}(h(\alpha)) = \psi(x)$ is irrational.   □

Finally, we may compare the statement of the J_A Imbedding Theorem with its classical counterpart.

45.2 Theorem (Classical Imbedding Theorem) *A metric space X of weight $|A| = \aleph_0$ is of dimension $\leq n$ if and only if it can be imbedded in the subset of I^{2n+1} whose tuples have at most n rational coordinates.*

§46 Comments

For an excellent account (along with relevant references and proofs) of the classical universal space theorems see Engelking [1978, Section 1.11, pages 118–133].

As discussed in §4.3, Karl Menger [1926a] showed that any compact metric space of dimension ≤ 1 may be imbedded in the unit cube I^3.

The universal space constructed in Menger [1926a] is a well-known fractal called the Menger sponge (Figure 49.2). Additional insight into Menger [1926a] may be obtained by reading Edgar's [1993] English translation, where one finds historical perspective, a color picture of the Sponge (Plate 3), and additional editorial comments. For example, at the end of the translation, Edgar points out that, "... any separable metric space is homeomorphic to a subset of a compact metric space with the same topological dimension." He also provides references to Menger [1928, Chapter IX, §1] along with Hurewicz [1927], Kuratowski [1937], and Hurewicz and Wallman [1948, page 65].

Edgar also states:

> Menger suggests – but does not prove – that the set R_{2n+1}^n is universal for (separable metric) spaces with topological dimension n. In Menger [1928], Chapter IX, Menger still provides only

a 'sketch' of the proof. The proof is carried out by S. Lefschetz [1931].

Four years after Menger's 1926 work on his "fractals as universal spaces," his student Georg Nöbeling [1931] proved Theorem 45.2, the Classical Imbedding Theorem. Nöbeling's universal space is the subspace of tuples in I^{2n+1} that have at most n rational coordinates, which is distinct from the "fractals" used by Menger.

Then three decades later, Jun-iti Nagata [1960] introduced a space that is universal for the class of general (not necessarily separable) metric spaces of dimension $\leq n$. In more detail, Nagata [1960] states:

> Although dimension theory for non-separable metric spaces has been greatly developed, it still seems that no universal n-dimensional set for non-separable metric spaces is known. Thus it will be of some interest to find a universal n-dimensional set for non-separable metric spaces in a generalized Hilbert space. This paper is devoted to this purpose.

Nagata then calls $f : \Omega \to \mathbb{R}$ a *finite function* if ord $f = |\{\alpha : f(\alpha) \neq 0\}|$ is finite; defines $F(\Omega) = \{f : f \text{ is finite}\}$ with metric

$$d(f,g) = \left[\Sigma_{\alpha \in \Omega} \left(f(\alpha) - g(\alpha)\right)^2\right]^{\frac{1}{2}};$$

and, for "f" now denoting "$(f_1, f_2, \ldots)$" where each $f_i \in F(\Omega)$ defines $H'(\Omega) = \{f : \Sigma_{i,\alpha} \left(f_i(\alpha)\right)^2 < +\infty\}$ with metric

$$\rho(f,g) = \sqrt{\Sigma_1^\infty \left(d(f_i, g_i)\right)^2}.$$

Nagata also specifies the *fundamental cube* $F'(\Omega) \subset H'(\Omega)$ as

$$F'(\Omega) = \{f : 0 \leq f_i \leq (1/i); \ \ \text{ord } f_i \geq \ \text{ord } f_{i+1}; i = 1, 2, \ldots\}$$

and uses $F'(\Omega)$ to construct the desired universal space.

46.1 Theorem (Nagata [1960]) *A metric space has dimension $\leq n$ if and only if it can be topologically imbedded in*

$$F_n(\Omega) \ \ = \{f \in F'(\Omega) : \text{at most } n \text{ of } f_i(\alpha), i = 1, 2, \ldots \text{and } \alpha \in \Omega, \\ \text{are rational and nonvanishing}\}$$

for some Ω.

Three years later, Nagata [1963] introduced another such universal space:

> Once [previously] we have constructed (Nagata [1960]) a universal n-dimensional set for *general* metric spaces which is a rather complicated subset of C. H. Dowker's (Dowker [1947]) generalized Hilbert space. In this brief note we shall show that we can find a simpler universal n-dimensional set in a countable product of H. J. Kowalsky's star-spaces.

Then, to demonstrate and compare the two applications of star spaces, he states Kowalsky's [1957] Theorem and Nagata's [1963] Theorem. (For precise statements of the Kowalsky and Nagata theorems, see Theorems 18.1 and 18.2 in Chapter 3, respectively.)

More than a decade later, the J_A^{n+1} Imbedding Theorem (which is also more simply called the "J_A Imbedding Theorem") appeared. Ironically, like the Menger-Nöbeling two-step, i.e., Menger [1926a] (dim ≤ 1) and Nöbeling [1931] (dim $\leq n$) publications, the corresponding J_A results occurred in two steps — Lipscomb [1973] (dim ≤ 1) and Lipscomb [1975] (dim $\leq n$).

There is also some history (and overdue thanks) regarding the term *ancestor*: It was Charles Alexander, my 1969–70 thesis advisor at the University of Virginia, who suggested the term *ancestor*. In fact, the reason for his suggestion was simply to intuitively convey the rather technical idea of the *ancestor map* $h : Z \to Y$. The context was his reading of the proof of the J_A^{n+1} Imbedding Theorem for $n \in \{0, 1\}$ that is contained in my University of Virginia thesis.

Looking back at the basics that led to the J_A^{n+1} Imbedding Theorem, I originally had the idea (in my minds eye) of somehow associating each point x in a metric space X with either one or at most two sequences of indices on members of appropriate decompositions of X. If there were two such sequences, then they were to differ at exactly one index (see the top of the right-side graphic in Figure 4.1). In such a case, these two sequences had to be mapped (via the appropriate component h^j of ancestor map h) to adjacent endpoints in $N(A)$ (see the bottom of the right-side graphic in Figure 4.1).

Eventually, the (at most two sequences)-to-(each point $x \in X$) idea was realized as a result of $(32)_i$, $(33)_i$, and $(36)_i$ in the Decomposition Theorem 39.1 and the Definition 42.1 of the decomposition map q. Then the change of two such sequences to adjacent endpoints in $N(A)$ was realized by the appropriate idea of "ancestors" in Definition 43.1. The rest of the development consisted of filling in the gaps and adjusting definitions.

CHAPTER 9

Minimal-Exponent Question

The J_A Imbedding Theorem tells us that any metric space X of weight $|A| \geq \aleph_0$ and dimension $n \geq 0$ may be imbedded in J_A^{n+1}. It is natural to ask, for $n \geq 1$ at least, "Could J_A^{n+1} be replaced with J_A^n?" And at first blush, since the 1-sphere S^1 may be imbedded in $J_3 = J_3^1$, and since Kuratowski's forbidden graphs may be imbedded in the 4-web $J_5 = J_5^1$ (see §50), one may be tempted to guess that the answer is yes. It turns out, however, that J_A^{n+1} cannot be replaced with J_A^n because the 2-sphere S^2 cannot be imbedded in the product of two one-dimensional spaces.[1]

In this chapter, after reviewing a few basics from Borsuk [1967] and Hocking and Young [1988] on Vietoris homology, we recall the homology group $H_2(S^2)$ and present the proofs in Borsuk [1975]. Borsuk's result was motivated by Nagata's [1965, page 163] statement of an open problem which, along with a brief review of the minimal-exponent question in the context of the Classical Imbedding Theorem, is detailed in §50.

§47 Vietoris Homology

Following a review of basic definitions and terminology presented in Borsuk [1967, pages 36–43], we provide a summary of Vietoris [1927] homology by quoting Hocking and Young [1988]. We learn that on a finite polytope, the Vietoris homology groups coincide with the simplicial homology groups. Throughout, we consider only integer coefficients. One of our main goals is that of precisely defining a *true cycle*, which is fundamental in Borsuk [1975].

47.1 THE $C_n(X, \mathbb{Z}, \varepsilon)$ GROUPS. Let $X = (X, \rho)$ be a metric space and let $\varepsilon > 0$. Then an *n-dimensional oriented ε-simplex* $\sigma = (a_0, \ldots, a_n)$ is a set of *vertices* $a_i \in X$ with indices $i \in N_n = \{0, 1, \ldots, n\}$ such that each $\rho(a_i, a_j) \leq \varepsilon$. Each oriented ε-simplex σ is a mapping from N_n into X, and these mappings determine either a "1σ" or a "-1σ" ε-simplex according to the following rules: If at least two of the vertices of σ are equal, then define $\sigma = 1\sigma = -1\sigma = -\sigma$ and call σ a *degenerate simplex*. Otherwise, for $n \geq 1$, $\sigma = (a_0, \ldots, a_n)$, and each permutation $i \mapsto \phi i$ of the set N_n, define

$$(a_{\phi 0}, \ldots, a_{\phi n}) = \begin{cases} \sigma & = 1\sigma \quad \text{if } i \mapsto \phi i \text{ is even} \\ -\sigma & = -1\sigma \quad \text{if } i \mapsto \phi i \text{ is odd}; \end{cases}$$

---

[1] This result is due to Borsuk [1975], where he remarks that Miss H. Patkowska observed that an obvious modification of his proof shows that the n-sphere S^n is not homeomorphic to any subset of the Cartesian product of n one-dimensional spaces.

S.L. Lipscomb, *Fractals and Universal Spaces in Dimension Theory*,
DOI 10.1007/978-0-387-85494-6_9, © Springer Science+Business Media, LLC 2009

for $n = 0$, define $-\sigma = (-1, a_0)$; and for the (-1)-dimensional case, define two (-1)-*dimensional oriented ε-simplexes* as the numbers "1" and "-1."

For each pair σ and $-\sigma$, select one member of $\{\sigma, -\sigma\}$ and call it *positively oriented*. The other simplex is then called *negatively oriented*. Further, denote the collection of all n-dimensional ε-simplexes of X as $\Sigma^n(X, \varepsilon)$; the subfamily of those that are positively oriented as $\Sigma^n_+(X, \varepsilon)$; the subfamily of those that are negatively oriented as $\Sigma^n_-(X, \varepsilon)$; and the subfamily of degenerate simplexes as

$$\Sigma^n_0(X, \varepsilon) = \Sigma^n_+(X, \varepsilon) \cap \Sigma^n_-(X, \varepsilon).$$

Let $\mathbb{Z} = (\mathbb{Z}, +)$ be the Abelian group of integers under the usual addition, and define an n-*dimensional ε-chain in X over* $\mathbb{Z}$ as a function

$$\chi : \Sigma^n(X, \varepsilon) \setminus \Sigma^n_0(X, \varepsilon) \;\rightarrow\; \mathbb{Z}; \quad \text{where}$$

$$\chi(-\sigma) = -\chi(\sigma), \quad \chi(\sigma) \neq 0 \text{ for finitely many } \sigma.$$

Those σ such that $\chi(\sigma) \neq 0$ are the *simplexes of* χ and the vertices of the simplexes of χ are called the *vertices of* χ.

Since all such chains have a common domain and common co-domain $\mathbb{Z}$, we define the *addition* "$\chi_1 + \chi_2$" *of chains* χ_1 and χ_2 by

$$(\chi_1 + \chi_2)(\sigma) = \chi_1(\sigma) + \chi_2(\sigma), \qquad \sigma \in \Sigma^n(X, \varepsilon) \setminus \Sigma^n_0(X, \varepsilon).$$

With respect to this addition, $C_n(X, \mathbb{Z}, \varepsilon)$ denotes the Abelian group of all n-dimensional ε-chains. Moreover, for $n < -1$, $C_n(X, \mathbb{Z}, \varepsilon)$ denotes the trivial group.

To develop representations of members of $C_n(X, \mathbb{Z}, \varepsilon)$, we introduce some concise notation: For a non-degenerate n-simplex σ and a $k \in \mathbb{Z}$, we shall use "$k\sigma$" to denote the chain $\chi : \Sigma^n(X, \varepsilon) \setminus \Sigma^n_0(X, \varepsilon) \to \mathbb{Z}$ given by

$$\chi(\tau) = \chi_{k\sigma}(\tau) = \begin{cases} k \in \mathbb{Z} & \text{if } \tau = \sigma; \\ -k \in \mathbb{Z} & \text{if } \tau = -\sigma; \\ 0 \in \mathbb{Z} & \text{if } \tau \neq \sigma. \end{cases}$$

And for a degenerate n-simplex σ and a $k \in \mathbb{Z}$, we use "$k\sigma$" to denote the zero of the group $C_n(X, \mathbb{Z}, \varepsilon)$.

In passing, note that the constraint $\chi(-\sigma) = -\chi(\sigma)$ is consistent with the "$k\sigma$" notation because

$$\chi_{k\sigma}(-\sigma) = -k = -\chi_{k\sigma}(\sigma).$$

The "$k\sigma$" notation allows us to represent any chain χ as a linear combination

$$(1) \qquad \chi = k_1\sigma_1 + \cdots + k_m\sigma_m \qquad \text{for } k_i \in \mathbb{Z} \text{ and } \sigma_i \in \Sigma^n(X, \varepsilon).$$

And in reverse, any such combination is a chain in $C_n(X, \mathbb{Z}, \varepsilon)$. The representation (1), however, is not unique, as $2\sigma = 1\sigma + 1\sigma$ demonstrates. Moreover, the notation (1) allows us to write $\sigma - \sigma = 1\sigma + (-1\sigma) = \chi_{1\sigma} + \chi_{(-1\sigma)} = 0$ where the "0" is the zero in $C_n(X, \mathbb{Z}, \varepsilon)$.

47.2 THE BOUNDARY MAP $\partial : C_n(X, \mathbb{Z}, \varepsilon) \to C_{n-1}(X, \mathbb{Z}, \varepsilon)$. For $n > 0$ and any n-simplex $\sigma = (a_0, \ldots, a_n)$, we define the *boundary* $\partial\sigma$ *of* σ by

$$(2) \qquad \partial\sigma = \sum_0^n (-1)^i (a_0, \ldots, a_{i-1}, a_{i+1}, \ldots, a_n).$$

So $\partial\sigma$ is an $(n-1)$-chain. In particular, when σ is a degenerate simplex, one may show that $\partial\sigma$ is the zero in $C_{n-1}(X, \mathbb{Z}, \varepsilon)$. For example, $\partial(a_0, a_1, a_2) = 1(a_1, a_2) + (-1)(a_0, a_2) + 1(a_0, a_1)$, and if (a_0, a_1, a_2) is degenerate, say $a_1 = a_2$, then $\partial(a_0, a_1, a_2) = 1(a_1, a_1)$ which is the zero in $C_{n-1}(X, \mathbb{Z}, \varepsilon)$ because (a_1, a_1) is degenerate.

For $n = 0$, we define $\partial(a_0) = 1$ and $\partial(-1, a_0) = -1$. And finally, for the case where $n \leq -1$, we define $\partial\sigma$ to be the zero of $C_{n-1}(X, \mathbb{Z}, \varepsilon)$.

To specify $\partial\chi$ for each chain $\chi \in C_n(X, \mathbb{Z}, \varepsilon)$, we let

$$(3) \qquad \partial\chi = \sum_0^\ell m_j(\partial\sigma_j)$$

where $m_j(\partial\sigma_j)$ is the chain derived by multiplying m_j by each of the coefficients in an expansion (2) with $\sigma = \sigma_j$.

It is straightforward to show that $\partial : C_n(X, \mathbb{Z}, \varepsilon) \to C_{n-1}(X, \mathbb{Z}, \varepsilon)$ is a homomorphism. The kernel $Z_n(X, \mathbb{Z}, \varepsilon)$ of ∂ is the set of *n-dimensional ε-cycles in X over* $\mathbb{Z}$; and the image $B_{n-1}(X, \mathbb{Z}, \varepsilon) = \partial C_n(X, \mathbb{Z}, \varepsilon)$ of ∂ is the set of the $(n-1)$-*dimensional ε-boundaries in X over* $\mathbb{Z}$.

One may also show that $B_{n-1}(X, \mathbb{Z}, \varepsilon) \subset Z_n(X, \mathbb{Z}, \varepsilon)$, which amounts to

$$\partial\partial\chi = 0 \qquad\qquad \chi \in C_n(X, \mathbb{Z}, \varepsilon).$$

For example,

$$\begin{aligned} \partial\partial(a_0, a_1, a_2) &= \partial(a_1, a_2) + \partial(-1)(a_0, a_2) + \partial(a_0, a_1) \\ &= (a_2) - (a_1) - (a_2) + (a_0) + (a_1) - (a_0) = 0. \end{aligned}$$

47.3 INFINITE CHAINS IN X. Let $\{\varepsilon_i\}$ be a sequence of positive numbers that converges to zero, and for each i, let $\chi_i \in C_n(X, \mathbb{Z}, \varepsilon_i)$. Then the sequence $\boldsymbol{\chi} = \{\chi_i\}$ is an *infinite n-dimensional chain in X* whenever there exists a compact subset $X_0 \subset X$ such that every vertex of every χ_i is an element of X_0.

The *majorant* and *carrier* of $\boldsymbol{\chi}$ is the sequence $\{\varepsilon_i\}$ and the X_0, respectively. Any convergent-to-zero sequence $\{\varepsilon_i'\}$ such that each $\varepsilon_i' \geq \varepsilon_i$ is also a majorant of $\boldsymbol{\chi}$. And any compact $X_0' \subset X$ that is a superset of X_0 is also a carrier of $\boldsymbol{\chi}$. We shall denote the infinite chain each of whose components is the zero chain as $\mathbf{0}$.

For infinite n-dimensional chains $\boldsymbol{\chi} = \{\chi_i\}$ and $\boldsymbol{\chi}' = \{\chi_i'\}$, the sequence $\boldsymbol{\chi} + \boldsymbol{\chi}' = \{\chi_i + \chi_i'\}$ is an infinite n-dimensional chain in X. Similarly, $\boldsymbol{\chi} - \boldsymbol{\chi}'$ denotes the infinite chain $\{\chi_i - \chi_i'\}$.

47.4 INFINITE, HOMOLOGOUS, ESSENTIAL, AND TRUE CYCLES IN X. An infinite chain $\boldsymbol{\gamma} = \{\gamma_i\}$ is an *infinite cycle* whenever each γ_i is a cycle. Thus, for any infinite n-dimensional chain $\boldsymbol{\chi} = \{\chi_i\}$, we have an infinite $(n-1)$-cycle $\partial\boldsymbol{\chi} = \{\partial\chi_i\}$. Two infinite cycles $\boldsymbol{\gamma}$ and $\boldsymbol{\gamma}'$ are *homologous in X*, written $\boldsymbol{\gamma} \sim \boldsymbol{\gamma}'$, when there exists a chain $\boldsymbol{\chi}$ in X such that $\partial\boldsymbol{\chi} = \boldsymbol{\gamma} - \boldsymbol{\gamma}'$.

An infinite cycle $\boldsymbol{\gamma}$ is an *essential cycle* if it has a carrier in which it is not homologous to the cycle $\mathbf{0}$. And an infinite cycle $\boldsymbol{\gamma} = \{\gamma_i\}$ in X is a *true cycle* if the infinite cycle $\boldsymbol{\gamma}' = \{\gamma_{i+1} - \gamma_i\}$ is not homologous to $\mathbf{0}$ in X.

47.5 THE HOMOLOGY (OR BETTI) GROUPS $H_n(X) = H_n(X, \mathbb{Z})$. The set $Z_n(X) = Z_n(X, \mathbb{Z})$ of all n-dimensional true cycles in X over $\mathbb{Z}$ under the addition given by

$$\{\gamma_i\} + \{\gamma_i'\} = \{\gamma_i + \gamma_i'\}$$

is an Abelian group. The elements in $Z_n(X)$ that are homologous to $\mathbf{0}$ form a subgroup $B_n(X) = B_n(X, \mathbb{Z})$ of $Z_n(X)$. The factor group $H_n(X) = Z_n(X)/B_n(X)$ is the *n-dimensional homology (or Betti) group of X over $\mathbb{Z}$.*

47.6 MAPS $X \xrightarrow{f} Y$ INDUCE HOMOMORPHISMS $H_n(X) \xrightarrow{f_*} H_n(Y)$. Let $\sigma = (a_0, \ldots, a_n)$ be an n-dimensional simplex in X. Then $f(\sigma)$, given by

$$f(\sigma) = (f(a_0), \ldots, f(a_n)),$$

is an n-dimensional simplex in Y. Since f may not be injective, $f(\sigma)$ may be degenerate even when σ is not degenerate. And the formula for $f(\sigma)$ induces an f-assignment of n-dimensional chains given by

$$\chi = k_1 \sigma_1 + \cdots + k_m \sigma_m \quad \mapsto \quad f(\chi) = k_1 f(\sigma_1) + \cdots + k_m f(\sigma_m).$$

Since f may not preserve distances, χ may be an ε-chain while $f(\chi)$ is not an ε-chain. Nevertheless, $f(\chi)$ does belong to $C_n(Y, \mathbb{Z}, \eta)$ where η depends on f and ε. And since f is uniformly continuous on any compact set, any sequence $\{f(\chi_i)\}$ associated by f with an infinite chain $\boldsymbol{\chi} = \{\chi_i\}$ is also an infinite chain $f(\boldsymbol{\chi})$ in Y and has a carrier $f(X_0)$ where X_0 is a carrier of $\boldsymbol{\chi}$.

One may show that $\partial f = f \partial$, which yields the fact that f maps infinite, true, and homologous cycles in X to, respectively, infinite, true, and homologous cycles in Y. It follows that f induces a homomorphism

$$f_* : H_n(X) \to H_n(Y).$$

Having reviewed the most basic concepts 47.1 – 47.6 of Vietoris homology theory, we turn to a short summary of the theory by quoting pages 346 and 347 of Hocking and Young [1988]. Our goal is to recall that we may calculate the Vietoris homology groups of the 2-sphere by using the simplicial homology theory on finite polytopes. (The reference numbers in the following quotation match those of this book.)

Vietoris homology theory. The Vietoris homology theory was the first of the Čech-type homology theories to appear. It was introduced by Vietoris [1927] and in this form applies only to metric spaces. While this theory has been used in many research papers, it has not been discussed so extensively as has the more general Čech theory. Again, for the sake of brevity, we consider only compact spaces in this presentation. ...

Let M be a compact metric space, and let ε be a positive number. We construct the simplicial complex $K_\varepsilon = \{\mathcal{V}, \Sigma\}$, where the vertices in $\mathcal{V}$ are the points of M and where a finite subcollection of vertices $p_0, \ldots, p_n$ forms an n-simplex in Σ if and only if the diameter of the set $\cup_{i=0}^n \{p_i\}[= \max d(p_i, p_j)]$ is less than ε. It is easy to prove that for each $\varepsilon > 0$, K_ε is a simplicial complex ... Therefore, for each $\varepsilon > 0$ and each integer $n \geq 0$, we may construct the simplicial homology $H_n(K_\varepsilon)$ of K_ε with integral coefficients.

Given $\varepsilon_1 > \varepsilon_2 > 0$ it is evident that each simplex of K_{ε_2} is also a simplex of K_{ε_1} and hence that there is an identity injection $j_{\varepsilon_1 \varepsilon_2}$ of K_{ε_2} into K_{ε_1}: This injection then induces a homomorphism $_*j_{\varepsilon_1 \varepsilon_2}$ of $H_n(K_{\varepsilon_2})$ into $H_n(K_{\varepsilon_1})$. Furthermore, if $\varepsilon_1 > \varepsilon_2 > \varepsilon_3 > 0$, then the induced homomorphisms satisfy the relation

$$_*j_{\varepsilon_1 \varepsilon_2}{}_*j_{\varepsilon_2 \varepsilon_3} = {}_*j_{\varepsilon_1 \varepsilon_3}.$$

Since the positive real numbers constitute a directed set, the collection $\{H_n(K_\varepsilon)\}$ together with the injection-induced homomorphisms $\{_*j_{\varepsilon\delta}\}$ form an inverse limit system of groups and homomorphisms. The inverse limit group of this system is the nth *Vietoris homology group* $V_n(M)$.

Clearly the complexes K_ε are much too large for convenient manipulation (they can certainly have a nondenumerable number of simplexes and infinite dimension). The usual technique in using Vietoris theory involves discussing the existence or, more often, the nonexistence of certain essential (nonbounding) cycles. In this way, one studies the connectivity properties of the space M without becoming involved with the complexes K_ε. It is known that the Vietoris groups, the singular groups, and the Čech groups coincide if the underlying space is sufficiently well-behaved. For instance, all these coincide with the simplicial homology groups on a finite polytope.

§48 The Vietoris Homology Group $H_2(S^2)$

Since the Vietoris homology groups coincide with the simplicial homology groups on a finite polytope, we may calculate the second Vietoris homology group of the 2-sphere S^2 by viewing S^2 as the finite polytope induced from the 2-skeleton (the union of all dimension ≤ 2 simplexes) of a 3-simplex, and then calculate the simplicial homology group $H_2(S^2)$.

To calculate $H_2(S^2)$ in the context of simplicial homology theory, we may consult almost any book on algebraic topology. For example, in Munkres [1984, §8], where a complex whose reduced homology vanishes in all dimensions is said to be *acyclic*, and where $\widetilde{H}_n(X)$ denotes the nth *reduced* homology group of X, we find the following theorem.

48.1 Theorem (simplicial reduced homology groups of spheres) *Let σ be an n-simplex. The complex K_σ consisting of σ and its faces is acyclic. If $n > 0$, let Σ^{n-1} denote the complex whose polytope is the boundary of σ. Orient σ. Then $\widetilde{H}_{n-1}(\Sigma^{n-1})$ is infinite cyclic and is generated by the chain $\partial\sigma$. Furthermore, $\widetilde{H}_i(\Sigma^{n-1}) = 0$ for $i \neq n - 1$.*

Using Munkres [1984, Theorem 7.2], which provides the exact relation between the *reduced* simplicial homology group $\widetilde{H}_0(K)$ and the nonreduced group $H_0(K)$ for the complex K, we may compare the *reduced* homology groups "$\widetilde{H}_q(S^n)$" of the n-sphere with those of the simplicial homology groups "$H_q(S^n)$." The binary relation symbol "$\cong$" denotes group isomorphism:

$$n > 0 \quad \widetilde{H}_q(S^n) \cong 0 \text{ if } q \neq 0, n \quad \widetilde{H}_0(S^n) \cong 0, \quad \widetilde{H}_n(S^n) \cong \mathbb{Z},$$
$$n = 0 \quad \widetilde{H}_q(S^0) \cong 0 \text{ if } q \neq 0, \qquad\qquad\qquad\qquad \widetilde{H}_0(S^0) \cong \mathbb{Z}.$$

$$n > 0 \quad H_q(S^n) \cong 0 \text{ if } q \neq 0, n \quad H_0(S^n) \cong \mathbb{Z}, \quad H_n(S^n) \cong \mathbb{Z},$$
$$n = 0 \quad H_q(S^0) \cong 0 \text{ if } q \neq 0, \qquad\qquad\qquad\qquad H_0(S^0) \cong \mathbb{Z} \oplus \mathbb{Z}.$$

It follows that the reduced homology sequence differs from the simplicial homology sequence only at dimension zero. And $H_2(S^2) \cong \mathbb{Z}$ because our coefficient group is $\mathbb{Z}$.

So the Vietoris homology group $H_2(S^2) \cong \mathbb{Z}$ is certainly nontrivial, which ensures that there exists a 2-dimensional true cycle γ that generates the Betti group $H_2(S^2)$.

§49 Borsuk's Theorem

In this section, we show that the 2-sphere S^2 is not topologically contained in the Cartesian product of two one-dimensional spaces. The presentation follows Borsuk [1975].

49.1 Lemma (Borsuk [1975]) *Let each of the polyhedra C and D be of dimension ≤ 1; let γ be a true 2-dimensional cycle that is a generator of the Betti group $H_2(S^2)$; and let f denote any continuous map $S^2 \to C \times D$. Then $f(\gamma) \sim 0$ in $C \times D$.*

PROOF. Since f is continuous, we may assume that each of C and D are connected. Further, since we are only considering homologous cycles in $C \times D$, we only need to consider the homotopy type of C and D. Thus we assume that C and D are finite bouquets of circles, i.e.,

$$C = C_1 \cup \cdots \cup C_m, \qquad D = D_1 \cup \cdots \cup D_n$$

where each C_i and each D_j is a circle and there exist points $c \in C$ and $d \in D$ such that

$$C_i \cap C_{i'} = \{c\} \quad \text{when} \quad i \neq i' \qquad D_j \cap D_{j'} = \{d\} \quad \text{when} \quad j \neq j'.$$

Now let α_i denote a 1-dimensional true cycle in C_i that generates $H_1(C_i)$ and let β_j denote a 1-dimensional true cycle in D_j that generates $H_1(D_j)$. It is well known (by a special case of a theorem of Künneth, see Alexandroff and Hopf [1935, page 308]) that the true 2-dimensional cycles $\alpha_i \times \beta_j$, where $i = 1, \ldots, m$ and $j = 1, \ldots, n$, generate the Betti group $H_2(C \times D)$. It follows that there exist integers k_{ij} such that

$$f(\gamma) \sim \sum_{i,j} k_{ij}(\alpha_i \times \beta_j) \quad \text{in} \quad C \times D.$$

Next, consider the retractions $\phi_i : C \to C_i$ and $\psi_j : D \to D_j$ such that

$$\phi_i(C_{i'}) = c \quad \text{for} \quad i \neq i' \qquad \text{and} \qquad \psi_j(D_{j'}) = d \quad \text{for} \quad j \neq j'.$$

Define

$$r_{ij}(x, y) = (\phi_i(x), \psi_j(y)) \qquad (x, y) \in C \times D.$$

Then each $r_{ij} : C \times D \to C_i \times D_j$ is a retraction such that

$$r_{ij}(\alpha_i \times \beta_j) = \alpha_i \times \beta_j \quad \text{and} \quad r_{ij}(\alpha_{i'} \times \beta_{j'}) \sim 0 \text{ for } (i, j) \neq (i'j').$$

It follows that for each pair (i_0, j_0), where $1 \leq i_0 \leq m$ and $1 \leq j_0 \leq n$, we have

$$r_{i_0 j_0} f(\gamma) \sim \sum_{i,j} k_{ij} r_{i_0 j_0}(\alpha_i \times \beta_j) \sim k_{i_0 j_0}(\alpha_{i_0} \times \beta_{j_0}).$$

But $r_{i_0 j_0} f$ maps S^2 into the surface $C_{i_0} \times D_{j_0}$, and it is well known that the degree of such a map is zero. Hence $k_{i_0 j_0} = 0$ and consequently $f(\gamma) \sim 0$ in $C \times D$, which finishes the proof. $\qquad \square$

In his proof of Theorem 49.5, Borsuk uses the *Menger sponge* M, which is universal for compact metric spaces of dimension ≤ 1.[2]

---

[2]For more background on M, and to view M as the attractor of an iterated function system, see §50.

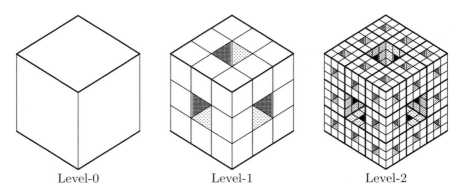

Level-0   Level-1   Level-2

Fig. 49.2 Levels of approximation to fractal known as Menger's Sponge.

In passing, we note that since the Menger sponge M is the intersection of the nested sequence level-0 $\supset$ level-1 $\supset$ level-2 $\supset \cdots$ of approximations, we have a sequence $P_0 \supset P_1 \supset P_2 \supset \cdots$ of 1-dimensional polyhedra that may be used to "approximate" M — for P_0, consider the union of the edges of the cube, for P_1, the union of the edges of all the subcubes at level-1, $\cdots$. With this sequence of 1-dimensional polyhedra, we may "approximate M" in the following sense: Using "$||x - y||$" to denote the usual distance between $x, y \in \mathbb{R}^3$, we see that for any $\varepsilon > 0$, an $N > 0$ exists such that for each $n > N$, the polyhedron P_n satisfies "$x \in M$ implies there exists $y \in P_n$ such that $||x - y|| < \varepsilon$."

In addition to Menger's sponge M, Borsuk applies a few concepts concerning continuous mappings: Recall that a continuous surjection $f : Y \to Y_0$ is a *retraction* (Borsuk [1967, page 10]) if $f(y) = y$ for each $y \in Y_0$. When a retraction $f : Y \to Y_0$ exists, we shall say that Y_0 *is a retract of* Y.

A closed subset Y_0 of a space Y is a *neighborhood retract in the space* Y (Borsuk [1967, page 14]) if there exists an open set $U \supset Y_0$ in Y such that Y_0 is a retract of U. Moreover, if a compact metric space X has the property that each homeomorphism $X \xrightarrow{h} h(X) = Y_0 \subset Y$ such that Y_0 is closed in Y produces a neighborhood retract Y_0 in the space Y, then X is called an *absolute neighborhood retract* (Borsuk [1967, page 100]), which we denote as "$X \in$ ANR."

For a proof of the following lemma, see Kuratowski [1968, page 354].

49.3 Lemma *Let X be a separable metric space. Then* $\dim X \leq n$ *if and only if for each closed $F \subset X$, each continuous mapping $F \to S^n$ has a continuous extension $X \to S^n$.*

49.4 Corollary *Let Z be a separable metric space where $\dim Z \leq 2$, and let $S \subset Z$ be homeomorphic to the 2-sphere S^2. Then there is a retraction $s : Z \to S$.*

PROOF. Let $F = S$ and consider the continuous identity map $F \to S$. Since

S^2 is compact, $F = S$ is closed in Z. An application of the previous lemma shows that there is a continuous extension $s : Z \to S$, which is the desired retraction. ☐

49.5 Theorem (Borsuk [1975]) *The Cartesian product $X \times Y$ of any two metric spaces X and Y where each of X and Y have dimension ≤ 1 does not contain any subset S homeomorphic to the 2-sphere S^2.*

PROOF. Suppose such an $S \subset X \times Y$ does exist. Setting

$$p(x, y) = x, \quad q(x, y) = y \qquad \text{for each } (x, y) \in X \times Y,$$

we get the two projections $p : X \times Y \to X$ and $q : X \times Y \to Y$. Then $X_0 = p(S)$ and $Y_0 = q(S)$ are continua of dimension ≤ 1 such that $S \subset X_0 \times Y_0$. Thus we may limit ourselves to the case where X and Y are continua. Since every continuum of dimension ≤ 1 is contained in the universal curve M of Menger [1926b], we may assume that $X = Y = M$. Now let us observe that to every $\varepsilon > 0$ we may assign a retraction

$$r_\varepsilon : M \to A^\varepsilon$$

where A^ε is a polyhedron and

$$\rho(x, r_\varepsilon(x)) < \varepsilon \qquad \text{for every} \quad x \in M = X = Y.$$

Since $\dim(X \times Y) \leq 2$, Corollary 49.4 shows that there exists a retraction $s : X \times Y \to S$. Letting, for each $\varepsilon > 0$,

$$f_\varepsilon(x, y) = (r_\varepsilon(x), r_\varepsilon(y)) \qquad (x, y) \in S$$

and

$$g_\varepsilon(x, y) = s f_\varepsilon(x, y) \qquad (x, y) \in S,$$

we obtain mappings $f_\varepsilon : S \to A^\varepsilon \times A^\varepsilon$ and $g_\varepsilon : S \to S$. It follows from Lemma 49.1 that the true cycle γ generating $H_2(S)$ is mapped by f_ε onto a true cycle homologous to zero in $A^\varepsilon \times A^\varepsilon$. So $g_\varepsilon(\gamma) \sim 0$ in S.

Now observe that for every $\eta > 0$ there is a positive number $\varepsilon < \eta$ such that

$$\rho(f_\varepsilon(x, y), s f_\varepsilon(x, y)) < \eta \qquad (x, y) \in S$$

because s is the identity on S and for ε sufficiently small and $(x, y) \in S$ the distance $f_\varepsilon(x, y)$ from S is arbitrarily small.

It follows that

$$\rho((x, y), g_\varepsilon(x, y)) \leq \rho((x, y), f_\varepsilon(x, y)) + \rho(f_\varepsilon(x, y), s f_\varepsilon(x, y)) < \varepsilon + \eta < 2\eta.$$

But $S \in \mathrm{ANR}$ and consequently for ε sufficiently small the map $g_\varepsilon : S \to S$ is homotopic to the identity map $i_S : S \to S$. Hence $g_\varepsilon(\gamma) \sim \gamma$ in S, which contradicts the relations $g_\varepsilon(\gamma) \sim 0$ in S and $\gamma \nsim 0$ in S. ☐

§50 Comments

Borsuk's Theorem (Theorem 49.5) was motivated not by the J_A Imbedding Theorem, but by an open problem cited by Nagata [1965, footnote on page 163]:

> ... it is an open problem whether every n-dimensional metric space can be topologically imbedded in the topological product of n one-dimensional metric spaces.

Nagata's motivation for citing this open problem was natural because of his proof (Nagata [1958]) that every n-dimensional metric space can be topologically imbedded in the topological product of $(n+1)$ one-dimensional metric spaces.

Borsuk's Theorem answered Nagata's question in the negative — the 2-sphere cannot be imbedded in any Cartesian product of two one-dimensional spaces. Nevertheless, Borsuk's result shows that the J_A Imbedding Theorem cannot be improved in the sense of reducing the index $(n + 1)$.

In the proof of his theorem, Borsuk used the Menger sponge. From the iterated function system (IFS) viewpoint, Menger's sponge is the 3-space generalization of Sierpiński's carpet in 2-space, which in turn is a generalization of Cantor's set in 1-space.

In detail, the Cantor set may be viewed as the attractor of the IFS $\{w_y\}$ where $y \in \{0, 1\}$ is on the boundary of the unit interval I and w_y is the $\frac{1}{3}$-contraction of I toward y. Analogously, the Sierpiński carpet (Menger sponge) is the attractor of the IFS $\{w_y\}$ where y lies in the boundary of the unit square I^2 (unit cube I^3) and ranges over the eight points in $\{0, \frac{1}{2}, 1\} \times \{0, \frac{1}{2}, 1\}$ (the 20 points in $\{0, \frac{1}{2}, 1\}^3$) that have at most one component equal to $\frac{1}{2}$; and each w_y denotes the $\frac{1}{3}$-contraction of I^2 (I^3) toward y, i.e.,

$$w_y(x) = y + \tfrac{1}{3}(x - y) = \tfrac{1}{3}x + \tfrac{2}{3}y.$$

Each of these three fractals, i.e., Cantor's set, Sierpiński's carpet, and Menger's sponge, are universal spaces for certain subclasses of separable metric spaces: Cantor's set is universal for zero-dimensional spaces. Sierpiński's carpet, as is shown in Sierpiński [1916], is universal for compact subspaces of the plane that have an empty interior; and, as shown in Sierpiński [1922], his "assumption of compactness" was not necessary. Menger's sponge, as is shown in Menger [1926a], is universal for the class of all compact metric spaces of dimension ≤ 1. For additional comments concerning various "universal space" developments, see Engelking [1978, Section 1.11], especially his "Historical and bibliographical notes" on his pages 128 and 129.

Nöbeling's [1931] Classical Imbedding Theorem shows that if X is an n-dimensional separable metric space, then X can be topologically imbedded in the product I^{2n+1} of $2n + 1$ copies of the one-dimensional unit interval I.

To address the "minimal-exponent question" for the Classical Imbedding Theorem, consider a related theorem concerning polytopes (see Hocking and Young [1988, page 215]).

50.1 Theorem (imbedding n-dimensional polytopes in $\mathbb{R}^{2n+1}$) *Let $|K|$ be an n-dimensional polytope with a triangulation K. Then $|K|$ may be imbedded rectilinearly in $\mathbb{R}^{2n+1}$.*

In the polytope case, Flores [1934] showed that the complex consisting of all faces of dimension $\leq n$ (the n-skeleton) of a $2n + 2$-simplex cannot be imbedded in $\mathbb{R}^{2n}$. (For a proof of Flores' result using the Borsuk-Ulam Antipodal Theorem, i.e., for every continuous mapping $g : S^n \to \mathbb{R}^n$ there exists a point $x \in S^n$ such that $g(x) = g(-x)$, see Problem 1.11.F on page 132 of Engelking [1978].)

As an obvious corollary to Flores' result, we see that the exponent $2n + 1$ of I^{2n+1} in the Classical Imbedding Theorem is minimal.

The $n = 1$ case is of interest in its own right: Flores' [1934] result implies that the 1-skeleton of the 4-simplex, which has five vertices, and is well known in graph theory as the *complete graph K_5 on five points*, cannot be imbedded in the plane $\mathbb{R}^{2n} = \mathbb{R}^2$. (For a proof using the Jordan Curve Theorem, see Example 1.11.8 on page 127 of Engelking [1978].)

Four years prior to Flores' [1934] result, however, Kuratowski [1930] showed that neither the complete graph K_5 nor the *complete bipartite graph $K_{3,3}$* can be imbedded in the plane. (Recall that the graph $K_{3,3}$ has as its vertex set a union of two disjoint size-three sets A and B whose size-nine edge set consists of all 1-simplexes with one endpoint in A and the other in B.) Any 1-dimensional metric space X, in particular any graph, that contains a topological copy of either K_5 or $K_{3,3}$ cannot be imbedded in the plane. In addition, however, Kuratowski [1930] also showed that a graph that cannot be imbedded in the plane must contain a topological copy of either K_5 or $K_{3,3}$. Today, these two graphs are often referred to as *Kuratowski's forbidden graphs*.

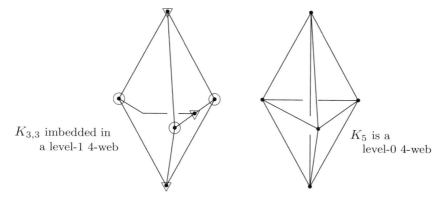

$K_{3,3}$ imbedded in
a level-1 4-web

K_5 is a
level-0 4-web

Fig. 50.2 Imbedding Kuratowski's Forbidden Graphs $K_{3,3}$, K_5 in J_5.

From Figure 50.2, it is clear that Kuratowski's forbidden graphs K_5 and K_{33} may be imbedded, respectively, in level-0 and level-1 4-webs.

CHAPTER 10

The J_A^∞ Imbedding Theorem

Every separable metric space may be imbedded in I^∞, and those of dimension $\leq n$ may be imbedded in I^{2n+1}. Moreover, every general (not necessarily separable) weight $|A|$ metric space may be imbedded in a countable product $S(A)^\infty$ of star spaces $S(A)$, and those of dimension $\leq n$ may be imbedded in an n-dimensional subspace of $S(A)^\infty$. Finally, as was shown in Chapter 8, every weight $|A| \geq \aleph_0$ metric space of dimension $\leq n$ may be imbedded in J_A^{n+1}. The remaining question is, "Can every metric space of weight $|A|$ be imbedded in a countably infinite product J_A^∞ of copies of J_A?"

In this chapter we show that J_A^∞ is indeed universal for the class of weight $|A|$ topological spaces that are metrizable. The proof shows how to view $S(A)$ as a subspace of J_A. The presentation follows Lipscomb [1976].

§51 Imbedding Theorems

In this section we provide detailed statements of six imbedding theorems that have appeared over approximately half a century — 1925 through 1976. Each theorem specifies a universal space, and the theorems fall naturally into three pairs. One member of each pair provides the description of a universal space in the n-dimensional case, while the other member provides a universal space for the general (not necessarily finite-dimensional) case.

As usual, "I" denotes the unit interval; "J_A" the image of the perfect mapping $p : N(A) \to J_A$ (as detailed in Chapter 1); and "$S(A)$" the star space (hedgehog with $|A|$ prickles).

We shall use the "standard metric" of the star space $S(A)$: A *star space* is a metric space $(S(A), d)$ where the set $S(A) = \cup_a I_a$ is the *star-shaped set* obtained by identifying the zeros of a disjoint union of $|A| \geq \aleph_0$ unit intervals I_a (the ath arm), and the metric d is given by

$$(1) \qquad d(x,y) = \begin{cases} |x - y| & \text{if } x \text{ and } y \text{ belong to the same arm} \\ |x + y| & \text{if } x \text{ and } y \text{ belong to distinct arms.} \end{cases}$$

The introduction of star spaces, as well as the imbedding theorems that use them as base spaces, predates the introduction of J_A, and the chronology is preserved in the following list — the three pairs of theorems given below appear in chronological order.

51.1 Theorem (Urysohn [1925a]) *A topological space of weight $\aleph_0$ is metrizable if and only if it can be imbedded in I^∞.*

S.L. Lipscomb, *Fractals and Universal Spaces in Dimension Theory*,
DOI 10.1007/978-0-387-85494-6_10, © Springer Science+Business Media, LLC 2009
   107

51.2 Theorem (Nöbeling [1931]) *A topological space of weight $\aleph_0$ is metrizable of dimension $\leq n$ if and only if it can be imbedded in the subspace of I^{2n+1} whose tuples contain at most n rational coordinates.*

51.3 Theorem (Kowalsky [1957]) *A topological space of weight $|A| \geq \aleph_0$ is metrizable if and only if it can be imbedded in $S(A)^\infty$.*

51.4 Theorem (Nagata [1963]) *A topological space of weight $|A| \geq \aleph_0$ is metrizable of dimension $\leq n$ if and only if it can be imbedded in the subspace of $S(A)^\infty$ whose tuples contain at most n nonvanishing rational coordinates.*

51.5 Theorem (Lipscomb [1975]) *A topological space of weight $|A| \geq \aleph_0$ is metrizable of dimension $\leq n$ if and only if it can be imbedded in the subspace of J_A^{n+1} whose tuples contain at most n rational coordinates.*

51.6 Theorem (Lipscomb [1976]) *A topological space of weight $|A| \geq \aleph_0$ is metrizable if and only if it can be imbedded in J_A^∞.*

§52 The Lemmas and Proof

From the hypothesis of Theorem 51.6, it is assumed throughout that A is an infinite set. And since A is infinite, for $z \in A$ fixed and $A' = A \setminus \{z\}$, it is clear that $S(A)$ and $S(A')$ are homeomorphic.

So to prove Theorem 51.6, it suffices to prove that $S(A')$ may be imbedded in J_A. It then follows that $S(A)^\infty$ may be imbedded in J_A^∞. Thus, the necessary part of Theorem 51.6 follows from the necessary part of Kowalsky's Theorem 51.3. The sufficiency part of Theorem 51.6 follows easily from the fact that a product of countably many metric spaces is metric. With these observations, it only remains to prove that $S(A') =_t X \subset J_A$.

To imbed $S(A')$ into J_A, we shall use the following theorem (see A4.3).

52.1 Theorem *Let $p : F \to X$ be surjective and quotient, $f : F \to S$ continuous, and $fp^{-1} : X \to S$ single valued, i.e., f is constant on each fiber $p^{-1}(x)$. Then fp^{-1} is continuous. Moreover, fp^{-1} is closed if and only if $f(H)$ is closed whenever H is a closed p-inverse set $(H = p^{-1}p(H))$.*

To apply Theorem 52.1, we begin by defining $F \subset N(A)$: For $\alpha = a_1 a_2 \cdots \in N(A)$, let $C(\alpha) = \{a_i : i = 1, 2, \ldots\} \subset A$. We call the members of $C(\alpha)$ the *characters* of α, and whenever $C(\alpha)$ is finite, we shall say that α *is of finite character*. With this terminology, we define

(2) $F = \{\alpha \in N(A) : |C(\alpha)| = 1, \text{ or }, |C(\alpha)| = 2 \text{ and } z \in C(\alpha)\}.$

And if each $F(a) = N(\{z, a\})$, we may show that $F = \cup\{F(a) : a \in A'\}$.

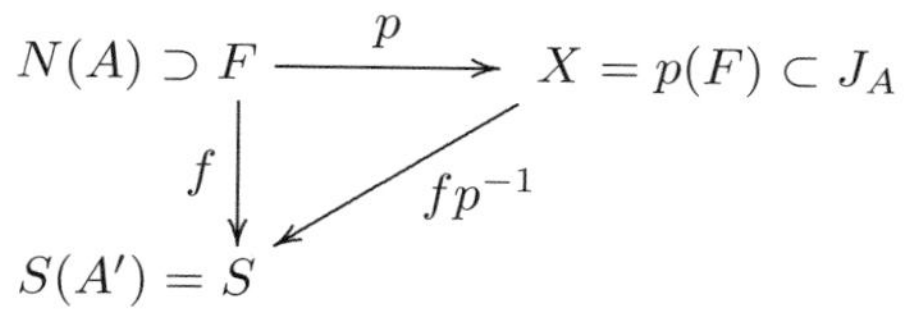

$$N(A) \supset F \xrightarrow{p} X = p(F) \subset J_A$$

$$f \downarrow \qquad \qquad \nearrow fp^{-1}$$

$$S(A') = S$$

Fig. 52.2 Diagram underlying the imbedding of $S(A')$ into J_A.

To define $f : F \to S = S(A')$ as indicated in Figure 52.2, let χ be the characteristic function of A' defined on A, i.e., $\chi(a) = 1$ if $a \in A'$, and $\chi(a) = 0$ if $a = z$. Then $f : F \to S(A') = \cup_{a \in A'} I_a$ is given by

$$(3) \quad f(\alpha) = \sum_{i=1}^{\infty} \chi(a_i)/2^i \in I_a \qquad \alpha = a_1 a_2 \cdots \in F(a) \text{ for some } a \in A'.$$

52.3 Lemma (f is continuous) *Let $f : F \to S = S(A')$ be the mapping specified in* (3). *Then f is continuous.*

PROOF. Then given the topologies on F and $S(A')$, it suffices to show that for any two points $\alpha = a_1 a_2 \cdots, \beta = b_1 b_2 \cdots \in F$ such that $a_i = b_i$ for $i = 1, \ldots, k$ we necessarily have $d(f(\alpha), f(\beta)) \leq 1/2^{k-1}$. The proof of this fact breaks into two cases: First, suppose there is an $a \in A'$ such that both $\alpha, \beta \in F(a)$. Then from the definition (3) of f and the definition (1) of the metric d, we see that

$$d(f(\alpha), f(\beta)) = \left| \sum_{i=1}^{\infty} \chi(a_i)/2^i - \sum_{i=1}^{\infty} \chi(b_i)/2^i \right| \leq \sum_{i=k+1}^{\infty} 1/2^i = 1/2^k < 1/2^{k-1}.$$

Second, suppose there is no $a \in A'$ such that both $\alpha, \beta \in F(a)$. Then $a_i = b_i$ for $1 \leq i \leq k$ implies that these first k characters satisfy $a_i = z = b_i$, which shows that the corresponding χ values satisfy $\chi(a_i) = 0 = \chi(b_i)$. Thus,

$$d(f(\alpha), f(\beta)) = \sum_{i=k+1}^{\infty} \chi(a_i)/2^i + \sum_{i=k+1}^{\infty} \chi(b_i)/2^i| \leq 1/2^k + 1/2^k = 1/2^{k-1}.$$

Therefore, from the first statement in this proof, we are finished. □

52.4 Lemma *Let $H \subset S(A')$, let "0" denote the "zero" in $S(A')$, and suppose that either $0 \notin \overline{H}$ or $0 \in H$. Then $H \cap I_a$ closed in I_a for each $a \in A'$ implies H is closed in $S(A')$.*

PROOF. Suppose H is not closed in $S(A')$. Then $x \in S(A')$ exists such that $x \in \overline{H} \setminus H$. Since either $0 \notin \overline{H}$ or $0 \in H$, it follows that $x \neq 0$. Thus,

$x \in I_a \setminus \{0\}$ for some $a \in A'$. However, since $I_a \setminus \{0\}$ is open in $S(A')$, we see that if $0 \notin \overline{H}$, then $0 \notin H$ and

$$x \in \text{cl}\,(H \cap (I_a \setminus \{0\})) = \text{cl}(H \cap I_a) = H \cap I_a$$

because $H \cap I_a$ is closed in I_a. Further, if $0 \in H$, then

$$x \in \text{cl}\,(H \cap (I_a \setminus \{0\})) \subset \text{cl}(H \cap I_a) = H \cap I_a$$

because (again) $H \cap I_a$ is closed in I_a. Thus $x \in H$ if either $0 \in \overline{H}$ or $0 \in H$, which contradicts the definition of x. $\square$

52.5 Lemma *Let $\overline{z} \notin K \subset F$ where K is closed in F. Then $f(\overline{z}) \notin \overline{f(K)}$.*

PROOF. If K is closed and $\overline{z} \notin K$, then an m exists such that whenever $\alpha \in F$ and $a_i = z$ for each $i \leq m$, then $\alpha \notin K$. It follows, since $\{\alpha : d(f(\overline{z}), f(\alpha) < 1/2^m\}$ is disjoint from K, that $f(\overline{z}) \notin \overline{f(K)}$. $\square$

52.6 Theorem (f is closed) *The continuous $f : F \to S(A')$ is also closed.*

PROOF. Let K be a closed subset of F. Then there are two cases: First, suppose $\overline{z} \notin K$. Then since $K \cap F(a)$ is compact and f is continuous (Lemma 52.3), $f(K \cap F(a)) = f(K) \cap I_a$ is a compact subset of I_a for each $a \in A'$. It follows that for $H = f(K)$, we have $H \cap I_a$ closed in I_a for each $a \in A'$. By Lemma 52.5, we have $f(\overline{z}) = 0 \notin \overline{H} = \overline{f(K)}$, and then an application of Lemma 52.4 shows that $H = f(K)$ is closed in $S(A')$. So the case $\overline{z} \notin K$ is finished.

Second, suppose $\overline{z} \in K$. Then $f(\overline{z}) = 0 \in H = f(K)$. Again, since $K \cap F(a)$ is compact and f is continuous we see that $f(K)$ is a compact subset of I_a for each $a \in A'$. Then again $H = f(K)$ is closed in $S(A')$ because $0 \in H$ allows an application of Lemma 52.4. $\square$

52.7 Theorem (J_A contains a star space $S(A)$) *Let $|A| \geq \aleph_0$. Then the star space $S(A)$ can be imbedded in J_A.*

PROOF. Since A is infinite, we may select and fix a point $z \in A$, define $A' = A \setminus \{z\}$, and obtain a homeomorphism $S(A) \to S(A')$. Thus, it suffices to show that $S(A')$ may be imbedded in J_A. For the proof, we shall apply Theorem 52.1 with the notation in the commutative diagram in Figure 52.2: We use (2) to define $F \subset N(A)$ and (3) to define $f : F \to S = S(A')$. We also let $p : F \to X = p(F) \subset J_A$ denote the restriction of the perfect (adjacent-endpoint identification) mapping $p : N(A) \to J_A$. Then from Lemmas 52.3 and 52.6, we see, respectively, that f is continuous and closed. Turning to the mapping $p : F \to X = p(F)$, we note that $F = p^{-1}(p(F))$, i.e., that F is a p^{-1}-inverse set. It follows, since $p : N(A) \to J_A$ is closed, that p restricted to an inverse set is also closed, and a fortiori quotient. Moreover, since $p : N(A) \to J_A$ is continuous, p restricted to F is continuous. It is also clear that f is constant on each fiber $p^{-1}(x)$. Thus, $fp^{-1} : p(F) \to S(A')$ is well defined and continuous. To see that fp^{-1} is also closed, we may invoke the equivalence stated in the last sentence of Theorem 52.1 because, by Lemma 52.6, f is closed.   □

§53 Comments

One should note that the six theorems (in §51) that span half a century (1925–1976) of constructions of universal spaces may be, from a topological viewpoint, unified. Indeed, the base spaces, i.e., the unit interval I, the star space $S(A)$, and the space J_A are, respectively, topological copies of $N(\{0,1\})/\sim$, $F/\sim$, and $N(A)/\sim$. Said differently, the adjacent-endpoint relation "$\sim$", as introduced in Lipscomb [1973], provides an "abstract method for constructing quotient spaces" that yields I, $S(A)$, and J_A as examples. (Recall that the relation "$\sim$" appears in Definitions §2.1 of Chapter 1.)

The proof, as given in this chapter, of the J^∞ Imbedding Theorem dodges the need for constructing "decompositions of metrizable spaces" and analogues of "ancestor maps" which were key to proving the J_A^{n+1} Imbedding Theorem. And it is an open problem to obtain a proof that J_A^∞ is universal for metrizable spaces of weight $|A| \geq \aleph_0$ using such decompositions and ancestor mappings.

In the finite-dimensional case, however, the decompositions of metrizable spaces that are fundamental to the proof of the J_A^{n+1} Imbedding Theorem were extended and modified in Ivanšić and Milutinović [2002] — for $|A| = \aleph_0$ the base space J_A in the J_A^{n+1} theorem may be replaced by J_3, yielding a J_3^{n+1} Imbedding Theorem for separable metric spaces.

One very nice aspect of J_3 serving as both a fractal and a base space for a universal space in dimension theory is that of visualization. The same can be said for the unit interval — it is certainly easy to visualize, and even though it is not a fractal, it is the attractor of the $\mathcal{F}_1$ IFS.

In the J_3 case, its most popular representation is the Sierpiński triangle. And J_4 has appeared in textbooks on fractals as the Sierpiński cheese. That brings us to J_5, the 4-web. In our attempts to visualize the 4-web, we have already presented several approximations (Chapter 2). All of those approximating representations, however, do not expose J_5 as one-dimensional.

In Figure 53.1, the one-dimensional aspect is exposed. The representation is a level-6 J_5. It has $5^6 \times 10 = 156{,}250$ "organized segments," each a cylindrical representation of the unit interval. And for $0 \leq i \leq 5$, the segments (pictured as cylinders) at level-i have a larger diameter than those that first appear at level-$(i + 1)$. So at the scale of the figure, the "forming clouds-of-segments at level-6" grow darker at level-7 (with 781,250 segments), still darker at level-8, ...

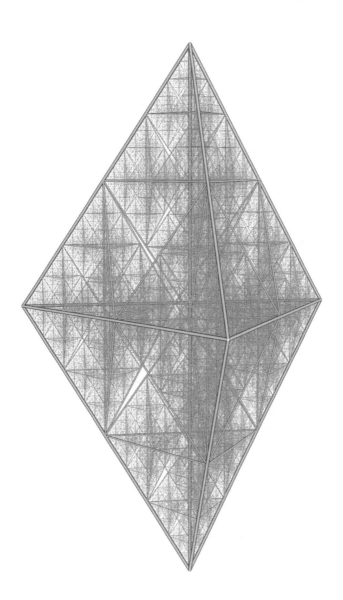

Fig. 53.1 Chris Dupilka's graphic representation of J_5.

CHAPTER 11

1992–2007 J_A-Related Research

The introduction of J_A (Chapter 1) and the first proof of the J_A^{n+1} Imbedding Theorem (Chapters 7 and 8) appeared during the first five years of the 1970s. Moreover, it was 1975 when, independently, both the J_A^{n+1} Imbedding Theorem and the term "fractal" first appeared in the literature. Slightly more than a decade later, following Hutchinson [1981] and Barnsley's [1988] popular book *Fractals Everywhere*, the notions of fractal, attractor, IFS, and Hutchinson operator became well known.

In this chapter we consider the emergence (over the 15 years 1992–2007) of mathematics that relates either directly or indirectly to J_A.

§54 Key Publications

From Cantor [1883b] to Miculescu and Mihail [2008],[1] certain publications have served to merge the mathematics of fractals and universal spaces as they relate to J_A. In this section, we begin with an overview of only a few — mainly those that directly relate to J_A and also contain new mathematics that has not previously appeared in any book.

54.1 Independently, Milutinović [1992][1993], and, Lipscomb and Perry [1992] imbedded J_A in $l^2(A)$ (generalized Hilbert space). Both approaches were centered around infinite IFSs, and Milutinović [1992] is discussed at length in this chapter.

54.2 Milutinović [1992] extended the adjacent-endpoint relation (defined on the infinite product $\times_1^\infty A_i$ where each $A_i = A$) to *finite products* $\times_1^n A_i$. He then used his (finite-product) "adjacency relation" to index the decompositions $\mathcal{W}_{ij}$ provided by the Decomposition Theorem 39.1, subsequently obtaining the second proof of the J_A Imbedding Theorem. In this chapter, the second proof is compared with the original.

54.3 Klavžar and Milutinović [1997] used the adjacency relation to introduce a new class of graphs that this author calls *Klavžar-Milutinović graphs*. These graphs represent a variant of the classical "Tower of Hanoi" problem.

54.4 Perry [1996] was the first to attempt to view J_A entirely within fractal theory — as a fixed point of an appropriate Hutchinson operator. While he was unable to show that J_A was such a fixed point, he did introduce an attractor ω_c^A of an infinite IFS (a fixed point of a Hutchinson operator). Perry's space ω_c^A was obtained by modifying the topology of J_A. In addition,

---

[1] Miculescu and Mihail's [2008] article was published in 2008, but was posted in 2007.

S.L. Lipscomb, *Fractals and Universal Spaces in Dimension Theory*,
DOI 10.1007/978-0-387-85494-6_11, © Springer Science+Business Media, LLC 2009
113

Perry's [1996] research set the stage for Miculescu and Mihail [2008], who showed that J_A is a fixed point of a Hutchinson operator. (Miculescu and Mihail's work is detailed in Chapter 5. For Perry's space ω_c^A see §31.)

54.5 Perry and Lipscomb [2003] showed that J_5, viewed as a subspace of the 4-simplex Δ_4 in 4-space, could be moved, via an isotopy that preserves fractal dimension, into 3-space. That is, we can see J_5. (Graphical approximations to J_5 appear in §7 and §10. The isotopy that moves J_5 into 3-space — graphically approximated in the color plates — is the topic of Chapter 12.)

54.6 Lipscomb [2005][2007] considers the problem that is inverse to constructing fractals: For fractals, we start with a manifold and then recursively cut (via an IFS) holes in the manifold. The inverse problem assumes that an IFS of a fractal is given. The problem is to extend the IFS to one that has the manifold as its attractor. The [2005] article extends the 2-web $\omega^2 =_t J_3$ IFS to a 2-simplex IFS, and the [2007] article extends the 3-web $\omega^3 =_t J_4$ IFS to a 3-simplex IFS (Chapters 13 and 14).

54.7 From 2002 to 2007 Ivanšić and Milutinović, and, Milutinović working alone produced a plethora of new J_A-related publications (§60). In their joint [2002] article we find the J_3^{n+1} *Imbedding Theorem*, i.e., the subspace of tuples in J_3^{n+1} with at most n rational coordinates is universal for the class of n-dimensional separable metric spaces.

§55 Chronological and Historical Context

Figure 55.1 provides context for the merging of fractals and universal spaces as they relate to J_A. The acronyms translate as follows:

B = Bing	K = Kowalsky	N = Nagata
Ba = Barnsley	Ka = Katětov	Nö = Nöbeling
C = Cantor	Kl = Klavžar	Os = Ostrand
D = Dowker	$l^2(A)$ = gen. Hilbert space	P = Perry
DT = Decomposition Theorem	L = Lipscomb	Pa = Parisse
E = Engelking	Le = Lefschetz	Pe = Pears
F = Falconer	M = Milutinović	Petr= Petr
H = Hutchinson	M_A = Milutinović's space	Po = Pontryagin
Hi = Hinz	Ma = Mandelbrot	S = Stone
Hu = Hurewicz	Me = Menger	Smi = Smirnov
I = Ivanšić	Mi = Miculescu	T = Tolstowa
IT = Imbedding Theorem	Mih = Mihail	US = universal space
IFS = Iterated Function System	Mo = Morita	W = Wallman

In addition, within Figure 55.1 a line segment indicates that the publication at the top endpoint depends on, or is related to, the one at the bottom.

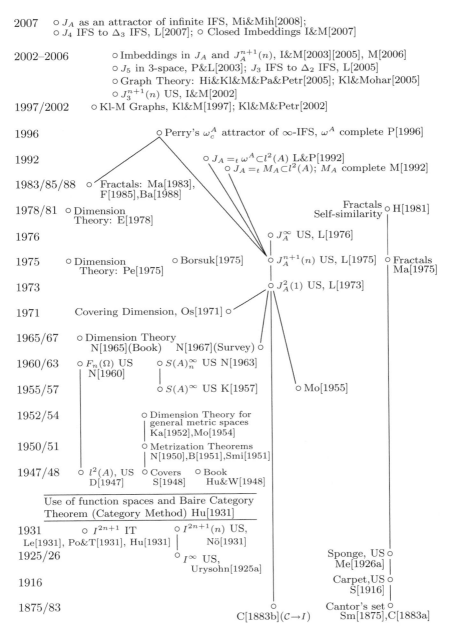

Fig. 55.1 References that provide context for J_A-related research.

For instance, Cantor's 1983 identification of adjacent endpoints $\mathcal{C} \to I$, a mapping from his set $\mathcal{C}$ onto the unit interval I, was fundamental motivation for this author's introduction of "adjacent endpoints" in $N(A)$, the goal being the construction of a one-dimensional analogue of the unit interval.

§56 Early History of J_A and M_A

To view J_A as a subspace of $l^2(A)$, Milutinović [1992] extended Sierpiński's original 2-space construction of Sierpiński's triangle to generalized Hilbert space $l^2(A)$. The result of the extension was the construction of a subspace M_A of the standard simplex Δ^A in $l^2(A)$.

He then showed that M_A is homeomorphic to J_A. His approach generalizes the fact that M_3, the Sierpiński triangle, and J_3 are homeomorphic.

In particular, for $A = \{0, 1, 2\}$ the homeomorphism $\chi : J_3 \to M_3$ is diagramed below, where, for each member w_{a_i} of the iterated function system W_A, the mapping $q(a_1 a_2 \cdots) = \cap_1^\infty w_{a_1} \circ \cdots \circ w_{a_n}(\Delta^2)$, the map $p : N(\{0, 1, 2\}) \to J_3$ is the natural mapping, and $\chi(p(a_1 a_2 \cdots)) = \cap_1^\infty w_{a_1} \circ \cdots \circ w_{a_n}(\Delta^2)$.[2]

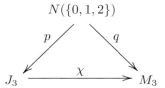

Milutinović's view of M_A was within dimension theory — he neither mentioned nor considered the mathematics required to solve the open (1992–2007) problem of showing that M_A is a fixed point of an appropriate Hutchinson operator. Nevertheless, guided by Sierpiński's [1915] recursive construction of the Sierpiński triangle, in 1992 Milutinović used the infinite IFS W_A to define his space M_A.[3]

Similarly, Lipscomb's [1973] introduction and view of J_A was within dimension theory. At that time, he was unaware of Sierpiński's triangle. Working within topology, he was motivated by Nagata's [1967] quote (§4.3), Morita's [1955] Theorem (Theorem 1.6), and Cantor's classical identification of adjacent endpoints mapping $\mathcal{C} \to I$. (For Cantor's classical work with references, see §4.3.)

Prior to 1973, following his formulation of the adjacent-endpoint relation in Baire spaces $N(A)$ for arbitrary non-empty A, Lipscomb was obviously very curious about geometric representations of the quotient spaces J_n for finite n. Beginning with J_3, a homeomorph of Sierpiński's triangle, Lipscomb deduced the J_3 structure *by using various (topological) views of the classical Cantor mapping* $\mathcal{C} \to I$ as indicated below:

The identification of adjacent endpoints in Cantor's space yields

---

[2]The corresponding general diagram (for $|A| \geq 1$) appears in Milutinović [1992]. For the corresponding "fractal diagram" (for $A = \{0, 1, 2\}$) where M_3 is replaced by its homeomorph ω^2 (i.e., the attractor ω^2 of the IFS $\mathcal{F}_2$), and q is replaced with the corresponding address map ϕ, see Theorem 8.5.

[3]For Sierpiński's construction see §21; for the construction of Δ^A see Appendix 2; and for the definitions of the Milutinović space M_A and the infinite IFS W_A see §22.

the unit interval I, i.e.,

$$I =_t \mathcal{C}/\sim \; =_t N(\{0,1\})/\sim \; = J_2.$$

The next problem? Find a geometrical representation of J_3. Lipscomb obtained the solution as follows: Since a representation of J_2 is a line segment $[u_0, u_1]$ with endpoints u_0 and u_1, for J_3 consider using three points u_0, u_1, and u_2 in the plane that are vertices of a triangle,

$$[u_i, u_j] =_t N(\{i,j\})/\sim \; = J_2, \quad |\{i,j\}| = 2 \text{ and } \{i,j\} \subset \{0,1,2\}.$$

So the edges of the triangle are obtained by identifying adjacent endpoints of character two — $a_1 a_2 \cdots \in N(\{0,1,2\})$ has *character two* if it is a nonconstant sequence in one of the sets $\{0,1\}$, $\{0,2\}$, or $\{1,2\}$. The next problem? Find a geometrical structure that corresponds to identifying adjacent endpoints in $\{i\} \times N(\{j,k\})$ where $\{i,j,k\} = \{0,1,2\}$. It is straightforward to show that the substructure consists of the geometric line segments (without endpoints) that connect the midpoints of the edges of the triangle $[u_0, u_1, u_2]$. And so on ad infinitum.

In fact, it was not until the 1980s, during a Michael Barnsley presentation on fractals, that this author became aware of the classical Sierpiński triangle — Professor Barnsley suddenly showed a slide of (an approximation to) the Sierpiński triangle. After the lecture, this author asked Barnsley if he had heard of the space J_A used in the theory of universal spaces in dimension theory. As I now recall, Barnsley's response was that he had not heard of the space J_A.

§57 Adjacency Relation

The adjacent-endpoint relation "$\sim$" (Definitions 2.1) is defined on the Baire space $N(A)$ whose underlying set is the countably infinite product set $\times_1^\infty A_i$ where each $A_i = A$. It is therefore natural to consider finite-product sets $A^n = \times_1^n A_i$ and an analogous "adjacency relation," which we shall also denote as "$\sim$".

57.1 Definition (Milutinović [1992]) (adjacency relation $\sim \; \subset \mathbf{A}^n \times \mathbf{A}^n$)
Let $|A| \geq 1$, let n be a positive integer, and let $A^n = \times_1^n A_i$ where each $A_i = A$. Then for points $a = a_1 \cdots a_n$ and $b = b_1 \cdots b_n$ in A^n, we may write "$a \sim b$" if either $a = b$, or, for distinct members $x \neq y$ of A, we have $a = a_1 \cdots a_{t-1} xyy \cdots y$ and $b = b_1 \cdots a_{t-1} yxx \cdots x$. If distinct a and b satisfy $a \sim b$, then the unique index $t \geq 1$ is called the *tail index* of a and b. The relation $\sim \; \subset A^n \times A^n$ given by "$a \sim b$" is called the *adjacency relation*, and whenever "$a \sim b$" we may say that a *is adjacent to* b.

57.2 Theorem   *Let $n \geq 1$, let $|A| \geq 1$, and let $A^n = \times_1^n A_i$ where each $A_i = A$. Then the adjacency relation $\sim$ is an equivalence relation on A^n with the property that each equivalence class contains at most two members. In addition, the only points in A^n that occupy singleton equivalence classes are the constant points, i.e., the length n strings $a = a_1 \cdots a_n$ of members of A such that $a_1 = \cdots = a_n$.*

PROOF. One may essentially copy the proof of Theorem 2.2. The only new aspect is the claim that only the constant points occupy singleton equivalence classes. But this follows because if $a \sim b$ where a is a constant, then distinct $x, y \in A$ cannot exist such that $a = a_1 \cdots a_{t-1} x y y \cdots y$. Thus $a = b$.   □

§58 Indexing the Decompositions

For an n-dimensional metric space X of weight $|A| \geq \aleph_0$, the decompositions of concern are the $\mathcal{W}_{ij} = \{W_a : a \in A\}$, $i \geq 1$, $1 \leq j \leq n+1$, whose existence and properties are specified by the Decomposition Theorem 39.1. In this section we provide an example of the first two steps ($i = 1, 2$) of Milutinović's [1992, §5] indexing scheme.

To demonstrate the basics, we shall use the decompositions $\mathcal{W}_{ij}$, $i, j \in \{1, 2\}$ described in Example 39.8 and illustrated in Figures 39.6 and 39.7. For $i = 1, 2$, these four decompositions are decompositions of the unit interval and they satisfy all of the properties listed in the Decomposition Theorem.

First two steps in indexing scheme (Milutinović [1992, §5]) Select a partition of $A = \cup_{ij} A_{ij}$ where each $|A_{ij}| = |A|$ and well order each A_{ij}. Then

(1)   let $\mathcal{W}_{1j} = \{W_a : a \in A_{1j}\} = \{W_{[a]} : [a] \in A_{1j}/\sim\}$;

(2α)   if $W \in \mathcal{W}_{2j}$ satisfies $W \cap \text{Bdry } \mathcal{W}_{1j} \neq \emptyset$,
          then for the unique distinct pair $W_a, W_b \in \mathcal{W}_{1j}$ that meet W,
          let $W_{[ab]} = W$ where $[ab] = \{ab, ba\} \in (A_{1j} \cup A_{2j})^2/\sim$; or

(2β)   if $W \in \mathcal{W}_{2j}$ satisfies $W \cap \text{Bdry } \mathcal{W}_{1j} = \emptyset$,
          then for the unique W_a such that $W \subset W_a \in \mathcal{W}_{1j}$, select (the smallest non-previously selected) $b \in A_{2j}$ and let $W_{ab} = W$.

The following example uses only finite index sets. For the transition to the infinite case and the recursive construction that yields an indexing for each $i = 1, 2, \ldots$, see Milutinović [1992].

58.1 EXAMPLE. For index sets, consider the following finite sets with their natural orderings:

$$A_{11} = \{a, b, c\}, \qquad A_{12} = \{d, e\},$$
$$A_{21} = \{g, h, i, j, k, l\}, \quad A_{22} = \{m, n, o, p\}.$$

Let $\mathcal{W}_{11}$, $\mathcal{W}_{12}$, $\mathcal{W}_{21}$, and $\mathcal{W}_{22}$ be the decompositions defined in Example 39.8. So $\mathcal{W}_{11}$ has three members, and we index $\mathcal{W}_{11} = \{W_a, W_b, W_c\}$ as illustrated at the top of Figure 58.2. Next, $\mathcal{W}_{12}$ has two members that we index as W_d and W_e as illustrated at the top of Figure 58.3.

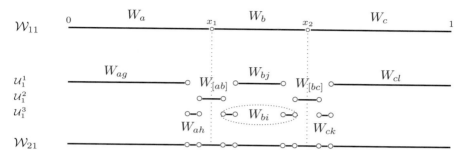

Fig. 58.2 Indexing the members of $\mathcal{W}_{ij}$ for $j = 1$.

For the $i = 2$ cases, note that the graphics in Figure 58.2 immediately below $\mathcal{W}_{11}$ display nine segments, two of which are interior to a dotted ellipse. The ellipse is used to indicate that the union of those two segments forms one member of $\mathcal{W}_{21} = \cup_1^3 \mathcal{U}_1^k$. So $|\mathcal{W}_{21}| = 8$. Thus, applying (2α) and (2β), we have

$$\mathcal{W}_{21} = \cup_1^3 \mathcal{U}_1^k = \{W_{ag}, W_{ah}, W_{[ab]}, W_{bi}, W_{bj}, W_{[bc]}, W_{ck}, W_{cl}\},$$

which is illustrated (within the unit interval) at the bottom of Figure 58.2.

Finally, turning to $\mathcal{W}_{22}$, we begin with the graphics in Figure 58.3 where $\mathcal{W}_{12}$ is indexed with the members of $A_{12} = \{d, e\}$, namely $\mathcal{W}_{12} = \{W_d, W_e\}$. At first blush, since the graphic shows the unit interval as three segments, one might think that three indices are required. However, the construction of $\mathcal{W}_{12}$ is graphically illustrated in Figure 39.3, where we see that

$$\mathcal{W}_{12} = \{[0, y_1) \cup (y_2, 1], (y_1, y_2)\},$$

so we may define $W_d = [0, y_1) \cup (y_2, 1]$ and $W_e = (y_1, y_2)$.

The construction of $\mathcal{W}_{22} = \cup_1^3 \mathcal{U}_2^k$ is graphically illustrated in Figure 39.5. The decomposition $\mathcal{W}_{22}$ is also illustrated in Figure 39.7, where the "dotted ellipses" appear. Further, detailed observations concerning $\mathcal{W}_{22}$ are provided in Example 39.8. With this background, the indexing scheme applied to $\mathcal{W}_{22}$ is given by

$$\mathcal{W}_{22} = \{W_{dm}, W_{[de]}, W_{en}, W_{eo}, W_{ep}\},$$

where the sets are illustrated in Figure 58.3.

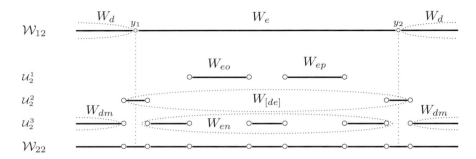

Fig. 58.3 Indexing the members of $\mathcal{W}_{ij}$ for $j = 2$.

That is, W_{dm} is the union of the two segments in $\mathcal{U}_2^3$ that contain the points "0" and "1", and W_{en} is the other member of $\mathcal{U}_2^3$. The set $W_{[de]}$ is the lone member of $\mathcal{U}_2^2$. And W_{eo} and W_{ep} are the two members of $\mathcal{U}_2^1$. These indexings of the $\mathcal{W}_{ij}$ for $i, j \in \{1, 2\}$ complete the example of Milutinović's indexing scheme.

In passing, note that in this section we are only attempting to provide an intuitive understanding behind some of the key concepts presented in Milutinović [1992]. The nuances that concern various modifications or extensions of the decompositions $\mathcal{W}_{ij}$, the notations used to intuitively describe the families $\mathcal{U}_j^k$, $k = 1, 2, 3$, and the detailed proofs are not addressed here.

For example, to complete the indexing scheme, Milutinović [1992] lists seven properties satisfied by the (1), (2α), and (2β) indexing. He then uses induction on $i = 1, 2, \ldots$ to prove the existence of an indexing of the $\mathcal{W}_{ij}$ that satisfies the seven properties.

§59 Proofs of the J_A^{n+1} Imbedding Theorem

In this section we provide an overview that compares and contrasts the approaches of Milutinović [1992] and Lipscomb [1975] to proving the J_A^{n+1} Imbedding Theorem.

From the beginning, as detailed in §56 "Early History of J_A and M_A," the spaces J_A and M_A were introduced for two distinct reasons — the former as a one-dimensional generalization of the unit interval, the latter as a generalization of Sierpiński's triangle.

As for proving the J_A^{n+1} Imbedding Theorem, however, both approaches apply the Decomposition Lemma 38.9 at the inductive step in a construction that yields an $\aleph_0 \times (n+1)$ matrix $[\mathcal{W}_{ij}]$ of decompositions.

59.1 Lipscomb's Approach. Lipscomb indexed the decompositions as in §42, namely $\mathcal{W}_{ij} = \{W_a : a \in A_{ij}\}$ where $\cup A_{ij}$ is a partition of A such that each $|A_{ij}| = |A|$. With this (nonspecial) indexing, Lipscomb focused on the $\aleph_0 \times (n+1)$ matrices $[\alpha_{ij}]$ where $W_{\alpha_{ij}} \in \mathcal{W}_{ij}$. That is, since the indexing of the members of the decompositions is rather arbitrary, each matrix $\alpha = [\alpha_{ij}]$, which is naturally a point in $N(A)^{n+1}$, may or may not be *of interest* — it is of interest if $\cap_{ij}\overline{W}_{\alpha_{ij}} \neq \emptyset$. The points $\alpha \in N(A)^{n+1}$ of interest define a set $Z \subset N(A)$, which is given in Definition 42.1, i.e.,

$$\alpha \in Z \subset N(A)^{n+1} \text{ if and only if } \{x_\alpha\} = \cap\{\overline{W}_{\alpha_{ij}} : i \geq 1; 1 \leq j \leq n+1\}.$$

The problem with Z is that it may happen that for distinct $\alpha, \beta \in Z$ we have $x_\alpha = x_\beta$ where corresponding columns α_j and β_j are not adjacent-endpoint related. In other words, Lipscomb does not consider the adjacent-endpoint indexing at the point in the proof where the non-empty $\cap_{ij}\overline{W}_{\alpha_{ij}}$ are first considered, i.e., when Z is defined. Then after Z is defined, Lipscomb addresses the "adjacent-endpoint indexing" by introducing the ancestor map h given in Definition 43.2 — the homeomorphism $h : Z \to h(Z) \subset N(A)^{n+1}$ essentially shows that the adjacent-endpoint relation is encoded in the members of Z such that, whenever $x_\alpha = x_\beta$, the columns $h(\alpha)_j$ and $h(\beta)_j$ are adjacent-endpoint related. In short, in the Lipscomb approach, the "adjacent-endpoint indexing" is addressed only in the final phase of constructing the imbedding, i.e., the adjacent-endpoint indexing may be viewed as the last piece of the puzzle.

59.2 MILUTINOVIĆ'S APPROACH. Similarly, for the given n-dimensional weight A metric space X, Milutinović begins with applications of the Decomposition Lemma 38.9 to define an $\aleph_0 \times (n+1)$ matrix of decompositions $\mathcal{W}_{ij}$. Then he applies his indexing scheme (the initial steps illustrated in Example 58.1), which is tantamount to the construction of Lipscomb's mapping $f = q \circ h^{-1}$ where h is the "ancestor map" and q the "decomposition mapping." (see Figure 41.1). That is, instead of creating something equivalent to q in the first step and then something equivalent to h in the second step, he presents an indexing scheme that dodges the need for the separate steps of constructing the "q" and "h" maps. Nevertheless, his indexing scheme also requires proofs for its construction and its properties. The observation here is that in contrasting the Lipscomb and Milutinović approaches, the major distinction occurs in the indexing schemes, and in both cases the proofs are rather technical.

For more details about Milutinović's approach, consider that he merges his two kinds of indices — those with brackets "$[a_1 \cdots a_i]$" and those without brackets "$a_1 \cdots a_i$." To use only bracketed strings, he adopts the dodge of saying that those indices that have brackets have *both members legitimate* — the "brackets" tell us that he is concerned with the finite adjacency relation. Otherwise, for an index "$a_1 \cdots a_i$" the class $[a_1 \cdots a_i]$ may contain another member $b_1 \cdots b_i \sim a_1 \cdots a_i$ which he calls *illegitimate*. He is careful to precisely keep track of his convention, and he describes several of its attributes.

Once the indexing scheme is fully understood and its properties detailed, for a fixed $j \in \{1, \ldots, n+1\}$ he uses the j superscript to indicate that he is only considering families from the jth column of the $\aleph_0 \times (n+1)$ matrix of decompositions. He then specifies the function $\phi_j : X \to M_A =_t J_A$ by

$$x \in \cap_1^\infty \overline{W}_{[a_1 \cdots a_i]}^j \quad \text{implies} \quad \phi_j(x) = [a_1 a_2 \cdots] = p(a_1 a_2 \cdots)$$

where $p : N(A) \to J_A$ is the natural mapping, and, most importantly (to quote Milutinović [1992]):

> ... with the left-side indices chosen in a coherent way, i.e., in such a way that the representatives of shorter ones are always initial segments of the representatives of longer ones.

He then proves that

$$\phi = (\phi_1, \ldots, \phi_{n+1}) : X \to J_A^{n+1}$$

is an imbedding.

§60 Ivanšić and Milutinović Theorems

In this section, we state and briefly discuss some of the main J_A-related theorems that were introduced by Ivanšić and Milutinović.

Ivanšić and Milutinović [2002] construct an $\aleph_0 \times (n+1)$ matrix of decompositions $\mathcal{V}_{ij}^*$. These decompositions are related to, but distinct from, the $\mathcal{W}_{ij}$ used in Milutinović [1992]. They obtain the $\mathcal{V}_{ij}^*$ decompositions by modifying other decompositions, which they overview as follows (with substitutions of mathematical notation and references used in this text):

> For a given metrizable separable space X of dimension $\leq n$, we shall construct $n + 1$ sequences of decompositions $\mathcal{V}_{ij}^*$, $i \geq 1$, $j = 1,\ldots,n+1$, of special type. These sequences of decompositions will mimic the behavior of finer and finer triangles in the Sierpiński curve — see Example 2 — then we shall use an indexing of their elements in order to describe an embedding of X into $J_3^{n+1}(n)$. That indexing will be a generalization of the standard coding of points in the Sierpiński curve M_A.

They construct the $\mathcal{V}_{ij}^*$ by modifying corresponding decompositions "$\mathcal{V}_{ij}$." The $\mathcal{V}_{ij}$ are constructed inductively (with respect to the index $i = 1, 2 \ldots$) using the Decomposition Lemma 38.9 at the inductive step:

$$\mathcal{V}_{(i-1)j} \xrightarrow{\text{Lemma 38.9}} \mathcal{W}_{ij} \longrightarrow \mathcal{V}_{ij}.$$

The technicalities are rather extensive (there are 14 properties that the desired decompositions must satisfy, and there are also four properties that the indexing satisfies); the result is impressive.

60.1 Theorem (Ivanšić and Milutinović [2002]) *Let $n \geq 0$, and let $J_3^{n+1}(n)$ denote the subspace of $(n+1)$-tuples in J_3^{n+1} that have at most n rational coordinates. Then $J_3^{n+1}(n)$ is universal for the class of separable metric spaces of dimension n.*

Thus *in the separable case, the classical fractal known as Sierpiński's triangle is the base space for a universal space.*

For the next result, we consider the question of *relative imbeddings,* "If X_0 is a closed subspace of X, and $f_0 : X_0 \to Y$ is an imbedding of X_0, then is there an extension $f : X \to Y$ of f_0 that is an imbedding of X?"

With respect to

$$J_A(0) = J_A^1(0) \quad = \{x \in J_A : x \text{ has at most zero rational coordinates}\}$$
$$= \{x \in J_A : x \text{ is irrational}\},$$

we have the following theorem.

60.2 Theorem (Ivanšić and Milutinović [2003]) *Let* $\dim X = 0$ *where* X *is a metric space of weight* $|A| \geq \aleph_0$. *Let* X_0 *be a compact subspace of* X. *Then any imbedding* $f_0 : X_0 \to J_A(0)$ *of* X_0 *has an extension* $f : X \to J_A(0)$ *that is an imbedding of* X.

Ivanšić and Milutinović [2003] pose the following open problem:

> For $X_0 \subset X$ and $f_0 : X_0 \to J_A(0)$, find other conditions on X_0, and perhaps on $f_0(X_0)$, that guarantee the existence of an embedding $f : X \to J_A(0)$ of X that extends f_0.

They show that "compact subspace of X" in Theorem 60.2 cannot be replaced with "closed subspace of X." In particular, let $X_0 = J_A(0)$ and let X_0' be a singleton set with trivial topology. Then consider $X = X_0 \vee X_0'$ as the disjoint union of X_0 and X_0', observe that $X_0 = J_A(0) \subset X$, and let $f_0 : X_0 \to J_A(0)$ be the identity mapping.

For the next result, we consider the case where X is now n-dimensional but the subspace X_0 is only finite and the co-domain $Y = J_A^{n+1}(n)$.

60.3 Theorem (Finitely Pointed Imbedding Theorem) (Ivanšić and Milutinović [2005]) *Let* X *be an* n-dimensional metric space of weight $|A| \geq \aleph_0$, and let $X_0 \subset X$ be finite. Then any imbedding $f_0 : X_0 \to J_A^{n+1}(n)$ of X_0 has an extension $f : X \to J_A^{n+1}(n)$ that is an imbedding of X.

And from Theorem 60.3, we have the following obvious corollary:

60.4 Corollary *Let* $|A| \geq \aleph_0$ *and let* $y_0 \in J_A^{n+1}(n)$. *Then the pointed space* $(J_A^{n+1}(n), y_0)$ *is a universal object in the category of pointed metrizable spaces of dimension* $\leq n$ *and weight* $|A|$.

In discussing their approach to the proof of their Finitely Pointed Imbedding Theorem, Ivanšić and Milutinović state the following (with substitutions of mathematical notation and references used in this text):

> This general strategy consists of constructing certain finer and finer sequences of decompositions and then indexing them in such

a way that the mappings defined by

$$x \mapsto [a_1 \cdots a_k xyy \cdots] = [a_1 \cdots a_k yxx \cdots]$$

(when x belongs to the boundaries of the sets indexed by the initial segments of the sequences $a_1 \cdots a_k xyy \cdots$ and $a_1 \cdots a_k yxx \cdots$)
or

$$x \mapsto [a_1 \cdots a_k \cdots]$$

(when x belongs to the sets indexed by the initial segments of the sequence $a_1 \cdots a_k \cdots$ and belongs to no boundary of elements of the decompositions) will be the $n+1$ *coordinate functions of an embedding into* M_A^{n+1}. This means that one may interpret the indexing of the decompositions as a sort of coordinatization of the space that mimics the coordinates of M_A^{n+1}.

For the next result, we consider the density of the set of imbeddings into J_A^{n+1}. In this case, there is a classical counterpart, "the density of the set of imbeddings into $\mathbb{R}^{2n+1}$" in the n-dimensional separable metric theory.[4]

At the beginning of Milutinović [2006], we find the following (with substitutions for mathematical notation and references used in this text):

> Results on density of the set of embeddings in the space of maps abound in topology. Recall the classic results on Nöbeling and Menger spaces (Engelking [1978], Hurewicz and Wallman [1948], Nagata [1983]). Most of the proofs in the literature are based on the Baire category theorem. Because of the topological completeness of Lipscomb's space (Milutinović [1992], Perry [1996]), it is possible to prove analogous results for Lipscomb's space using the same approach.

> Our proof is different. It is based on geometric properties of M_A (which is homeomorphic to Lipscomb's space). It is done in such a way that a similar geometric structure is imposed on any metric space of appropriate dimension and weight (via certain sequences of decompositions of the space), thus obtaining a more explicit and graphic description of the approximation.

In the following theorem, J_A is identified with M_A, which has the metric "d" inherited from the generalized Hilbert space $l^2(A)$ that contains M_A. Thus, we may assume that J_A^{n+1} is equipped with the metric

$$d(x, y) = \max \{d(x_j, y_j) : j = 1, 2, \ldots, n+1\}.$$

---

[4]See Figure 55.1 where Hurewicz [1931], Hurewicz and Wallman [1948], and the "Category Method" are listed. The method involves applications of function spaces and the Baire Category Theorem.

60.5 Theorem (Milutinović [2006])   *Let X be an n-dimensional metrizable space of weight $\leq |A|$. Let $f : X \to J_A^{n+1}$ be a continuous mapping and let ε be a positive number. Then there exists an embedding $\psi : X \to J_A^{n+1}(n) \subset J_A^{n+1}$ such that $d(f, \psi) \leq \varepsilon$.*

For Ivanšić and Milutinović's most recent result, we consider *closed imbeddings*. The topic and corresponding theorem are introduced by Ivanšić and Milutinović in their 2007 article (with substitutions of mathematical notation and references used in this text):

> If a topological space is embedded into a topologically complete metrizable space (i.e., into a space that can be endowed by a complete metric) as a closed subset, it must be topologically complete metrizable itself. On the other hand, if a topologically complete metrizable space is embedded into another such space, the embedding need not be closed (embedding $\mathbb{R}$ as an open interval in itself, or $\mathbb{N}$ as $\{1/m : m \in \mathbb{N}\}$ into $\mathbb{R}$, are easy examples for this claim). The problem of the existence of closed embeddings of topologically complete metrizable spaces has been extensively treated in the theory of universal spaces. Tsuda [1985a][1985b], Waśko [1986], Hattori [1989], Olszewski and Piątkiewicz [1992], and Nagórko [2006] have proved results about existence of closed embeddings of complete metric spaces into several universal spaces. This often required special modifications of the previously known universal spaces. Also, in all cases the proofs were obtained by the use of the Baire category theorem. For Lipscomb's universal space no results on existence of closed embeddings have appeared yet. In this paper we prove that the direct approach of obtaining embeddings into Lipscomb's universal space, developed in Milutinović [1992][1993] and later exploited in Ivanšić and Milutinović [2002][2003][2005] and Milutinović [2006] yields closed embeddings with no further changes made, in the case when the embedded space is topologically complete.

For the following theorems, keep in mind that the set of tuples in J_A^{n+1} that have at most n rational coordinates is denoted $J_A^{n+1}(n)$.

60.6 Theorem (Ivanšić and Milutinović [2007])   *Let (X, ρ) be a complete n-dimensional metric space of weight $|A| \geq \aleph_0$. Then there is a closed embedding of X into $J_A^{n+1}(n)$.*

60.7 Theorem (Ivanšić and Milutinović [2007])   *Let (X, ρ) be a complete n-dimensional metric space of weight $\leq |A|$, and let $f : X \to J_A^{n+1}$ be a continuous mapping. Then for any $\varepsilon > 0$ there is a closed embedding $\psi : X \to J_A^{n+1}(n)$ such that for each $x \in X$ we have $d(f(x), \psi(x)) \leq \varepsilon$.*

§61 Comments

Because the goal in this chapter is that of surveying the J_A-related mathematics that has expanded the theories of fractals or universal spaces in dimension theory either by merging the theories or extending the individual theories, the J_A-induced graph theory was not discussed.

As for the Klavžar-Milutinović graphs, they first appeared as a result of Milutinović's knowledge of the adjacency relation ($\sim$ on finite products) together with Klavžar's knowledge of graph theory (i.e., Klavžar proposed that together they apply $\sim$ to study an induced class of graphs). The class was introduced in 1997, and its members were defined as follows:

For $k \geq 1$ and any $n \geq 1$, the *Klavžar-Milutinović graph* KM_{nk} is the graph that has *vertex set* $V_{nk} = \times_{i=1}^{n}\{1,\ldots,k\}_i$ where each factor equals $\{1,\ldots,k\}$; and *edge set* E_{nk}, whose members are given by

$$[a_1 \cdots a_{t-1}xyy \cdots y,\ a_1 \cdots a_{t-1}yxx \cdots x] \quad \text{or}$$
$$[a_1 \cdots a_{n-1}x,\ a_1 \cdots a_{n-1}y] \quad \text{or}$$
$$[xy \cdots y,\ yx \cdots x]$$

where $x, y \in \{1,\ldots,k\}$ and $x \neq y$. For example, in $KM_{2,4}$ we see that each of $[11, 12]$, $[11, 13]$, and $[11, 14]$ is an edge of $KM_{2,4}$, while $[11, 22]$ is not an edge. The $KM_{2,4}$ graph is pictured in Figure 61.1:

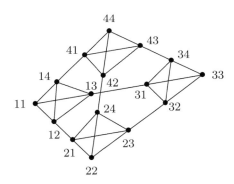

Fig. 61.1 The $KM_{2,4}$ graph.

It turns out, for example, that the graphs $KM_{n,3}$ are isomorphic to the graphs of the Tower of Hanoi Problem, and that for all values of n and k, the graphs KM_{nk} are Hamiltonian. (See Klavžar and Milutinović [1997].)

The four research articles, listed at the top of the "tree" in Figure 55.1 — Hinz, Klavžar, Milutinović, Parisse, and Petr [2005], Klavžar and Mohar [2005], Klavžar, Milutinović, and Petre [2002], and Klavžar and Milutinović [1997] — represent a natural and relatively recent newly constructed class of graphs.[5]

---

[5]It is interesting to note, however, that the Klavžar and Milutinović [1997] article was

Illustration of Chapter 12 Isotopy
That Moves J_5 into 3-space[1]

The J_4 subspace of J_5 is the Sierpiński cheese, which lives inside of 3-space and is illustrated below at the start $(t = 0)$ of the isotopy $(0 \leq t \leq 1)$. The variation in color, ranging from red to magenta, is a coloring of the points of J_5 in 4-space at the start of the isotopy. The coloring scheme serves to indicate distance from 3-space — red-colored points are at distance zero from 3-space, while those with the magenta color are at a maximum distance from 3-space. With the colors fixed, note that the isotopy gradually moves the nonred points from 4-space into 3-space, and that the magenta-colored points only enter 3-space when $t \approx 1$.

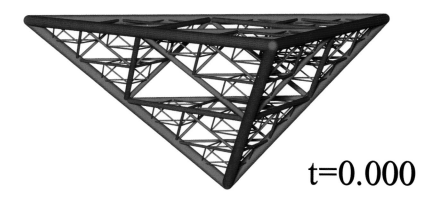

t=0.000

---

[1]Chris Dupilka generated these color plates using Pov-Ray software to encode the mathematics of Chapter 12.

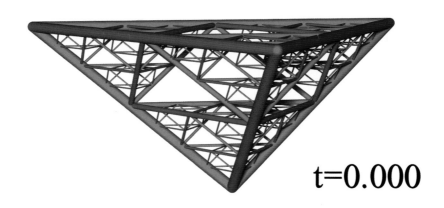

t=0.000

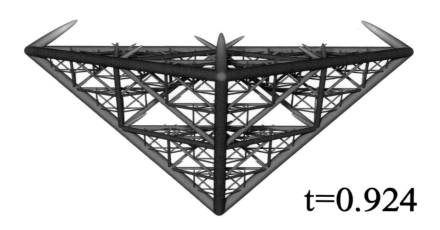

t=0.924

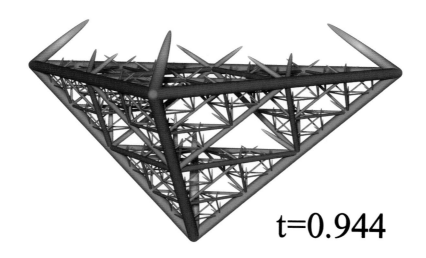

t=0.944

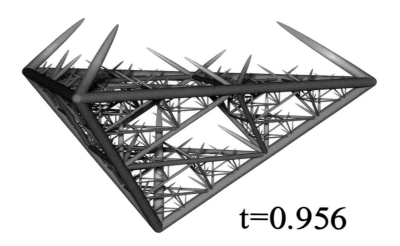

t=0.956

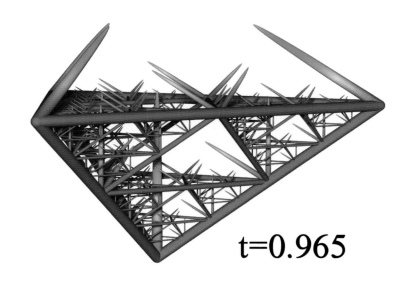

t=0.965

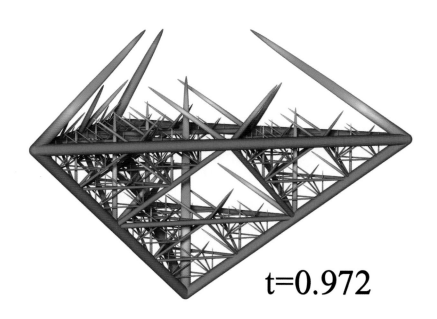

t=0.972

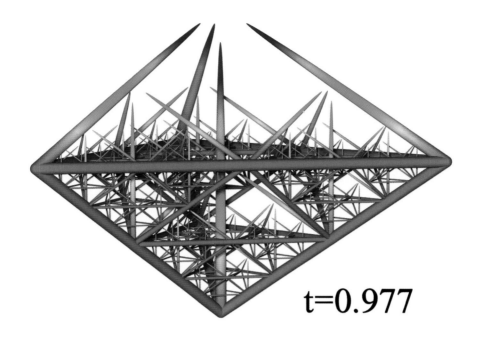

t=0.977

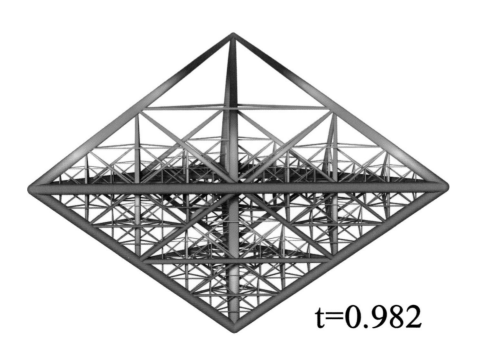

t=0.982

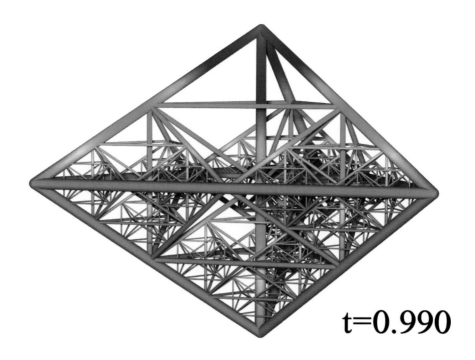

t=0.990

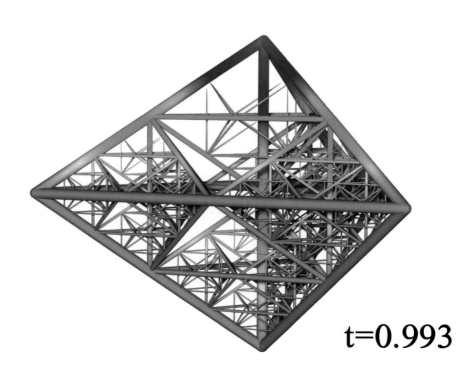

t=0.993

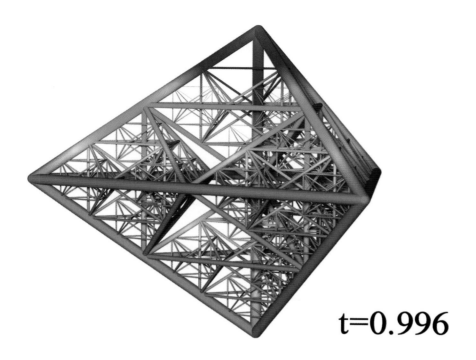

t=0.996

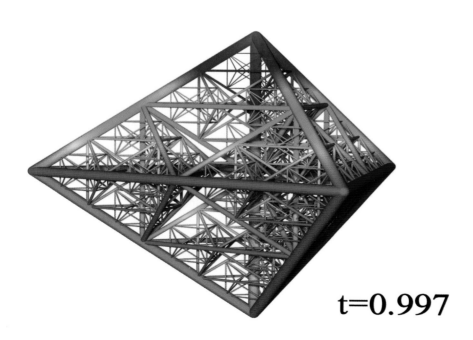

t=0.997

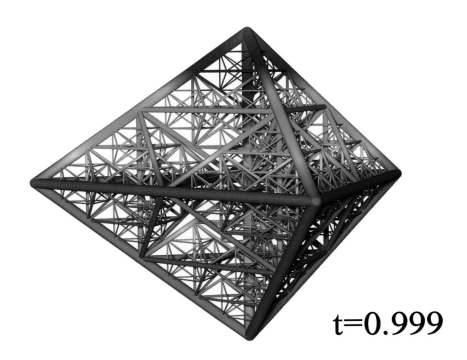

t=0.999

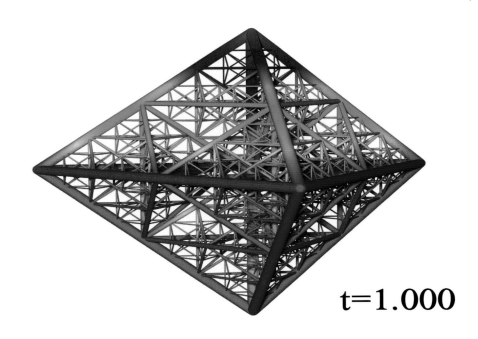

t=1.000

So as we climb the tree in Figure 55.1, looking at the dates from 2005 through 2007, we see two other articles (Lipscomb [2005] and Lipscomb [2007]) that pose yet another general problem, the problem that is inverse to creating fractals from manifolds. Indeed, fractals were viewed historically as the residual of an infinite process of cutting holes in manifolds. The prime example is the Sierpiński triangle obtained by cutting holes in a 2-simplex manifold.

The reverse problem is that of starting with a fractal, and then (in a sense) reconstructing the manifold. The idea is that the fractal is the attractor of an IFS acting on a manifold. Can we extend the given IFS to one whose attractor is the containing manifold? Special cases of this problem are the topics of Chapters 13 and 14. The mathematics follows Lipscomb [2005] and [2007].

Another piece of mathematics listed in Figure 55.1 that has not been addressed in this chapter is the mathematics of moving J_5 from 4-space into 3-space with its fractal dimension preserved. That result is due to Perry and Lipscomb [2003] and is the topic of Chapter 12.

To motivate the material of Chapter 12, however, we first present eight pages of color plates that show what one would see if he were watching J_5 move into 3-space.

---

received by the Czechoslovak Mathematical Journal in 1994. So one may rightly say that this new contribution to graph theory, whose roots date back to Cantor's [1883b] identification of adjacent endpoints, actually sprouted in 1994.

CHAPTER 12

Isotopy Moves J_5 into 3-Space

For finite $n = |A|$, the space $J_A = J_n$ is a one-dimensional separable metric space. And for small n, it is natural to represent J_{n+1} inside of 3-space as a union of $n + 1$ congruent just-touching scaled-by-$\frac{1}{2}$ copies of itself. For example, we know that J_2 is represented as the unit interval $\omega^1 = [0, 1/2] \cup [1/2, 1] = f_0(\omega^1) \cup f_1(\omega^1)$, J_3 as Sierpiński's triangle $\omega^2 = \cup_0^2 f_i(\omega^2)$, and J_4 as the Sierpiński cheese (3D-gasket) $\omega^3 = \cup_0^3 f_i(\omega^3)$.

In 2003, the homeomorph of J_5 that lives in 4-space, namely the 4-web ω^4, was moved into 3-space via an isotopy that preserved the ω^4 5-fold self-similarity. In other words, the isotopy preserves the fractal dimension $D(\omega^4) = \ln(4+1)/\ln(2)$ of ω^4. Intuitively, this allows us to "see" J_5, just as we "see" J_2, J_3, and J_4.[1]

In this chapter we construct the desired isotopy. The mathematics also yields the fact that for each $n \geq 4$, the n-web ω^n may be represented in $(n-1)$-space. The presentation follows Perry and Lipscomb [2003].

§62 Representing J_{n+1} in 3-Space

From the Classical Imbedding Theorem it is clear that the one-dimensional separable-metric space J_{n+1} may be topologically imbedded in 3-space. But an arbitrary imbedding may not shed light on the self-similarity feature of J_{n+1} — the natural map $p : N(\{0, 1, \ldots, n\}) \rightarrow J_{n+1}$ induces $n + 1$ just-touching copies $p(\langle 0 \rangle), \ldots, p(\langle n \rangle)$ of J_{n+1} (Lemma 5.1).

Nevertheless, from Theorem 8.5 it is clear that J_{n+1} is homeomorphic to the n-web ω^n, which is the attractor of the IFS $\mathcal{F}_n = \{f_0, \ldots, f_n\}$ that resides in the n-simplex Δ_n in n-space. In more detail, consider the following $n + 1$ vectors $\mathbf{u}_i$ in n-space $\mathbb{R}^n$:

$$\mathbf{u}_0 = (0, \ldots, 0), \quad \mathbf{u}_1 = (1, 0, 0, \ldots, 0), \quad \ldots, \quad \mathbf{u}_n = (0, 0, \ldots, 0, 1).$$

Then $\Delta_n \subset \mathbb{R}^n$ is the n-simplex with vertices $\mathbf{u}_0, \ldots, \mathbf{u}_n$, and each member $f_i(\mathbf{x}) = \mathbf{x}/2 + \mathbf{u}_i/2$ of $\mathcal{F}_n$ is the $\frac{1}{2}$-contraction toward $\mathbf{u}_i$.

Since $\omega^n \subset \mathbb{R}^n$, we see that $n \geq 4$ implies that we can "see" neither ω^n nor any of its scaled copies $f_i(\omega^n)$. In particular, to "see" $\omega^4 =_t J_5$, the ideal imbedding of J_5 into 3-space would be an isotopy that moves ω^4 from 4-space into 3-space while preserving its fractal dimension. Such an isotopy would "show" J_5 as a union of its five $f_i(\omega^4)$ just-touching self-similar copies.

---

[1] The dimension function D used here is the *self-similarity dimension* (§A14). For examples, motivation, and a discussion of the self-similarity dimension, see Peitgen, Jürgens, and Saupe [1992].

S.L. Lipscomb, *Fractals and Universal Spaces in Dimension Theory*,
DOI 10.1007/978-0-387-85494-6_12, © Springer Science+Business Media, LLC 2009

To imbed ω^n in m-space $\mathbb{R}^m$ with its fractal dimension preserved, it is necessary that the fractal dimension $D(\omega^n) \leq \dim \mathbb{R}^m = m$ (Barnsley [1988, Theorem 2, page 202]). So let us look at the fractal dimension $D(\omega^n) = \ln(n+1)/\ln(2)$ of the n-webs for small values of n:

$$D(\omega^1) = 1 \quad < \quad D(\omega^2) \approx 1.58 \quad < \quad D(\omega^3) = 2 \quad < \quad D(\omega^4) \approx 2.32 \quad <$$
$$< \quad D(\omega^5) \approx 2.58 \quad < \quad D(\omega^6) \approx 2.81 \quad < \quad D(\omega^7) = 3 \quad < \cdots$$

Thus, $D(\omega^2) > 1$ implies that ω^2 cannot be viewed on the real line. However, $D(\omega^3) = 2 = \dim(\mathbb{R}^2)$ sheds no light on the fact that ω^3 cannot be viewed in the plane, as was discussed in §9.

For $n = 4, 5, 6, 7$, and 8 we see that

$$D(\omega^4) \approx 2.32 < 3, \quad D(\omega^5) \approx 2.58 < 3, \quad D(\omega^6) \approx 2.81 < 3,$$
$$D(\omega^7) = 3 \leq 3, \qquad D(\omega^8) \approx 3.16 > 3.$$

Since $D(\omega^8) > 3$, we know that the 8-web ω^8 cannot be viewed (with fractal dimension preserved) in $\mathbb{R}^3$. Prior to 2003, it had been an open question as to whether any n-web ω^n for $n = 4, 5, 6, 7$ can be viewed in $\mathbb{R}^3$.

In this chapter, we show that the 4-web can indeed be "viewed" in 3-space. The self-similarity of ω^4 makes the key observation combinatorial: Roughly, the 2-web resides in the plane because there exist *three* congruent triangles that may be positioned such that each *just touches the other two*; the 3-web resides in 3-space because there exist *four* congruent tetrahedra that may be positioned such that each *just touches the other three*; and the 4-web may be viewed in 3-space because there exist *five* congruent hexahedra (two tetrahedra pasted along a face) that may be positioned in 3-space such that each *just touches the other four*.

§63 The IFS and Five Points in 3-Space

In addition to the definitions given above, for 3-space we let $\mathbf{v}_i = \mathbf{u}_i$ ($i = 0, 1, 2, 3$); and for 4-space, we use the insertion $(x, y, z) \mapsto (x, y, z, 0) \in \mathbb{R}^4$ of 3-space into 4-space so that we may also think of each $\mathbf{v}_i \in \mathbb{R}^4$. We also define $\mathbf{v}_4$ as either $(2/3, 2/3, 2/3)$ or $(2/3, 2/3, 2/3, 0)$, the choice will be clear from the context. At other times, we denote either the $\mathbf{v}$- or $\mathbf{u}$-vectors as $\mathbf{w}$-vectors, i.e., $\mathbf{w}_i \in \{\mathbf{u}_i, \mathbf{v}_i\}$. And we shall also use the notation "$\mathbf{w}_i$" for the *terminal point* of the vector $\mathbf{w}_i$, i.e., the discussion/figures may concern/illustrate either vectors $\mathbf{w}_i$ or points $\mathbf{w}_i$.

The vectors $\mathbf{v}_0, \mathbf{v}_1, \mathbf{v}_2, \mathbf{v}_3, \mathbf{v}_4$ are the vertices of the *hexahedron* Λ^3. And given the hexahedron $\Lambda^3 \subset \mathbb{R}^3$ as illustrated in Figure 63.1, we associate the IFS $\mathcal{G}_3 = \{g_0, g_1, \ldots, g_4\}$, where $g_i(\mathbf{x}) = \mathbf{x}/2 + \mathbf{v}_i/2$. The attractor of $\mathcal{G}_3$ is denoted ω_3^4. Our goal is to show that ω_3^4 is homeomorphic to ω^4 and that $D(\omega_3^4) = D(\omega^4)$. Since several arguments are essentially the same for $\mathcal{F}_n$ and $\mathcal{G}_3$, we also use a variable IFS $\mathcal{H} \in \{\mathcal{F}_n, \mathcal{G}_3\}$ that contains functions $h_i(\mathbf{x}) = \mathbf{x}/2 + \mathbf{w}_i/2$.

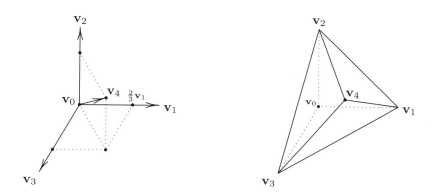

Fig. 63.1 The **v**-vectors and the hexahedron Λ^3.

§64 The Isotopy

Recall (from §17) the no-carry characterization of ω^n. In particular, ω^4 consists of those points $\mathbf{x} = x_1\mathbf{u}_1 + x_2\mathbf{u}_2 + x_3\mathbf{u}_3 + x_4\mathbf{u}_4$ in Δ_4 such that there exist binary representations of x_1, x_2, x_3, and x_4 where the sum of any two representations induces a "no carry." In other words, $\mathbf{x} \in \Delta_4$ is also in ω^4 if and only if there exists a $4 \times \aleph_0$ matrix $[a_{ij}]$ where the ith row is a binary expansion of x_i and where each column contains at most one "1".

It follows from this no-carry characterization of ω^4, and, since the 3-simplex Δ_3 is the face of Δ_4 opposite $\mathbf{u}_4$, that the 4-web ω^4 contains the 3-web ω^3. Intuitively, the isotopy that we construct fixes this 3-web ω^3 while moving (the terminal point of) the vector $\mathbf{u}_4 \in \mathbb{R}^4$ along the line segment $[\mathbf{u}_4, \mathbf{v}_4] = \{(1 - t)\mathbf{u}_4 + t\mathbf{v}_4 \mid 0 \le t \le 1\} \subset \mathbb{R}^4$ to (the terminal point of) the vector $\mathbf{v}_4 \in \mathbb{R}^4$. More precisely, let

$$H : \omega^4 \times I \to \mathbb{R}^4$$

be given by

$$H(\mathbf{x}, t) = H_t(\mathbf{x}) = \begin{bmatrix} 1 & 0 & 0 & \frac{2}{3}t \\ 0 & 1 & 0 & \frac{2}{3}t \\ 0 & 0 & 1 & \frac{2}{3}t \\ 0 & 0 & 0 & (1-t) \end{bmatrix} \mathbf{x} \quad (\mathbf{x} \in \omega^4 \subset \mathbb{R}^4; \ t \in I = [0,1]).$$

We shall show that H is an isotopy rel ω^3 (homotopy with each H_t a homeomorphism that is the identity on ω^3). Since each $H_t : \mathbb{R}^4 \to \mathbb{R}^4$ is a linear transformation with an upper-left 3×3 identity submatrix, it is clear that H is a homotopy and that H fixes Δ^3, a fortiori, fixes ω^3. When $t < 1$, then H_t is nonsingular and H_t^{-1} exists. It follows that H_t and H_t^{-1} are bounded, and we may conclude (Rudin [1966, Theorem 5.10]) that

$$\frac{1}{||H_t^{-1}||} \cdot ||\mathbf{x} - \mathbf{y}|| \le ||H_t(\mathbf{x}) - H_t(\mathbf{y})|| \le ||H_t|| \cdot ||\mathbf{x} - \mathbf{y}||.$$

Thus, for $t < 1$, $H_t : \mathbb{R}^4 \to \mathbb{R}^4$ induces an equivalent metric on $\mathbb{R}^4$ (Barnsley [1988, page 13, Definition 3]), and from Barnsley [1988, page 180, Theorem 3], H_t preserves fractal dimension, i.e., $D(H_t(\omega^4)) = D(\omega^4)$. So linear algebra and metric space theory suffice to prove that each H_t $(t < 1)$ is a homeomorphism that both fixes ω^3 and preserves fractal dimension.

The proof that H_1 is one-to-one (on ω^4) and respects fractal dimension, however, involves $\mathcal{F}_4$ and $\mathcal{G}_3$, and that is where the choice of "2/3" becomes critical. For example, if we replace "2/3" with "1", i.e., if we define

$$H'(\mathbf{x}, t) = \begin{bmatrix} 1 & 0 & 0 & 1t \\ 0 & 1 & 0 & 1t \\ 0 & 0 & 1 & 1t \\ 0 & 0 & 0 & (1-t) \end{bmatrix} \mathbf{x} \qquad (\mathbf{x} \in \omega^4 \subset \mathbb{R}^4; \ t \in I = [0,1]),$$

then for $t < 1$, an argument similar to the one used in the "2/3" case above would show that each H'_t $(t < 1)$ is a homeomorphism that both fixes ω^3 and preserves fractal dimension. But for $t = 1$, consider the two distinct points $\mathbf{x}, \mathbf{y} \in \omega^4$ whose components are expressed in binary as follows:

$$\mathbf{x} = (.00111\cdots)\mathbf{u}_1 + (.0000\cdots)\mathbf{u}_2 + (.0000\cdots)\mathbf{u}_3 + (.0100\cdots)\mathbf{u}_4$$
$$\mathbf{y} = (.10000\cdots)\mathbf{u}_1 + (.0100\cdots)\mathbf{u}_2 + (.0011\cdots)\mathbf{u}_3 + (.0000\cdots)\mathbf{u}_4.$$

Then clearly $\mathbf{x} \neq \mathbf{y}$, but since the binary expansions $.b_1 b_2 \cdots b_k 0111 \cdots$ and $.b_1 b_2 \cdots b_k 1000 \cdots$ represent the same number, $H'_1(\mathbf{x}) = \mathbf{y} = H'_1(\mathbf{y})$, showing that H'_1 is not one-to-one.

§65 The Hexahedron

The 3-*simplex*

$$\Delta_3 = \{\mathbf{x} \in \mathbb{R}^3 \ : \ \mathbf{x} = x_0 \mathbf{u}_0 + x_1 \mathbf{u}_1 + x_2 \mathbf{u}_2 + x_3 \mathbf{u}_3; \ x_i \geq 0; \ \Sigma_i x_i \leq 1\}$$
$$= \{\mathbf{x} \in \mathbb{R}^3 \ : \ \mathbf{x} = x_1 \mathbf{u}_1 + x_2 \mathbf{u}_2 + x_3 \mathbf{u}_3; \ x_i \geq 0; \ \Sigma_i x_i \leq 1\}.$$

Recall that the *face* T_0 *opposite* the zero vertex $\mathbf{u}_0 = 0$ is the 2-simplex

$$T_0 = \{\mathbf{x} \in \Delta_3 \ : \ \mathbf{x} = x_1 \mathbf{u}_1 + x_2 \mathbf{u}_2 + x_3 \mathbf{u}_3; \ x_i \geq 0; \ \Sigma_i x_i = 1\}.$$

The scalars "x_i" that define an $\mathbf{x} \in T_0$ are sometimes called the *barycentric coordinates of* $\mathbf{x}$. For $n = 3$, the face T_0 opposite the origin $\mathbf{u}_0 = \mathbf{v}_0$ is a closed equilateral triangle that contains the point $p = (1/3, 1/3, 1/3)$ at its *barycenter* (p is the unique point whose barycentric coordinates are all the same). By reflecting the 3-simplex Δ_3 through T_0, we obtain its mirror image Δ_3^R. Thus, $\Delta_3 \cap \Delta_3^R = T_0$, and we note that $p \in T_0$ and that $p = \frac{1}{2}\mathbf{v}_4$. It follows that $\mathbf{v}_4$ is the mirror image of the origin $\mathbf{v}_0$ and consequently a vertex of the simplex Δ_3^R. Thus, Δ_3^R is just the *cone* consisting of T_0 together with all line segments that have one endpoint in T_0 and the other at the point $\mathbf{v}_4$. The union $\Delta_3 \cup \Delta_3^R$ as a subspace of $\mathbb{R}^3$ will be called the *hexahedron* Λ^3.

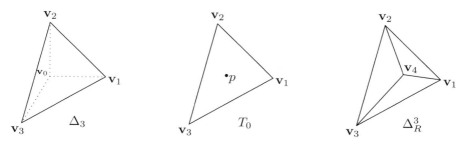

Fig. 65.1 Tetrahedron Δ^3, its face T_0 opposite $\mathbf{v}_0$, its mirror image Δ_3^R.

An algebraic description of Λ^3 will be needed. Our approach is standard and uses the representation of closed convex subsets of $\mathbb{R}^n$ as intersections of supporting hyperplanes. In short, a *hyperplane* $P = \{\mathbf{x} \in \mathbb{R}^n : \mathbf{a} \cdot \mathbf{x} = c\}$, where $\mathbf{a} \in \mathbb{R}^n$ and $c \in \mathbb{R}^1$ are given, and where "$\mathbf{a} \cdot \mathbf{x}$" is the usual inner product $a_1 x_1 + \cdots + a_n x_n$. Each such P induces two closed *half spaces*, obtained by replacing the equality in the definition of P by either "$\leq$" or "$\geq$." A hyperplane P is called *supporting for* a closed convex set K if $P \cap K$ is not empty and K is contained in one of the two closed half-spaces that is bounded by P.

Thus, since the three planes that contain, respectively, the three triangles with vertices $\{\mathbf{v}_0, \mathbf{v}_i, \mathbf{v}_j\}$ (where i and j are distinct members of $\{1, 2, 3\}$) are supporting for Λ^3, we see that $(x_1, x_2, x_3) \in \Lambda^3$ implies the three inequalities $x_1, x_2, x_3 \geq 0$. Using the other three planes that contain, respectively, the three triangles with vertices $\{\mathbf{v}_4, \mathbf{v}_i, \mathbf{v}_j\}$, we obtain another three inequalities, namely, for distinct $i, j, k = 1, 2, 3$, we have $x_i + x_j - (1/2)x_k \leq 1$. Thus, for $\mathbf{x}$ denoting $(x_1, x_2, x_3) \in \mathbb{R}^3$, we have

(1) $\Lambda^3 = \{\mathbf{x} \in \mathbb{R}^3 : x_i \geq 0; \text{ distinct } i, j, k = 1, 2, 3, \ x_i + x_j - (1/2)x_k \leq 1\}.$

65.2 Lemma *Let* $\mathbf{x} = (x_1, x_2, x_3) \in \Lambda^3$. *Then* $0 \leq x_1 + x_2 + x_3 \leq 2$, *where* $x_1 + x_2 + x_3 \leq 1$ *implies* $\mathbf{x} \in \Delta_3$, *and* $1 \leq x_1 + x_2 + x_3 \leq 2$ *implies* $\mathbf{x} \in \Delta_3^R$.

PROOF. The statement "$0 \leq x_1 + x_2 + x_3 \leq 2$" follows from the fact that the linear functional $x_1 + x_2 + x_3$ takes on its extreme values at the extreme points $\mathbf{v}_0, \mathbf{v}_1, \mathbf{v}_2, \mathbf{v}_3, \mathbf{v}_4$ of Λ^3. The other two inequalities follow from the same observation with $x_1 + x_2 + x_3$ restricted to Δ_3 and Δ_3^R, respectively. □

65.3 Lemma *Let* $\mathbf{x} = (x_1, x_2, x_3) \in \Delta_3^R$ *be such that there exist distinct indices* i *and* j *for which* $x_i + x_j \geq 4/3$. *Then* $\mathbf{x} = \mathbf{v}_4 = (2/3, 2/3, 2/3)$.

PROOF. From the representation of Λ^3 given in (1), we see that

$$4/3 - (1/2)x_k \leq x_i + x_j - (1/2)x_k \leq 1$$

and it follows that $(2/3) \leq x_k$. But since one of either $x_i \geq 2/3$ or $x_j \geq 2/3$

is true, a similar argument shows all three components are $\geq 2/3$. Then, since $2/3 + 2/3 + 2/3 = 2$, Lemma 65.2 provides the desired result. $\qquad \square$

65.4 Lemma *Let* $\mathbf{x} = (x_1, x_2, x_3) \in \Lambda^3$. *Then* $x_i = 1$ *implies* $\mathbf{x} = \mathbf{u}_i$.

PROOF. The two inequalities $x_i + x_k - (1/2)x_j \leq 1$ and $x_i + x_j - (1/2)x_k \leq 1$ with $x_i = 1$ yield $2x_j \leq x_k$ and $2x_k \leq x_j$, showing that $x_j = 0 = x_k$. $\qquad \square$

65.5 Lemma   *Let* $\mathbf{x} = (x_1, x_2, x_3) \in \Lambda^3$. *Then each* $0 \leq x_i \leq 1$.

PROOF. The inequality $0 \leq x_i$ follows from the definition of Λ^3. To see that $x_i \leq 1$ for each i, consider $\mathbf{x} \mapsto x_i$ as a linear functional, and recall that x_i is maximum at the extreme points of Λ^3, i.e., at $(0,0,0)$, $(1,0,0)$, $(0,1,0)$, $(0,0,1)$, and $(2/3, 2/3, 2/3)$. $\qquad \square$

§66 IFSs and the Just-Touching Property

At the beginning of §62 and the end of §63 we defined, respectively, $\mathcal{F}_n$ and $\mathcal{G}_3$. Here we consider their "just-touching property."

66.1 Theorem *The IFS* $\mathcal{F}_n$ *acting on* Δ_n *satisfies the just-touching property.*

PROOF. Let i and j denote distinct indices in $\{0, 1, \ldots, n\}$, and let $\mathbf{x}, \mathbf{y} \in \Delta_n$ be such that $f_i(\mathbf{x}) = f_j(\mathbf{y})$. Then $f_i(\mathbf{x}) = \frac{1}{2}\mathbf{x} + \frac{1}{2}\mathbf{u}_i = \frac{1}{2}\mathbf{y} + \frac{1}{2}\mathbf{u}_j = f_j(\mathbf{y})$. Thus, $\mathbf{x} - \mathbf{y} = \mathbf{u}_j - \mathbf{u}_i$, i.e., in terms of components,

$$(x_1 - y_1, ..., x_j - y_j, ..., x_i - y_i, ..., x_n - y_n) = (0, ..., 0, \delta_j, 0, ..., 0, -\delta_i, 0, ..., 0)$$

where $\delta_j, \delta_i \in \{0, 1\}$ ($\delta_k = 0$ iff $k = 0$). We have three cases: (a) $\delta_j = 0 \neq \delta_i$; (b) $\delta_j \neq 0 = \delta_i$; (c) $\delta_j \neq 0 \neq \delta_i$. If (a), then $x_j - y_j = \delta_j = 0$ and $x_i - y_i = -\delta_i = -1$. So $x_j = y_j$, and, since $y_i \leq \Sigma_k y_k \leq 1$ and $x_i \geq 0$, we have $x_i = 0 \neq y_i = 1$. The definition of Δ_n shows that $\mathbf{y} = \mathbf{u}_i$. It follows that $\mathbf{x} = \mathbf{u}_0 = \mathbf{u}_j = 0$. Thus,

$$(2) \qquad f_i(\mathbf{x}) = f_i(\mathbf{u}_j) = \mathbf{u}_j/2 + \mathbf{u}_i/2 = f_j(\mathbf{u}_i) = f_j(\mathbf{y}),$$

showing that the point $\mathbf{u}_j/2 + \mathbf{u}_i/2 \in f_i(\Delta_n) \cap f_j(\Delta_n)$. Turning to case (b), we see that $x_j - y_j = \delta_j = 1$ and $x_i - y_i = -\delta_i = 0$. So $x_i = y_i$, and, since $x_j \leq \Sigma_k x_k \leq 1$ and $y_j \geq 0$, we have $x_j = 1$ and $y_j = 0$. And by definition of Δ_n, we have $\mathbf{x} = \mathbf{u}_j$. It follows that $\mathbf{y} = \mathbf{0} = \mathbf{u}_0 = \mathbf{u}_i$. Clearly, (2) holds in this case also. Finally, for case (c), we have $x_j - y_j = \delta_j = 1$ and $x_i - y_i = -\delta_i = -1$. As before, it follows that $x_j = 1$ and $y_j = 0$, while $x_i = 0$ and $y_i = 1$. Then Lemma 65.4 implies both $\mathbf{x} = \mathbf{u}_i$ and $\mathbf{y} = \mathbf{u}_j$, which in turn shows that (2) holds in this final case. Thus, (2) holds for every $\mathbf{x}, \mathbf{y} \in \Delta_n$ such that $f_i(\mathbf{x}) = f_j(\mathbf{y})$, i.e., $\{\mathbf{u}_j/2 + \mathbf{u}_i/2\} = f_i(\Delta_n) \cap f_j(\Delta_n)$. $\qquad \square$

Note that the previous theorem provides existence as well as uniqueness of the "just-touching" points $\frac{1}{2}(\mathbf{u}_i + \mathbf{u}_j)$. It turns out, in the context of addressing, that these points have dual addresses, namely $ijjj\cdots$ and $jiii\cdots$.

The idea of constructing five "just touching and congruent" hexahedra is basic to building a model of the 4-web in 3-space. An example of such a combinatorial construction is pictured in Figures 7.1, 7.2, and 7.3.

66.2 Theorem *The IFS $\mathcal{G}_3$ has the just-touching property on Λ^3.*

PROOF. Consider distinct indices $i, j \in \{0, 1, 2, 3, 4\}$, and let $\mathbf{x}$ and $\mathbf{y}$ be points in Λ^3 such that $g_i(\mathbf{x}) = g_j(\mathbf{y})$. Then

$$g_i(\mathbf{x}) = \mathbf{x}/2 + \mathbf{v}_i/2 = \mathbf{y}/2 + \mathbf{v}_j/2.$$

Thus, $\mathbf{x} - \mathbf{y} = \mathbf{v}_j - \mathbf{v}_i$, i.e.,

$$(x_1 - y_1, x_2 - y_2, x_3 - y_3) = (\delta_1, \delta_2, \delta_3)$$

where each $\delta_k \in \{0, \pm 1, \pm 1/3, \pm 2/3\}$. We break the possibilities into those where $j = 0$ (Case I) and those where $j \neq 0$ (Case II).

Case I ($j = 0$). In this case $\mathbf{x} - \mathbf{y} = -\mathbf{v}_i$ for some $i \in \{1, 2, 3, 4\}$. And the possibilities are either exactly one component $x_i - y_i = -1$ with the others equal to 0, or $\mathbf{x} - \mathbf{y} = (-2/3, -2/3, -2/3)$. In the former subcase $i \in \{1, 2, 3\}$, and then Lemma 65.5 shows that $x_i = 0$ and $y_i = 1$. Then Lemma 65.4 gives $\mathbf{y} = \mathbf{v}_i$. It follows that $\mathbf{x} = \mathbf{v}_0$. Thus, $g_i(\mathbf{x}) = g_i(\mathbf{v}_0) = \mathbf{v}_0/2 + \mathbf{v}_i/2 = \mathbf{v}_j/2 + \mathbf{v}_i/2 = g_j(\mathbf{y})$ is the unique point $\mathbf{v}_j/2 + \mathbf{v}_i/2$ that depends only on i and j, i.e.,

(3) $$g_i(\mathbf{x}) = \mathbf{v}_j/2 + \mathbf{v}_i/2 = g_j(\mathbf{y}).$$

In the latter subcase $i = 4$, and $x_k - y_k = -2/3$ for each $k = 1, 2, 3$, showing that $y_k = x_k + 2/3 \geq 2/3$ for each $k = 1, 2, 3$. It follows from Lemma 65.3 that $\mathbf{y} = \mathbf{v}_4$ and (consequently) that $\mathbf{x} = \mathbf{v}_0$. And so (3) also holds when $j = 0$ and $i = 4$. This finishes Case I.

Case II ($j \neq 0$). The possibilities correspond to three subcases, namely $i = 0$, $1 \leq i \leq 3$, and $i = 4$. If $i = 0$ and $j \in \{1, 2, 3\}$, then $x_j - y_j = 1$ and $x_k - y_k = 0$ when $k \neq j$. Lemma 65.5 then shows that $\mathbf{x} = \mathbf{v}_j$, and (consequently) $\mathbf{y} = \mathbf{v}_0 = \mathbf{v}_i$. So (3) also holds for $j \in \{1, 2, 3\}$, and $i = 0$. Next, if $i = 0$ and $j = 4$, then $\mathbf{x} - \mathbf{y} = (2/3, 2/3, 2/3)$, showing that $x_k = y_k + 2/3 \geq 2/3$ for each $k = 1, 2, 3$. Lemma 65.3 shows that $\mathbf{x} = \mathbf{v}_4 = \mathbf{v}_j$ and (consequently) that $\mathbf{y} = \mathbf{v}_0 = \mathbf{v}_i$. So (3) also holds for $i = 0$ and $j = 4$. This finishes the subcase $i = 0$. We turn to the subcase $1 \leq i \leq 3$. If j is also such that $1 \leq j \leq 3$, then we may assume that $x_j - y_j = 1$, $x_i - y_i = -1$, and $x_k - y_k = 0$ where $i \neq k \neq j$. Lemmas 65.4 and 65.5 show that the first of these equations yields $\mathbf{x} = \mathbf{v}_j$. Then Lemma 65.5 shows that $y_i = 1$, and then Lemma 65.4 shows $\mathbf{y} = \mathbf{v}_i$. So (3) also holds for distinct $i, j \in \{1, 2, 3\}$. For $1 \leq i \leq 3$ and $j = 4$, we have $x_i - y_i = -1/3$, and $x_k - y_k = 2/3$ for $k \neq i$. Since there are two values of k that satisfy the last equation, we again deduce that the sum of two components of $\mathbf{x}$ is $\geq 4/3$, and then Lemma 65.3 shows that $\mathbf{x} = \mathbf{v}_4 = \mathbf{v}_j$. It follows that $\mathbf{y} = \mathbf{v}_i$, and so (3) also holds when

$1 \le i \le 3$ and $j = 4$. Finally, consider the subcase $i = 4$. In this instance $1 \le j \le 3$, and we see that $x_j - y_j = 1/3$ and $x_k - y_k = -2/3$ for two indices $k \ne j$. Thus, the sum of two components of $\mathbf{y}$ is $\ge 4/3$, showing (via Lemma 65.3) that $\mathbf{y} = \mathbf{v}_4 = \mathbf{v}_i$. It follows that $\mathbf{x} = \mathbf{v}_j$. So (3) also holds for $j = 1, 2, 3$ and $i = 4$. This finishes Case II.

Thus, for every $\mathbf{x}, \mathbf{y} \in \Lambda^3$ such that $g_i(\mathbf{x}) = g_j(\mathbf{y})$, equation (3) holds. It follows that $\{\mathbf{v}_j/2 + \mathbf{v}_i/2\} = g_i(\Lambda^3) \cap g_j(\Lambda^3)$ for each $i \ne j$.   □

§67 Addressing and the Isotopy

In this section, we let $A = \{0, 1, 2, 3, 4\}$, and use $\mathcal{H} = \{h_i : i \in A\}$ to denote one of the iterated function systems $\mathcal{F}_4$ or $\mathcal{G}_3$. So for each $i \in A$,

$$h_i(\mathbf{x}) = \mathbf{x}/2 + \mathbf{w}_i/2 = \begin{cases} \mathbf{x}/2 + \mathbf{u}_i/2 & \text{when } \mathcal{H} = \mathcal{F}_4; \\ \mathbf{x}/2 + \mathbf{v}_i/2 & \text{when } \mathcal{H} = \mathcal{G}_3. \end{cases}$$

For the code space $N(\{0, 1, 2, 3, 4\})$, each $\sigma \in N(A)$ determines a sequence $\{\sigma^n\}$, where $\sigma^n = \sigma_1 \cdots \sigma_n 000 \cdots$, that obviously converges to σ. And the continuity of the address map $\phi : N(A) \to K$ shows that the corresponding sequence $\{p^n = p_{\sigma^n}\}$ in the attractor K of $\mathcal{H}$ converges to p_σ.

With each address $\sigma = \sigma_1 \sigma_2 \cdots \in N(A)$, we also associate an infinite matrix $M_\sigma = [a_{ij}]_\sigma$ $(1 \le i \le 4; 1 \le j)$ of zeros and ones via the *nth-column formula*: If $\sigma_n = 0$, then let the nth column contain only zeros; and if $\sigma_n = k \ne 0$, then let $a_{kn} = 1$ be the only 1 in the nth column. Clearly, then, the rows of M_σ induce binary representations that satisfy the no-carry condition.

67.1 Theorem *Let $\{\mathbf{w}_i\} \in \{\{\mathbf{u}_i\}, \{\mathbf{v}_i\}\}$, the choice conforming to the choice of $\mathcal{H} \in \{\mathcal{F}_4, \mathcal{G}_3\}$. Let K denote the attractor of $\mathcal{H}$. Let $\sigma \in N(\{0, 1, 2, 3, 4\})$, and let p_σ be the image of σ under the address map. Then p_σ may be written as a linear combination*

(4) $$p_\sigma = a_1 \mathbf{w}_1 + a_2 \mathbf{w}_2 + a_3 \mathbf{w}_3 + a_4 \mathbf{w}_4$$

where each coefficient a_i, $1 \le i \le 4$, has the binary expansion $.a_{i1} a_{i2} \cdots$ where $a_{i1} a_{i2} \cdots$ is the ith row of $M_\sigma = [a_{ij}]_\sigma$.

PROOF. Let $\{\sigma^n\}$ and $\{p^n\}$ denote the sequences defined in the next-to-last paragraph preceding this theorem. Since the m-fold composition $(h_0 \circ \cdots \circ h_0)(K) = K/(2^m)$, since $h_{\sigma_1} \circ \cdots \circ h_{\sigma_n}$ is one-to-one, since p^n is the only point in

$$\cap_{m=1}^\infty (h_{\sigma_1} \circ \cdots \circ h_{\sigma_n}(K/(2^m))),$$

and since $\{\mathbf{w}_0\} = \cap_{m=1}^\infty K/(2^m)$, it follows that $p^n = h_{\sigma_1} \circ \cdots \circ h_{\sigma_n}(\mathbf{w}_0)$. Now certainly the origin $p_0 = 0$ corresponds to the sequence $000 \cdots \in N(A)$ of zeros and satisfies the conclusion of the theorem. So inductively, let $\sigma' = \sigma_2 \sigma_3 \cdots \sigma_n 000 \cdots$ and suppose that $p_{\sigma'}$ satisfies the conclusion of the theorem. We show that the image p^n of $\sigma^n = \sigma_1\sigma_2\cdots\sigma_n 000\cdots$ also satisfies the conclusion of the theorem: Letting $\sigma_1 = k \in \{0,1,2,3,4\}$, we have, in binary, $\mathbf{w}_k/2 = (.1)\mathbf{w}_k$, and so

$$p^n = h_{\sigma_1} \circ \cdots \circ h_{\sigma_n}(\mathbf{w}_0) = h_k(p_{\sigma'})$$

$$= (1/2)\left[(.a'_{11}a'_{12}\cdots)\mathbf{w}_1 + \cdots + (.a'_{41}a'_{42}\cdots)\mathbf{w}_4\right] + (.1)\mathbf{w}_k$$

$$= \begin{cases} (.0a'_{11}a'_{12}\cdots)\mathbf{w}_1 + \cdots + (.1a'_{k1}a'_{k2}\cdots)\mathbf{w}_k + \cdots + (.0a'_{41}a'_{42}\cdots)\mathbf{w}_4 & \text{if } k \neq 0; \\ (.0a'_{11}a'_{12}\cdots)\mathbf{w}_1 + \cdots + (.0a'_{41}a'_{42}\cdots)\mathbf{w}_4 & \text{if } k = 0. \end{cases}$$

That is, we may obtain p^n and $M_{\sigma^n} = [a_{ij}]_{\sigma^n}$ using $p_{\sigma'}$ and $M_{\sigma'} = [a'_{ij}]$. Indeed, for each $i \in \{1,2,3,4\}$,

$$a_i^n \mathbf{w}_i = (.a_{i1}^n a_{i2}^n \cdots)\mathbf{w}_i = \begin{cases} (.0a'_{i1}a'_{i2}\cdots)\mathbf{w}_i & \text{when } \sigma_1 = k \neq i; \\ (.1a'_{i1}a'_{i2}\cdots)\mathbf{w}_i & \text{when } \sigma_1 = k = i. \end{cases}$$

So these "shifted and possibly one added" binary expansions represent the scalar coefficients a_i^n of the vectors $\mathbf{w}_i$ that appear in the sum

$$p^n = \sum_{i=1}^4 a_i^n \mathbf{w}_i$$

which corresponds to (4). Moreover, since the matrix M_{σ^n} is obtained from $M_{\sigma'}$ by shifting the columns of the latter to the right by one index, and then determining the first column via the value of σ_1, it is clear that this representation of p^n satisfies the conclusion of the theorem.

Turning to p_σ, we see that $\sigma^n \to \sigma$ in $N(A)$ (term-by-term) implies that $M_{\sigma^n} \to M_\sigma$ (column-by-column), and so the binary representations $.a_{i1}^n a_{i2}^n \cdots a_{in}^n 000$ induced by the rows of the M_{σ^n} matrices converge (term-by-term) to the binary representations $.a_{i1}a_{i2}\cdots a_{in}a_{i(n+1)}\cdots$ induced by the corresponding rows of M_σ. That is, $a_i^n \to .a_{i1}a_{i2}\cdots$. Now suppose $\mathcal{H} = \mathcal{F}_4$: Let $p_\sigma = \sum_{i=1}^4 b_i\mathbf{u}_i$ where the "b_i" are the unique scalars that define the point p_σ relative to the basis $\{\mathbf{u}_i\}$ of $\mathbb{R}^4$. Then $(p^n = p_{\sigma^n}) \to p_\sigma$ implies $a_i^n \to b_i$. But $a_i^n \to .a_{i1}a_{i2}\cdots$, and thus the b_i scalars have the corresponding binary representations $.a_{i1}a_{i2}\cdots = a_i$ when $\mathcal{H} = \mathcal{F}_4$. Finally, suppose $\mathcal{H} = \mathcal{G}_3$: Let $p_\sigma = \sum_{i=1}^3 b_i\mathbf{v}_i$ where the b_i are the unique scalars that define p_σ relative to the basis $\{\mathbf{v}_1, \mathbf{v}_2, \mathbf{v}_3\}$ of $\mathbb{R}^3$. Then $p_{\sigma^n} \to p_\sigma$ together with

$$p^n = p_{\sigma^n} = \sum_{i=1}^4 a_i^n \mathbf{w}_i = \sum_{i=1}^3 a_i^n \mathbf{v}_i + \sum_{i=1}^3 a_4^n(2/3)\mathbf{v}_i = \sum_{i=1}^3 (a_i^n + (2/3)a_4^n)\mathbf{v}_i$$

implies $(a_i^n + (2/3)a_4^n) \to b_i$ for each $i = 1,2,3$. But $a_i^n \to .a_{i1}a_{i2}\cdots = a_i$ for each $i = 1,2,3,4$, showing that $p_\sigma = a_1\mathbf{v}_1 + a_2\mathbf{v}_2 + a_3\mathbf{v}_3 + a_4\mathbf{v}_4$ where the coefficients a_i $(i = 1,2,3,4)$ have the desired binary expansions. $\qquad\Box$

67.2 Theorem   *Let $\mathcal{H}$ be given as in the hypothesis of Theorem 67.1. Let $p \in K$ where K is the attractor of $\mathcal{H}$, let $O \in \{\Delta^4, \Lambda^3\}$ correspond to the choice of $\mathcal{H}$, and let each of two distinct sequences σ and τ address p, i.e., $p = p_\sigma = p_\tau$. Then there is an index $l \geq 0$ such that*

$$\sigma = j_1 j_2 \cdots j_l i k k k \cdots \quad and \quad \tau = j_1 j_2 \cdots j_l k i i i \cdots$$

where $k \neq i$ and $i, k \in A$.

PROOF. Let l be either the $\max\{j \ : \ \sigma_1 = \tau_1, \cdots, \sigma_j = \tau_j\}$ or 0 (if $\sigma_1 \neq \tau_1$). Then for $\sigma' = \sigma_{l+1} \sigma_{l+2} \cdots$ and $\tau' = \tau_{l+1} \tau_{l+2} \cdots$, we see that

$$p = p_\sigma = h_{j_1} h_{j_2} \cdots h_{j_l}(p_{\sigma'}) = p_\tau = h_{j_1} h_{j_2} \cdots h_{j_l}(p_{\tau'}).$$

We let $i = \sigma_{l+1}$ and $k = \tau_{l+1}$. Then $k \neq i$, and $p_{\sigma'} \in h_i(O)$ while $p_{\tau'} \in h_k(O)$. But since $h_{j_1} \circ \cdots \circ h_{j_l}$ is one-to-one, $p' = p_{\sigma'} = p_{\tau'}$, showing that $p' \in h_i(O) \cap h_k(O)$, and therefore must be the "just-touching" point $(\mathbf{w}_i + \mathbf{w}_k)/2$. For example, if $i = 1$ and $k = 2$

$$M_{p_{\sigma'}} = \begin{bmatrix} 1 & 0 & 0 & \cdots \\ 0 & 1 & 1 & \cdots \\ 0 & 0 & 0 & \cdots \\ 0 & 0 & 0 & \cdots \end{bmatrix} \quad \text{and} \quad M_{p_{\tau'}} = \begin{bmatrix} 0 & 1 & 1 & \cdots \\ 1 & 0 & 0 & \cdots \\ 0 & 0 & 0 & \cdots \\ 0 & 0 & 0 & \cdots \end{bmatrix}$$

By evaluating $h_{j_1} \circ \cdots \circ h_{j_l}$ at p', we, in effect, move the columns of $M_{p_{\sigma'}}$ and $M_{p_{\tau'}}$, l places to the right and then fill in the first l columns as prescribed by the h_js. The results are M_σ and M_τ, respectively. Thus, by comparing these two matrices, the desired relation between σ and τ holds. □

67.3 Theorem   *The homotopy H defined in §64 above is an isotopy that preserves fractal dimension.*

PROOF.   It follows from Theorem 67.2 that the equivalence relation "$\sim$" induced on $N(A)$ via the address map (in either case where $\mathcal{H} = \mathcal{F}_4$ or $\mathcal{H} = \mathcal{G}_3$) is the one of *identifying adjacent endpoints* in $N(A)$ (Lipscomb and Perry [1992]). Since $N(A)$ is compact, the continuous surjective addressing map from $N(A)$ to the attractor of $\mathcal{H}$ is closed, and hence quotient. It follows that since each of the attractors ω^4 and ω_3^4 is homeomorphic to the quotient $N(A)/\sim$, they are homeomorphic to each other. More precisely, the homeomorphisms (induced from the address maps) $\alpha : N(A)/\sim \ \to \omega^4$ and $\beta : N(A)/\sim \ \to \omega_3^4$ map an equivalence class $[\sigma] \in N(A)/\sim$ to $a_1\mathbf{u}_1 + a_2\mathbf{u}_2 + a_3\mathbf{u}_3 + a_4\mathbf{u}_4$ in the former case, and to $a_1\mathbf{v}_1 + a_2\mathbf{v}_2 + a_3\mathbf{v}_3 + a_4\mathbf{v}_4$ in the latter case (the unique 4-tuple (a_1, a_2, a_3, a_4) being specified as in Theorem 67.3). It follows that $\beta \circ \alpha^{-1} : \omega^4 \to \omega_3^4$ is a homeomorphism that is given by

$$a_1\mathbf{u}_1 + a_2\mathbf{u}_2 + a_3\mathbf{u}_3 + a_4\mathbf{u}_4 \ \mapsto \ a_1\mathbf{v}_1 + a_2\mathbf{v}_2 + a_3\mathbf{v}_3 + a_4\mathbf{v}_4$$

Rewriting $\mathbf{v}_4 = (2/3)\mathbf{v}_1 + (2/3)\mathbf{v}_2 + (2/3)\mathbf{v}_3$ and substituting, we have $\mathbf{x} = (a_1, a_2, a_3, a_4) \in \omega^4$ implies $\beta \circ \alpha^{-1}(\mathbf{x}) = H_1(\mathbf{x})$. In other words, H_1 is one-to-one on ω^4, and so by previous remarks, it follows that H is an isotopy. To see that H_1 respects the fractal dimension, we recall that $\mathcal{F}_4$ and $\mathcal{G}_3$ contain the same number "5" of affine transformations, each with the scale factor of $1/2$. It follows that both fractals have dimension $\ln(5)/\ln(2) \approx 2.2319$. □

67.4 Corollary *Let $n \geq 4$. Then the n-web $\omega^n \subset \mathbb{R}^n$ can be imbedded in $\mathbb{R}^{n-1}$ with fractal dimension preserved.*

PROOF. For $n = 4$, Theorem 67.3 provides the linear transformation $H_1 : \mathbb{R}^4 \to \mathbb{R}^4$ whose restriction to ω^4 is a homeomorphism into $\mathbb{R}^3 \subset \mathbb{R}^4$ that preserves fractal dimension. The matrix representation of H_1 relative to the standard basis $\{\mathbf{u}_1, \mathbf{u}_2, \mathbf{u}_3, \mathbf{u}_4\}$ is

$$H_1 = \begin{bmatrix} 1 & 0 & 0 & 2/3 \\ 0 & 1 & 0 & 2/3 \\ 0 & 0 & 1 & 2/3 \\ 0 & 0 & 0 & 0 \end{bmatrix}.$$

So we only need to consider the $n > 4$ case. In this case, let $\mathbf{u}_1, \mathbf{u}_2, \ldots, \mathbf{u}_n$ denote the standard basis vectors of $\mathbb{R}^n$, and let $\mathbb{R}^{n-4}$ denote the subspace of $\mathbb{R}^n$ that has basis $\{\mathbf{u}_5, \mathbf{u}_6, \ldots, \mathbf{u}_n\}$. Identify $\mathbb{R}^n$ with $\mathbb{R}^4 \oplus \mathbb{R}^{n-4}$, and define $L_n : \mathbb{R}^n \to \mathbb{R}^n$ as the product map $L_n = H_1 \times 1_{n-4}$ where 1_{n-4} is the identity (linear) transformation on $\mathbb{R}^{n-4}$. That is, L_n is given by

$$H_1 \times 1_{n-4} : \mathbb{R}^4 \oplus \mathbb{R}^{n-4} \to \mathbb{R}^4 \oplus \mathbb{R}^{n-4}.$$

It follows that relative to the basis $\{\mathbf{u}_i\}_{i=1}^n$, the "matrix" H_1 and the $(n-4) \times (n-4)$ identity matrix I_{n-4} may serve as blocks in a block-matrix representation of L_n, namely

$$\begin{bmatrix} H_1 & 0 \\ 0 & I_{n-4} \end{bmatrix}.$$

With this matrix representation, clearly $L_n(\omega^n) \subset \{(x_1, x_2, x_3, 0, x_5, \ldots, x_n) \in \mathbb{R}^n\}$, i.e., essentially, $L_n(\omega^n) \subset \mathbb{R}^{n-1}$. Since L_n is linear, it is continuous; and since ω^n is compact, it suffices to show that L_n is one-to-one on ω^n. So suppose that $\mathbf{x}, \mathbf{y} \in \omega^n$ are such that $L_n(\mathbf{x}) = L_n(\mathbf{y})$. Then $L_n = H_1 \times 1_{n-4}$ implies that $H_1(x_1, \ldots, x_4) = H_1(y_1, \ldots, y_4)$ and that $1_{n-4}(x_5, \ldots, x_n) = 1_{n-4}(y_5, \ldots, y_n)$. Since each of $\mathbf{x}$ and $\mathbf{y}$ satisfies the no-carry condition, each of $(x_1, \ldots, x_4)$, $(y_1, \ldots, y_4) \in \mathbb{R}^4$ satisfies the no-carry condition. Hence, they are members of $\omega^4 \subset \mathbb{R}^4$. But since H_1 is one-to-one on ω^4, it follows that $\mathbf{x} = \mathbf{y}$. To see that L_n preserves the fractal dimension of ω^n, first extend $\mathcal{G}_3$ to the IFS $\mathcal{G}_n = \{g_0, g_1, \ldots, g_n\}$ by letting $\mathbf{v}_i = \mathbf{u}_i$ for $i > 4$. Second, define $\omega_{n-1}^n = L_n(\omega^n)$. Then show that $L_n(f_i(\omega^n)) = g_i(\omega_{n-1}^n)$ for each $i = 0, \ldots, n$ where $f_i \in \mathcal{F}_n$. Deduce that $\omega_{n-1}^n = \cup_i g_i(\omega_{n-1}^n)$, i.e., that ω_{n-1}^n is the unique attractor of the IFS $\mathcal{G}_n$. As in the last paragraph of Theorem 67.3, it follows that $D(\omega^n) = D(\omega_{n-1}^n)$. □

§68 Comments

From the approach used in this chapter, it appears that to solve the open problems (§62) of determining whether one can *visualize* (within 3-space) the self-similarity of the 5-, 6-, or 7-web, respectively J_6, J_7, or J_8, one would need to construct many candidate polyhedra within 3-space. For example, in the 5-web ω^5 case the desired polyhedron would be the convex hull of six points in 3-space. Suppose P_6 denotes such a polyhedron. Then using a computer and appropriate software, one would contract P_6 by $1/2$ toward each of its six vertices and "visually check" for the "just-touching property." But even if a candidate P_6 were found whose level-1 iterates "visually" appear to satisfy the just-touching property, a mathematical proof would still be required. And in such a case, the approach in this chapter could serve as a guide.

CHAPTER 13

From 2-Web IFS to 2-Simplex IFS
2-Space and the 1-Sphere

Sierpiński's classical construction of his triangle (gasket) begins with a 2-simplex Δ^2 (manifold) and ends with the 2-web ω^2 (fractal) subspace. It is therefore natural (inverse of moving from manifolds to fractals) to seek a minimal code-space and address-map extension of the n-web system to an n-simplex system (a fractals-to-manifolds problem).

In this chapter we consider the $n = 2$ case, and extend the 2-web IFS system $\mathcal{F}_2$ to a 2-simplex IFS system $\mathcal{F}_2^*$. The extension, when viewed as identification of certain sequences in code space, yields a representation of 2-space and the 1-sphere. The following chapter provides a solution for the $n = 3$ case. Here, however, we follow the presentation in Lipscomb [2005].

§69 Overview

The IFS of interest is $\mathcal{F}_2 = \{w_0, w_1, w_2\}$ whose affine transformations

$$w_k(\mathbf{x}) = \mathbf{u}_k + (1/2)(\mathbf{x} - \mathbf{u}_k) = 1/2(\mathbf{x} + \mathbf{u}_k) \qquad (\mathbf{x} \in \Delta^2; k = 0, 1, 2)$$

are contractions by $1/2$ toward the $\mathbf{u}_k$ where $\mathbf{u}_0 = (1, 0, 0)^T$, $\mathbf{u}_1 = (0, 1, 0)^T$, and $\mathbf{u}_2 = (0, 0, 1)^T$ are the standard basis vectors in $\mathbb{R}^3$ and also the vertices of Δ^2. (This definition is a variant of the one given (for $n = 2$) at the beginning of §8. In this case we use the *standard 2-simplex* $\Delta^2 \subset \mathbb{R}^3$ instead of $\Delta_2 \subset \mathbb{R}^2$.)

This IFS $\mathcal{F}_2$ has ω^2 as the attractor, and the corresponding address map $\phi : N(\{0, 1, 2\}) \to \omega^2$ may also be viewed as the natural mapping $p : N(\{0.1, 2\}) \to J_3$ (see Theorem 8.4). From the fractal viewpoint, $N(\{0, 1, 2\})$ is a *code space*, and ω^2 is *Sierpiński's triangle/gasket*. In topological terms, $N(\{0, 1, 2\})$ is a Baire space and ω^2 is the 2-*web* (Definition 8.3), which is homeomorphic to J_3.

For $A = \{0, 1, 2\}$, we seek a "minimal" extension of both Baire's space $N(A)$ and the adjacent-endpoint relation on $N(A)$ that yields, as a quotient structure, the entire 2-simplex Δ^2. In particular, we extend $N(\{0, 1, 2\})$ to $N(\{0, 1, 2, 3\})$, and, $\sim$ on $N(\{0, 1, 2\})$ to a relation $R \supset \sim$ on $N(\{0, 1, 2, 3\})$. It turns out that the new relation R induces equivalence classes of cardinalities 1, 2, 3, and 6 only. *One of the goals here is to emulate the adjacent-endpoint approach*: In a relatively simple way, recognize the basic forms of the sequences that when identified yield the 2-simplex Δ^2. From the fractal viewpoint, we shall define an IFS $\mathcal{F}_2^* = \{w_0, w_1, w_2, w_3\}$ with attractor Δ^2.

S.L. Lipscomb, *Fractals and Universal Spaces in Dimension Theory*,
DOI 10.1007/978-0-387-85494-6_13, © Springer Science+Business Media, LLC 2009

As an application, using the address map $\phi : N(\{0, 1, 2, 3\}) \to \Delta^2$ we specify ϕ-inverse sets $F, G = (N(\{0, 1, 2, 3\}) \setminus F) \subset N(\{0, 1, 2, 3\})$ such that $\phi|_F : F \to \partial \Delta^2$ and $\phi|_G : G \to (\Delta^2 \setminus \partial \Delta^2)$ are quotient maps, the former onto a copy of the 1-sphere and the latter onto a copy of 2-space.

§70 The $\mathcal{F}_2^*$ IFS

The intuitive idea behind the desired iterated function system $\mathcal{F}_2^*$ is that of using one extra digit "3" to encode an *iterated pasting* — at each level "w_3" serves to "fill the holes" in Sierpiński's gasket.[1]

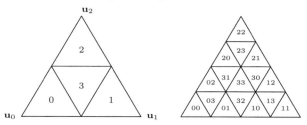

Fig. 70.1 Decompositions of Δ^2; scaled by $\frac{1}{2}$ and $\frac{1}{4}$ subtriangles.

Recall that when we identify a point $\mathbf{x}$ in Δ^2 with its column vector $[x_0, x_1, x_2]^T$ in 3-space, we may view the components x_k as either barycentric coordinates or Cartesian coordinates — our 2-simplex Δ^2 is the one where barycentric coordinates and Cartesian coordinates are equal.

With this background, we extend the IFS $\mathcal{F}_2$ by considering the barycenter $\mathbf{u}_3 = (1/3)(\mathbf{u}_0 + \mathbf{u}_1 + \mathbf{u}_2)$ of Δ^2, and the affine transformation[2]

$$w_3(\mathbf{x}) = (1/2)L\mathbf{x} + (1/2)\mathbf{u}_3$$

where

$$L = \begin{bmatrix} -1/3 & 2/3 & 2/3 \\ 2/3 & -1/3 & 2/3 \\ 2/3 & 2/3 & -1/3 \end{bmatrix}.$$

Since L is not the identity matrix, the affine transformation "w_3" is not a contraction. Nevertheless, w_3 is the composition of a 180° rotation L (an isometry) followed by a contraction by $1/2$ toward $\mathbf{u}_3$. That is, the linear transformation L is a rotation of 180° about the line containing $\mathbf{u}_3$ that is

---

[1]The phrase *iterated pasting* is the author's attempt to intuitively describe what $w_3 \in \mathcal{F}_2^* \setminus \mathcal{F}_2$ contributes in the context of the iteration process. The iterations of the members of $\mathcal{F}_2$ serve to "iteratively cut holes" in the manifold Δ^2, while on the other hand, the addition of the function w_3 "surgically repairs" the "cuts" at each level of iteration by pasting just the right size triangles in the holes. For example, the left-side graphic in Figure 70.1 shows the first level: The hole $\Delta^2 \setminus (w_0(\Delta^2) \cup w_1(\Delta^2) \cup w_2(\Delta^2))$ appears. But $w_3(\Delta^2)$, the triangle labeled "3" fills the hole. The process is repeated at the second level where the iterates $w_i \circ w_j(\Delta^2)$ appear in the right-side graphic of Figure 70.1. Again, those triangles whose label contains the digit "3" fill all holes.

[2]Also note that $w_3(\mathbf{x}) = \frac{1}{2}(\mathbf{v} - \mathbf{x})$ for $\mathbf{v} = [1, 1, 1]^T$.

perpendicular to the plane containing Δ^2, and as a consequence, the affine w_3 maps $\mathbf{u}_k$ to the midpoint of the edge opposite $\mathbf{u}_k$.

The extension $\mathcal{F}_2^* = \{w_0, w_1, w_2, w_3\}$ of $\mathcal{F}_2 = \{w_0, w_1, w_2\}$ is an IFS with *contractivity factor* $1/2$. Furthermore, from the left-side graphic of Figure 70.1 we see that the 2-simplex Δ^2 satisfies

$$\Delta^2 = w_0(\Delta^2) \cup w_1(\Delta^2) \cup w_2(\Delta^2) \cup w_3(\Delta^2),$$

showing that the attractor of $\mathcal{F}_2^*$ is Δ^2. (The 2-simplex Δ^2 is the fixed point of the Hutchinson operator.) Our goal is to show that the point-inverse sets of the address map ϕ

$$\alpha_1\alpha_2 \cdots \mapsto \cap_{j=1}^{\infty} \left(w_{\alpha_1} \circ w_{\alpha_2} \circ \cdots \circ w_{\alpha_j}(\Delta^2) \right)$$

from code space $N(\{0, 1, 2, 3\})$ onto Δ^2 define the classes of a relation R that is a superset of the adjacent-endpoint relation $\sim$ on $N(\{0, 1, 2\})$.

As a result, we obtain representations of the 2-space and the 1-sphere.

§71 The Quotient/Address Map

For a sequence $\alpha_1\alpha_2 \cdots$ in $\{0, 1, 2, 3\}$, define the subtriangle $T_{\alpha_1} = w_{\alpha_1}(\Delta^2)$, and then, for $j > 1$, recursively define the subtriangle

$$T_{\alpha_1\alpha_2\cdots\alpha_j} = w_{\alpha_1}(T_{\alpha_2\alpha_3\cdots\alpha_j}) = w_{\alpha_1} \circ w_{\alpha_2} \circ \cdots \circ w_{\alpha_j}(\Delta^2).$$

With this notation, the quotient map (address map) ϕ is given by $\phi(\alpha) = x$ where $\{x\} = \cap_{j=1}^{\infty} T_{\alpha_1\alpha_2\cdots\alpha_j}$.

Sequences in $\{0, 1, 2\}$. For $i \in \{0, 1, 2\}$, consider $\mathbf{y} \in T_i$. Then there exists a $\mathbf{z} \in \Delta^2$ such that $\mathbf{y} = w_i(\mathbf{z})$. Represent the barycentric coordinates $z_k = .z_{k1}z_{k2}\cdots$ of $\mathbf{z}$ in binary, i.e., each z_{kj} is a binary digit. Then since multiplication by $1/2$ is a "right-shift of these digits," $y_k = .\delta_{ki}z_{k1}z_{k2}\cdots$ where $\delta_{ki} = 1$ when $k = i$ and zero otherwise.

Applying this observation to sequences α in $\{0, 1, 2\}$, we see that $\mathbf{x} \in T_{\alpha_1}$ implies that there exists $\mathbf{y} \in \Delta^2$ such that $\mathbf{x} = w_{\alpha_1}(\mathbf{y})$. Thus,

$$(1) \qquad x_k = .\delta_{k\alpha_1}y_{k1}y_{k2}\cdots$$

where the second binary digit x_{k2} in this expansion is the first digit y_{k1} in that of y_k. Moreover, since $\mathbf{x} \in T_{\alpha_1\alpha_2} = w_{\alpha_1}(w_{\alpha_2}(\Delta^2))$, there also exists a $\mathbf{z} \in \Delta^2$ such that $\mathbf{y} = w_{\alpha_2}(\mathbf{z})$ and $\mathbf{x} = w_{\alpha_1}(\mathbf{y})$. It follows that $y_k = .\delta_{k\alpha_2}z_{k1}z_{k2}\cdots$, yielding $x_k = .\delta_{k\alpha_1}\delta_{k\alpha_2}z_{k1}z_{k2}\cdots$. In general, if $\mathbf{x} \in \cap_{j=1}^{\infty}T_{\alpha_1\cdots\alpha_j}$, then $x_k = .\delta_{k\alpha_1}\delta_{k\alpha_2}\delta_{k\alpha_3}\cdots$ where $k \in \{0, 1, 2\}$. This line of reasoning yields the following proposition (see Milutinović [1992, Corollary 8]).

71.1 Proposition *Let* $\alpha_1\alpha_2 \cdots \in N(\{0, 1, 2\})$. *Let* $\{\mathbf{x}\} = \cap_{j=1}^{\infty}T_{\alpha_1\alpha_2\cdots\alpha_j}$, *and, for each* $k \in \{0, 1, 2\}$ *and each* $j = 1, 2, \ldots$, *let* $x_{kj} = \delta_{k\alpha_j}$. *Then the barycentric coordinates* x_k *of* $\mathbf{x}$ *are given by* $x_k = \Sigma_{j=1}^{\infty}x_{kj}/2^j$.

Sequences in $\{0, 1, 2, 3\}$. Let α be a sequence in $\{0, 1, 2, 3\}$. In particular, it is illustrative to consider the constant sequence $333\cdots$. Then $\mathbf{x} \in T_3$ implies $\mathbf{x} = w_3(\mathbf{y})$ for some $\mathbf{y} = [y_0, y_1, y_2]^T \in \Delta^2$. So for $\{k, \ell, m\} = \{0, 1, 2\}$ and $y_k' = 1 - y_k$, we may use $w_3(\mathbf{x}) = \frac{1}{2}(\mathbf{v} - \mathbf{x})$ for $v = [1, 1, 1]^T$ to deduce that

$$x_k = (1/2)(y_k').$$

Thus, w_3 maps $y_k = .y_{k1}y_{k2}\cdots$ to

(2) $$x_k = .0y_{k1}'y_{k2}'\cdots$$

where each binary digit $y_{kj}' = 1 - y_{kj}$. Continuing, since $\mathbf{x} \in T_{33}$ there is also a $\mathbf{z} \in \Delta^2$ such that $w_3(\mathbf{z}) = \mathbf{y}$, showing that $\mathbf{x} = 1/4(\mathbf{v} + \mathbf{z})$. Thus $z_k = .z_{k1}z_{k2}\cdots$ maps to

(3) $$x_k = .01z_{k1}z_{k2}\cdots.$$

And $\mathbf{x} \in T_{333}$ provides a $\mathbf{t} \in \Delta^2$ such that $w_3(\mathbf{t}) = \mathbf{z}$; and then $t_k = .t_{k1}t_{k2}\cdots$ maps to

(4) $$x_k = .010t_{k1}'t_{k2}'\cdots.$$

The upshot? Equations (2), (3), and (4) expose the alternating pattern of "primed" and "non-primed." That is, for a composition involving an odd number of w_3s, the right-shift is followed by an application of "primes," while an even number involves only the right-shift, i.e., $(y_{km}')' = y_{km}$.

Moreover, when α is the constant sequence of 3s, the containment $\mathbf{x} \in \cap_{j=1}^\infty w_{\alpha_1} \circ \cdots w_{\alpha_j}(\Delta^2)$ yields each $x_k = .010101\overline{01} = \Sigma_{j=1}^\infty 1/4^j = 1/3$: The quotient map maps the constant sequence $333\ldots$ to the barycenter of the original triangle.

It follows that any algorithm for calculating the jth binary digit x_{kj} of the coordinate x_k must account for the parity of the number of w_3s appearing in $w_{\alpha_1} \circ \cdots \circ w_{\alpha_j}(\Delta^2)$.

The parity of the w_3s also appears in the basic geometry: While each subtriangle T_0, T_1, T_2, and T_3 in Figure 70.1 has a horizontal edge and corresponding opposite vertex, only T_3 has the corresponding vertex positioned *below* its horizontal edge. (The other triangles "point up.") However, for two "3s", the subtriangle T_{33} points up. Indeed, a subtriangle $T_{\alpha_1\alpha_2\cdots\alpha_j}$ *points up* if and only if the number of α_n (in the list $\alpha_1, \alpha_2, \ldots, \alpha_j$) equaling 3 is an even integer. In particular, since "zero" is an even integer, all subtriangles indexed via sequences in $\{0, 1, 2\}$ "point up," allowing us, in the Sierpiński gasket case, to dodge the need for "primes."

§72 The x_{kj}-Algorithm

Relative to a sequence α in $\{0, 1, 2, 3\}$, we shall say that the subscript j (of α_j) is *up* whenever the number of α_n, $1 \le n \le j$, satisfying $\alpha_n = 3$ is even. Otherwise, j is *down*. Thus, "j is up" if and only if $T_{\alpha_1\cdots\alpha_j}$ "points up."

In addition, we introduce the "x_{kj}-algorithm," which allows us to view the address map as the composition $\phi = \eta \circ \vartheta$: The map $\vartheta : N(0, 1, 2, 3) \rightarrow N(0, 1)^3$ maps α to the ordered triple $[x_{01}x_{02} \cdots, x_{11}x_{12} \cdots, x_{21}x_{22} \cdots]^T \in N(0, 1)^3$ calculated via the algorithm, while $\eta : N(0, 1)^3 \rightarrow \Delta^2$ maps each such triple to the point $[.x_{01}x_{02} \cdots, .x_{11}x_{12} \cdots, .x_{21}x_{22} \cdots]^T \in \Delta^2$ whose components have the indicated binary expansions.

72.1 Proposition (x_{kj}-algorithm) *Let α be a sequence in $\{0, 1, 2, 3\}$. Let $\{\mathbf{x}\} = \cap_{j=1}^{\infty} T_{\alpha_1 \alpha_2 \cdots \alpha_j}$, and, for each $k \in \{0, 1, 2\}$ and each $j = 1, 2, \ldots,$ let*

$$x_{kj} = \begin{cases} \delta_{k\alpha_j} & j \text{ is up and } \alpha_j \neq 3 \\ \delta'_{k\alpha_j} & j \text{ is down and } \alpha_j \neq 3 \\ 1 & j \text{ is up and } \alpha_j = 3 \\ 0 & j \text{ is down and } \alpha_j = 3 \end{cases}$$

Then the barycentric coordinates x_k of $\mathbf{x}$ are given by $x_k = \Sigma_{j=1}^{\infty} x_{kj}/2^j$.

PROOF. We begin with an induction argument that shows whenever $\mathbf{x} = w_{\alpha_1} \circ \cdots \circ w_{\alpha_j}(\mathbf{y})$, then the first j values $x_{k1}, \ldots, x_{kj}$ output by the x_{kj}-algorithm are the first j digits in a binary expansion of x_k, i.e.,

$$x_k = \begin{cases} .x_{k1} \cdots x_{kj} y_{k1} y_{k2} \cdots & j \text{ is up; } y_k = y_{k1} y_{k2} \cdots \\ .x_{k1} \cdots x_{kj} y'_{k1} y'_{k2} \cdots & j \text{ is down; } y_k = y_{k1} y_{k2} \cdots. \end{cases}$$

So we begin with $j = 1$. Let $\mathbf{x} = w_{\alpha_1}(\mathbf{y})$ for some $\mathbf{y} \in \Delta^2$. There are two cases according to $\alpha_1 = 3$ or $\alpha_1 \neq 3$: In the latter case, "j is up and $\alpha_j \neq 3$," so the x_{kj}-algorithm output is $x_{k1} = \delta_{k\alpha_1}$, which is the first digit x_{k1} in the binary expansion $.\delta_{k\alpha_1} y_{k1} y_{k2} \cdots$ of x_k. (see equation (1)). In the former case, "j is down and $\alpha_j = 3$," so the algorithm output is $x_{k1} = 0$, which is the first digit in the binary expansion $.0y'_{k1} y'_{k2} \cdots$ of x_k (see equation (2)). Thus, the first step in the induction is complete.

Now let $j > 1$. Let $\mathbf{x} = w_{\alpha_1} \circ \cdots w_{\alpha_j}(\mathbf{z})$, i.e.,

$$\mathbf{x} = w_{\alpha_1} \circ \cdots \circ w_{\alpha_{j-1}}(\mathbf{y}) \qquad \text{and} \qquad \mathbf{y} = w_{\alpha_j}(\mathbf{z}).$$

Then by the inductive hypothesis,

$$y_k = \begin{cases} .\delta_{k\alpha_j} z_{k1} z_{k2} \cdots & \alpha_j \neq 3 \\ .0z'_{k1} z'_{k2} \cdots & \alpha_j = 3. \end{cases}$$

The inductive hypothesis also tells us that the first $j - 1$ values $x_{k1}, \ldots, x_{k(j-1)}$ output by the x_{kj}-algorithm are also the first $(j - 1)$ digits in the binary expansion

$$x_k = \begin{cases} .x_{k1} \cdots x_{k(j-1)} y_{k1} y_{k2} \cdots & j - 1 \text{ is up} \\ .x_{k1} \cdots x_{k(j-1)} y'_{k1} y'_{k2} \cdots & j - 1 \text{ is down.} \end{cases}$$

Substituting into these two possible expansions of x_k the two possible expansions for y_k given above, we obtain

$$
x_k = \begin{cases}
.x_{k1} \cdots x_{k(j-1)} \delta_{k\alpha_j} z_{k1} z_{k2} \cdots & j-1 \text{ is up and } \alpha_j \neq 3 \\
.x_{k1} \cdots x_{k(j-1)} \delta'_{k\alpha_j} z'_{k1} z'_{k2} \cdots & j-1 \text{ is down and } \alpha_j \neq 3 \\
.x_{k1} \cdots x_{k(j-1)} 0 z'_{k1} z'_{k2} \cdots & j-1 \text{ is up and } \alpha_j = 3 \\
.x_{k1} \cdots x_{k(j-1)} 0' z_{k1} z_{k2} \cdots & j-1 \text{ is down and } \alpha_j = 3.
\end{cases}
$$

So there are four possible cases for expansions of x_k, depending on whether α_j equals or does not equal 3 and whether "$j-1$" is "up" or "down." In each case, we show that the first j algorithm output-values are the first j digits in the corresponding expansion: Recall that the induction hypothesis ensures that the first $j-1$ digits are the algorithm output-values, so we only need to show that the algorithm output for x_{kj} is the jth digit in each of the four cases: First, suppose "$j-1$ is up and $\alpha_j \neq 3$." Then $\alpha_j \neq 3$ specifies that $y_{k1} = \delta_{k\alpha_j}$. Also, "$j-1$ is up" and "$\alpha_j \neq 3$" imply that "j is up". Thus, "j is up and $\alpha_j \neq 3$," so the algorithm output is $x_{kj} = \delta_{k\alpha_j} = y_{k1}$, which agrees with the first expansion listed above. Second, suppose "$j-1$ is down and $\alpha_j \neq 3$." Then again $y_{k1} = \delta_{k\alpha_j}$. Also, "$j-1$ is down" and "$\alpha_j \neq 3$" imply that "j is down." Thus, "j is down and $\alpha_j \neq 3$," so the algorithm output is $x_{kj} = \delta'_{k\alpha_j} = y'_{k1}$, which agrees with the second expansion listed above. Third, suppose "$j-1$ is up and $\alpha_j = 3$." Then $y_{k1} = 0$. Also, "$j-1$ is up" and "$\alpha_j = 3$" imply that "j is down." Thus, "j is down and $\alpha_j = 3$," so the algorithm output is $x_{kj} = 0 = y_{k1}$, which agrees with the third expansion listed above. Fourth and finally, suppose "$j-1$ is down and $\alpha_j = 3$." Then again $y_{k1} = 0$. Also, "$j-1$ is down" and "$\alpha_j = 3$" imply that "j is up." Thus, "j is up and $\alpha_j = 3$," so the algorithm output is $x_{kj} = 1 = 0' = y'_{k1}$, which agrees with the fourth and final expansion listed above. This finishes the induction step. It follows that there is a *constant* sequence of binary expansions of x_k whose jth term has its first j digits x_{k1}, $\ldots$, x_{kj} calculated via the x_{kj}-algorithm. Since this constant sequence clearly converges to $.x_{k1} x_{k2} \cdots$ where *all* digits are calculated via the x_{kj}-algorithm, the proof is complete.   □

With Proposition 72.1 and $\phi(\alpha) = \mathbf{x} = [x_0, x_1, x_2]^T$, we may calculate each x_k using α as input to the x_{kj}-algorithm. For example, consider the sequence $\alpha = 2, 1, 0, 0, 0, \cdots$. Then $\phi(\alpha) = [1/4, 1/4, 1/2]^T$, and the x_{kj}-algorithm with input α has output $x_0 = .0011 \cdots$, $x_1 = .0100 \cdots$, and $x_2 = .1000 \cdots$. But as we shall see, as an application of Proposition 74.1(v), there is no input sequence that will allow the x_{kj}-algorithm to output the given equivalent binary expansions $x_0 = x_1 = .0100 \cdots$ and $x_2 = .1000 \cdots$. It follows that even though the quotient map ϕ and the x_{kj}-algorithm have the same domain $N(0, 1, 2, 3)$ and they are equal when viewed as functions onto Δ^2, the x_{kj}-algorithm cannot produce all "binary representations" of all points in Δ^2.

§73 Binary Representations

To understand the limitations of the x_{kj}-algorithm, we present several propositions, the first of which may be deduced from the observation that the equality $\Sigma_{j=i}^{\infty} 1/2^j = 1/2^{i-1}$ yields the equality $.x_{k1} \cdots x_{k(i-2)} 0111 \cdots = .x_{k1} \cdots x_{k(i-2)} 1000 \cdots .$

73.1 Proposition   *Let $x_k \in [0,1]$ have a binary expansion $x_k = .x_{k1} x_{k2} \cdots$. That is, $x_k = \Sigma_{j=1}^{\infty} x_{kj}/2^j$ where $x_{k1} x_{k2} \cdots$ is a sequence in $\{0,1\}$. Then x_k has another binary expansion $.y_{k1} y_{k2} \cdots$ if and only if the sequences $x_{k1} x_{k2} \cdots$ and $y_{k1} y_{k2} \cdots$ are adjacent endpoints in $N(0,1)$ if and only if $x_{k1} x_{k2} \cdots$ has a tail index.*

Thus, since each $x_k \in [0,1]$ has at most two binary expansions, there exist at most $8 = 2^3$ distinct representations $[x_{01} x_{02} \cdots, x_{11} x_{12} \cdots, x_{21} x_{22} \cdots]^T \in N(0,1)^3$ whose components are *strings* of binary digits (in contrast to the corresponding triple $\mathbf{x} = [x_0, x_1, x_2]^T \in \Delta^2$ whose components are the *values* of the corresponding binary expansions). We shall refer to any such representation as a *binary representation of* $\mathbf{x} \in \Delta^2$, and say that for a given sequence $\alpha \in N(0,1,2,3)$, the x_{kj}-algorithm produces a binary representation of $\phi(\alpha) = \mathbf{x}$. To count binary representations of a given point $\mathbf{x} \in \Delta^2$, we may use the previous proposition and a simple counting argument.

73.2 Proposition   *Let $\mathbf{x} \in \Delta^2$, and let χ denote the number of barycentric coordinates x_k that have a binary expansion whose sequence of digits has a tail index. Then the number of binary representations of $\mathbf{x}$ is 2^{χ}.*

Our next proposition shows that the x_{kj}-algorithm maps $N(0,1,2,3)$ one-to-one onto the binary representations that it can produce.

73.3 Proposition (ϑ is an injection)   *Let $\phi : N(0,1,2,3) \to \Delta^2$ be the quotient map, and let $\phi(\alpha) = \phi(\beta) = \mathbf{x} = [x_0, x_1, x_2]^T$. If the x_{kj}-algorithm applied to α and β gives the same binary representation of $\mathbf{x}$, then $\alpha = \beta$.*

PROOF. For the input sequence α, we let $.x_{k1}^{\alpha} x_{k2}^{\alpha} \cdots$ denote the output. Then the jth digit of the output is "x_{kj}^{α}." Likewise, "x_{kj}^{β}" has the obvious meaning. Note that among the components of the ordered triple $(x_{0j}^{\alpha}, x_{1j}^{\alpha}, x_{2j}^{\alpha})$ there is exactly "one 1" in the case "j is up and $\alpha_j \neq 3$," or exactly "two 1s" in the case "j is down and $\alpha_j \neq 3$," or exactly "three 1s" in the case "j is up and $\alpha_j = 3$," or exactly "zero 1s" in the case "j is down and $\alpha_j = 3$." Likewise, we may consider the corresponding ordered triple produced with input β. It follows, since both inputs α and β produce the same binary representation, that these ordered triples have the "same number of 1s," and consequently, by exhaustive analysis, that $\alpha_j = \beta_j$. This finishes the proof.   □

It follows that Propositions 73.1, 73.2, and 73.3 yield the following.

73.4 Corollary   *Let* $\mathbf{x} = [x_0, x_1, x_2]^T$ *be such that each coordinate* x_k *has a binary expansion whose sequence of digits in* $\{0, 1\}$ *has no tail index. Then for some input sequence* α, *the* x_{kj}-*algorithm yields the unique binary representation of* $\mathbf{x}$, *and* $\{\alpha\} = \phi^{-1}(\mathbf{x})$.

§74 Associated Matrices

Let α and β be distinct members of $\phi^{-1}(\mathbf{x})$. Then the x_{kj}-algorithm, with input α and then input β, will output distinct binary representations of $\mathbf{x}$ (Proposition 73.3). It follows that $\phi^{-1}(\mathbf{x})$ can contain at most the number of distinct binary representations of $\mathbf{x}$ that the x_{kj}-algorithm can produce, which, according to Proposition 73.2, is at most 8. To count the exact numbers of representations for various points $\mathbf{x} \in \Delta^2$, however, we shall use certain matrices that will help us understand the algorithm:

Let α be an input sequence to the x_{kj}-algorithm, which then yields $x_0 = .x_{01}x_{02} \cdots$ and $x_1 = .x_{11}x_{12} \cdots$ and $x_2 = .x_{21}x_{22} \cdots$. A matrix M_α *is associated with* α if

$$M_\alpha = \begin{bmatrix} x_{\ell 1} & x_{\ell 2} & \cdots \\ x_{m1} & x_{m2} & \cdots \\ x_{n1} & x_{n2} & \cdots \end{bmatrix}$$

where $\{\ell, m, n\} = \{0, 1, 2\}$. (The ordering of the rows is not important.) It is the properties of the columns of these matrices that allow us to better understand the x_{kj}-algorithm, and subsequently the quotient map itself.

Observe that while there is no restriction on a particular column, i.e., a column may contain no "1s", one "1", two "1s", or three "1s", the x_{kj}-algorithm places constraints on which columns can be adjacent. The proof of the following proposition is straightforward and therefore omitted.

74.1 Proposition   *Let* α *be input for the* x_{kj}-*algorithm, and let* M_α *be associated with* α. *Then the columns of* M_α *satisfy the following:*

   (i) When α *is a sequence in* $\{0, 1, 2\}$, *each column contains exactly one* 1.
   (ii) When r *is the smallest index such that* $\alpha_r = 3$, *each column preceding the* rth *column contains exactly one* 1.
   (iii) When $\alpha_r = 3$ *and the smallest index* $s > r$ *such that* $\alpha_s = 3$ *also satisfies* $s > r + 1$, *then each column between the* rth *and* s *columns contains exactly two* 1s *when* r *is down, and exactly one* 1 *when* r *is up.*
   (iv) When $\alpha_r = 3$, *then the* rth *column contains only zeros when* r *is down and contains only ones when* r *is up.*
   (v) Columns with exactly one 1 *cannot be adjacent to columns with exactly two* 1s, *i.e., these two distinct kinds of columns are separated by columns of all zeros or columns of all ones.*
   (vi) Two columns whose entries are all zeros cannot be adjacent, and two columns whose entries are all ones cannot be adjacent.

With the aid of Proposition 74.1 and the "matched sequences" introduced in the following section, we shall be in a position to understand the point-inverse sets $\phi^{-1}(\mathbf{x})$.

In passing, note that Proposition 74.1(i) tells us that for each α in the code space $N(0, 1, 2)$ of Sierpiński's gasket, the associated matrix M_α has exactly one 1 in each column.

§75 Matched Sequences

For a given doubleton subset K of $\{0, 1, 2\}$, we use "a_i" and "b_i", for each $i = 1, 2, \ldots$, to denote the elements of K with the constraint that $\{a_i, b_i\} = K$. In addition, for any such i, we use "c_i" to denote the lone element in $\{0, 1, 2\} - K$, and we use "d_i" to denote any member of $\{0, 1, 2, 3\}$. The following definitions concern sequences in $\{0, 1, 2, 3\}$:

Singleton sequences: Constant sequences, sequences with the constant subsequence $33 \cdots$, sequences with three constant subsequences, or sequences in a doubleton set $K \subset \{0, 1, 2\}$ with no tail index.

Matched doubleton sequences: Two sequences α, β with no tail index, but with an index $r \geq 1$ and a doubleton subset K of $\{0, 1, 2\}$ such that

$$
\begin{aligned}
\alpha &= d_1 \cdots d_{r-1} c_r a_{r+1} a_{r+2} \cdots \\
\beta &= d_1 \cdots d_{r-1} 3\, b_{r+1} b_{r+2} \cdots.
\end{aligned}
$$

Matched tripleton sequences: Three sequences α, β, γ with a common tail index $t \geq 1$, where a doubleton subset K of $\{0, 1, 2\}$ exists such that

$$
\begin{aligned}
\alpha &= a_1 a_2 \cdots a_{t-1} a_t \overline{a_{t+1}} \\
\beta &= a_1 a_2 \cdots a_{t-1} a_{t+1} \overline{a_t} \\
\gamma &= a_1 a_2 \cdots a_{t-1} 3 \overline{c_{t+1}}.
\end{aligned}
$$

Matched hexeton sequences: Six sequences $\alpha, \beta, \gamma, \delta, \epsilon, \zeta$ with a common tail index $t > r \geq 1$ where a doubleton subset K of $\{0, 1, 2\}$ exists such that

$$
\begin{aligned}
\alpha &= d_1 \cdots d_{r-1} c_r a_{r+1} \cdots a_{t-1} a_t \overline{a_{t+1}} \\
\beta &= d_1 \cdots d_{r-1} c_r a_{r+1} \cdots a_{t-1} a_{t+1} \overline{a_t} \\
\gamma &= d_1 \cdots d_{r-1} c_r a_{r+1} \cdots a_{t-1} 3 \overline{c_{t+1}} \\
\delta &= d_1 \cdots d_{r-1} 3\, b_{r+1} \cdots b_{t-1} a_t \overline{a_{t+1}} \\
\epsilon &= d_1 \cdots d_{r-1} 3\, b_{r+1} \cdots b_{t-1} a_{t+1} \overline{a_t} \\
\zeta &= d_1 \cdots d_{r-1} 3\, b_{r+1} \cdots b_{t-1} 3 \overline{c_{t+1}}.
\end{aligned}
$$

(To avoid confusion, note that the form as presented is clear when $t > r + 1$, but for the lone case when $t = r + 1$, we need to remove each instance of "$a_{r+1} \cdots a_{t-1}$" and each instance of "$b_{r+1} \cdots b_{t-1}$". For example, when $t = r + 1$, then α is simply $d_1 \cdots d_{r-1} c_r a_t \overline{a_{t+1}}$.)

To illustrate and motivate these classes of sequences, first consider rational points in the 2-web ω^2. In this case, matched tripleton sequences address those on the boundary $\partial \Delta^2$ of Δ^2 (Figure 75.1):

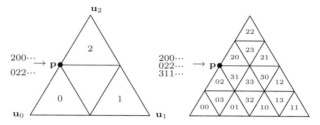

Fig. 75.1 Matched tripleton sequences address a rational $\mathbf{p}$ in $\omega^2 \cap \partial \Delta^2$.

And as illustrated in Figure 75.2, matched hexeton sequences address points in $\omega^2 \cap (\Delta^2 - \partial \Delta^2)$:

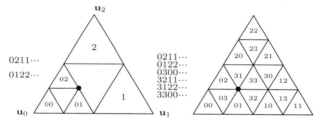

Fig. 75.2 Matched hexeton sequences — a rational in $\omega^2 \cap (\Delta^2 - \partial \Delta^2)$.

Second, consider the irrational points in ω^2. Such a point p on the edge $[\mathbf{u}_0, \mathbf{u}_2]$ of $\partial \Delta^2$ can lie on neither an edge nor a vertex of any triangle $T_{\beta_1 \beta_2 \cdots \beta_j}$ where any $\beta_i = 3$. Intuitively, it follows that the lone address of p relative to ω^2 is also the lone address of p relative to Δ^2. (So in this case, a (singleton) sequence in a doubleton set $K = \{0, 2\} \subset \{0, 1, 2\}$ with no tail index is the address of p.) But an irrational point p on say the edge common to triangles T_0 and T_3 will have a matched doubleton sequence of addresses where $r = 1$ and $K = \{1, 2\}$.

§76 Point-Inverse Sets

Using mostly Proposition 74.1 and matched sequences, we prove the following propositions:

76.1 Proposition *A singleton sequence α has the property that $\phi^{-1}\phi(\alpha) = \alpha$.*

PROOF. First, consider $\alpha = kk \cdots$ where $k \neq 3$. Then Proposition 74.1(i) shows that M_α contains one row of ones and two rows of zeros. Each induced sequence of binary digits therefore has no tail index, and Corollary 73.4 then yields the desired result. Second, if α contains the subsequence $333\cdots$, then Proposition 74.1(iv) shows that the columns of M_α containing all zeros are infinite in number, as are those containing all ones. Again, apply Corollary 73.4. Third, suppose that the sequence α does not contain the constant subsequence $33\cdots$, but it does contain the constant subsequences $kk\cdots$ for each $k \neq 3$. Then eventually all indices are up or all indices are down. In either case, each sequence corresponding to a row of M_α has no tail index. Again, apply Corollary 73.4. Fourth and finally, suppose $\alpha = a_1 a_2 \cdots$ is a two-valued sequence in $\{0,1,2\}$ with no tail index. Then Proposition 74.1(i) shows that each column of M_α contains exactly one 1. And since α is twovalued, M_α has a row of zeros and the corresponding sequence of digits thus has no tail index. Moreover, since α has no tail index, each of the other two induced sequences has no tail index. And once again and finally, apply Corollary 73.4. □

76.2 Proposition *Let α, β be matched doubleton sequences. Then $\phi(\alpha) = \phi(\beta)$, and if $\phi(\alpha) = \phi(\beta) = \mathbf{x}$, then $\phi^{-1}(\mathbf{x}) = \{\alpha, \beta\}$.*

PROOF. Let $K = \{\ell, m\}$, and let $\{n\} = \{0,1,2\} - K$. Since α and β have the same first $r - 1$ values $d_1, \ldots, d_{r-1}$, the first $(r-1)$ columns of the associated matrices of the matched doubleton sequences define the same $3 \times (r-1)$ submatrix D. Moreover, we assume (for illustrative purposes, i.e., the general case is similar) that $a_{r+1} = \ell$, $a_{r+2} = m$, and $a_{r+3} = \ell$. So first, with this assumption, we suppose that $r - 1$ is up relative to α (note that $r - 1$ is up relative to β). Then r is up relative to α and down relative to β. Thus,

$$M_\alpha = \begin{bmatrix} & 0 & 1 & 0 & 1 & \cdots \\ D & 0 & 0 & 1 & 0 & \cdots \\ & 1 & 0 & 0 & 0 & 00\cdots \end{bmatrix} \quad \text{and} \quad M_\beta = \begin{bmatrix} & 0 & 1 & 0 & 1 & \cdots \\ D & 0 & 0 & 1 & 0 & \cdots \\ & 0 & 1 & 1 & 1 & 11\cdots \end{bmatrix}.$$

From inspection of the binary expansions induced by these matrices it is clear that ϕ maps the matched doubleton sequence to the same point, say $\mathbf{x}$. Moreover, we see that rows of M_α that correspond to ℓ and m have no tail index and only the row corresponding to n represents a binary sequence with a tail index r. It follows (Proposition 73.2) that $\mathbf{x}$ has only two binary representations. Thus, Proposition 73.3 shows that $\phi^{-1}(\mathbf{x})$ has size two and the desired result follows. So second, under the same assumption ($a_{r+1} = \ell$, $a_{r+2} = m$, and $a_{r+3} = \ell$), we now suppose that $r - 1$ is down relative to both α and β: Then r is down relative to α and up relative to β. Thus,

$$M_\alpha = \begin{bmatrix} & 1 & 0 & 1 & 0 & \cdots \\ D & 1 & 1 & 0 & 1 & \cdots \\ & 0 & 1 & 1 & 1 & 11\cdots \end{bmatrix} \quad \text{and} \quad M_\beta = \begin{bmatrix} & 1 & 0 & 1 & 0 & \cdots \\ D & 1 & 1 & 0 & 1 & \cdots \\ & 1 & 0 & 0 & 0 & 00\cdots \end{bmatrix}.$$

An argument similar to the one used in the first case shows that the desired result is again true. Thus, we are finished. □

76.3 Proposition *Let α, β, γ be matched tripleton sequences. Then $\phi(\alpha) = \phi(\beta) = \phi(\gamma)$, and if $\phi(\alpha) = \phi(\beta) = \phi(\gamma) = \mathbf{x}$, then $\phi^{-1}(\mathbf{x}) = \{\alpha, \beta, \gamma\}$.*

PROOF. Let $K = \{\ell, m\}$. Since α, β, and γ have the same first $t - 1$ values $a_1, \ldots, a_{t-1}$, the first $(t - 1)$ columns of their associated matrices (with the same ordering ℓ, m, n of rows) contain the same $3 \times (t - 1)$ submatrix U (a mnemonic for up) whose entries in the nth row are all zero. Moreover, we assume (for illustrative purposes, i.e., the general case is similar) that $a_t = \ell$. Since neither a_t nor a_{t+1} equals 3, the indices t and $t + 1$ are up relative to both α and β, while both of t and $t + 1$ are down relative to γ. Thus,

$$M_\alpha = \begin{bmatrix} & 1 & 0 & 0 & \cdots \\ U & 0 & 1 & 1 & \cdots \\ & 0 & 0 & 0 & \cdots \end{bmatrix}, \; M_\beta = \begin{bmatrix} & 0 & 1 & 1 & \cdots \\ U & 1 & 0 & 0 & \cdots \\ & 0 & 0 & 0 & \cdots \end{bmatrix}, \; M_\gamma = \begin{bmatrix} & 0 & 1 & 1 & \cdots \\ U & 0 & 1 & 1 & \cdots \\ & 0 & 0 & 0 & \cdots \end{bmatrix}.$$

From the binary expansions induced by these matrices, we see that ϕ is constant on the matched tripleton sequences. And if we let $\mathbf{x} = \phi(\alpha) = \phi(\beta) = \phi(\gamma)$, then since exactly two rows of any of these associated matrices induce sequences with tail indices, we see (Proposition 73.2) that the image point $\mathbf{x}$ has $4 = 2^2$ binary representations. Three of the four representations are given by the associated matrices. The fourth representation (matrix) is

$$\begin{bmatrix} & 1 & 0 & 0 & \cdots \\ U & 1 & 0 & 0 & \cdots \\ & 0 & 0 & 0 & \cdots \end{bmatrix}.$$

But Proposition 74.1(vi) shows that the x_{kj}-algorithm cannot produce this matrix. It follows that the size of $\phi^{-1}(\mathbf{x})$ is three, and this finishes the proof. $\square$

We shall need the following matrices in the proof of the next proposition.

$$M_\alpha = \begin{bmatrix} & 0 & 1 & 0 & 1 & 1 & 0 & 0 \cdots \\ D & 0 & 0 & 1 & 0 & \mathbf{0} & 1 & 1 \cdots \\ & 1 & 0 & 0 & 0 & \mathbf{0} & 0 & 0 \cdots \end{bmatrix}, M_\delta = \begin{bmatrix} & 0 & 1 & 0 & 1 & 1 & 0 & 0 \cdots \\ D & 0 & 0 & 1 & 0 & \mathbf{0} & 1 & 1 \cdots \\ & 0 & 1 & 1 & 1 & \mathbf{1} & 1 & 1 \cdots \end{bmatrix}$$

$$M_\beta = \begin{bmatrix} & 0 & 1 & 0 & 1 & \mathbf{0} & 1 & 1 \cdots \\ D & 0 & 0 & 1 & 0 & \mathbf{1} & 0 & 0 \cdots \\ & 1 & 0 & 0 & 0 & \mathbf{0} & 0 & 0 \cdots \end{bmatrix}, M_\epsilon = \begin{bmatrix} & 0 & 1 & 0 & 1 & \mathbf{0} & 1 & 1 \cdots \\ D & 0 & 0 & 1 & 0 & \mathbf{1} & 0 & 0 \cdots \\ & 0 & 1 & 1 & 1 & \mathbf{1} & 1 & 1 \cdots \end{bmatrix}$$

$$M_\gamma = \begin{bmatrix} & 0 & 1 & 0 & 1 & \mathbf{0} & 1 & 1 \cdots \\ D & 0 & 0 & 1 & 0 & \mathbf{0} & 1 & 1 \cdots \\ & 1 & 0 & 0 & 0 & \mathbf{0} & 0 & 0 \cdots \end{bmatrix}, M_\zeta = \begin{bmatrix} & 0 & 1 & 0 & 1 & 1 & 0 & 0 \cdots \\ D & 0 & 0 & 1 & 0 & \mathbf{1} & 0 & 0 \cdots \\ & 0 & 1 & 1 & 1 & \mathbf{1} & 1 & 1 \cdots \end{bmatrix}.$$

76.4 Proposition *Let α, β, γ, δ, ϵ, ζ be matched hexeton sequences. Then $\phi(\alpha) = \phi(\beta) = \phi(\gamma) = \phi(\delta) = \phi(\epsilon) = \phi(\zeta)$, and if $\phi(\alpha) = \phi(\beta) = \phi(\gamma) = \phi(\delta) = \phi(\epsilon) = \phi(\zeta) = \mathbf{x}$, then $\phi^{-1}(\mathbf{x}) = \{\alpha, \beta, \gamma, \delta, \epsilon, \zeta\}$.*

PROOF. Let $K = \{\ell, m\}$. Similar to the "doubleton sequence case" above, each of the six associated matrices $M_\alpha, \ldots, M_\zeta$ whose rows are ordered ℓ, m, n contain the same $3 \times (r-1)$ submatrix D. We assume (for illustrative purposes, i.e., the general case is similar) that $a_{r+1} = \ell$, $a_{r+2} = m$, $a_{r+3} = a_{t-1} = \ell$, and $a_t = \ell$. So first suppose that $r - 1$ is up relative to α. Then, using a boldface font to indicate the tth column, we have the six matrices M_α, M_β, M_γ, M_δ, M_ϵ, and M_ζ whose entries were detailed prior to the statement of the theorem.

From the binary expansions given by these matrices, we see that ϕ is constant on the matched hexeton sequences. And if we let $\mathbf{x} = \phi(\alpha) = \phi(\beta) = \phi(\gamma) = \phi(\delta) = \phi(\epsilon) = \phi(\zeta)$, then since all three rows of any of these associated matrices induce sequences with tail indices, we see (Proposition 73.2) that the image point $\mathbf{x}$ has $8 = 2^3$ binary representations, while the associated matrices account for only six of the eight. Indeed, the two remaining representations (matrices) are

$$\begin{bmatrix} 0 & 1 & 0 & 1 & \mathbf{1} & 0 & 0 \cdots \\ D\ 0 & 0 & 1 & 0 & \mathbf{1} & 0 & 0 \cdots \\ 1 & 0 & 0 & 0 & \mathbf{0} & 0 & 0 \cdots \end{bmatrix} \quad \text{and} \quad \begin{bmatrix} 0 & 1 & 0 & 1 & \mathbf{0} & 1 & 1 \cdots \\ D\ 0 & 0 & 1 & 0 & \mathbf{0} & 1 & 1 \cdots \\ 0 & 1 & 1 & 1 & \mathbf{1} & 1 & 1 \cdots \end{bmatrix}.$$

But Proposition 74.1(vi) shows that the x_{kj}-algorithm cannot produce these representations. If follows that the size of $\phi^{-1}(\mathbf{x})$ is six, and thus we are finished with the first case. So second, suppose that $r - 1$ is down relative to α. In this case we may calculate the associated matrices by simply changing each "1" to "0" and each "0" to "1" in the formula for the six matrices listed above (in the "up" case). (This property is imbedded in the algorithm, e.g., when $\alpha_j \neq 3$, the $\delta_{k\alpha_j}$ output may be obtained from the output $\delta'_{k\alpha_j}$ by permuting the zeros and ones.) Thus, in the "down case" the two "missing representations" will once again exhibit either adjacent columns of all zeros or adjacent columns of all ones, and such matrices do not lie in the range of the x_{kj}-algorithm. □

§77 The Relation R

The following theorem makes the definition of R obvious.

77.1 Theorem *Let $\phi : N(0, 1, 2, 3) \to \Delta^2$ be the quotient/address map, and let $\mathbf{x} \in \Delta^2$. Then $\phi^{-1}(\mathbf{x})$ is a singleton set containing a singleton sequence, or a doubleton set containing matched doubleton sequences, or a tripleton set containing matched tripleton sequences, or a hexeton set containing matched hexeton sequences.*

PROOF. If $\mathbf{x}$ is the ϕ-image of a singleton sequence, then by Proposition 76.1 we are finished. So suppose $\mathbf{x}$ is not the image of a singleton sequence. Since ϕ is onto Δ^2, there exists a sequence α in $\{0, 1, 2, 3\}$ such that $\phi(\alpha) = \mathbf{x}$. Since α is not a singleton sequence, it is non constant and eventually in a doubleton subset $K \subset \{0,1,2\}$, and whenever *all* values of α are in K, then α has a tail index. Consider the last case first. Then $\alpha = a_1 a_2 \cdots a_t \overline{a_{t+1}}$, showing that α is one among three matched tripleton sequences; and an application of Proposition 76.3 finishes the proof. So we are left with the case where a doubleton subset $K \subset \{0,1,2\}$ and a smallest index r exist such that $\alpha_i \in K$ when $i > r \geq 1$. There are two subcases, according to whether α has or does not have a tail index. If α has a tail index $t > r$, then

$$\alpha = d_1 \cdots d_{r-1} c_r a_{r+1} \cdots a_{t-1} a_t \overline{a_{t+1}} \qquad \text{or}$$
$$\alpha = d_1 \cdots d_{r-1} 3 \, a_{r+1} \cdots a_{t-1} a_t \overline{a_{t+1}}$$

when $t > r + 1$, and

$$\alpha = d_1 \cdots d_{r-1} c_r a_t \overline{a_{t+1}} \qquad \text{or}$$
$$\alpha = d_1 \cdots d_{r-1} 3 \, a_t \overline{a_{t+1}}$$

when $t = r + 1$. In other words, when α has a tail index, then α is one among six matched hexeton sequences; and an application of Proposition 76.4 finishes the proof. So, turning to the last subcase where our α has no tail index, we see that

$$\alpha = d_1 \cdots d_{r-1} c_r a_{r+1} a_{r+2} \cdots \qquad \text{or}$$
$$\alpha = d_1 \cdots d_{r-1} 3 \, a_{r+1} a_{r+2} \cdots .$$

In other words, when α has no tail index, then α is one among two matched doubleton sequences; and an application of Proposition 76.2 finishes the proof. Since all possibilities for the form of α have been exhausted, and in every case the theorem was true, the proof is finished. □

Since the relation R may be defined as the equivalence relation induced by the partition $\{\phi^{-1}(\mathbf{x}) | \mathbf{x} \in \Delta^2\}$, Theorem 77.1 shows that the classes of R contain one, two, three, or six elements.

§78 Representations of 2-Space and the 1-Sphere

The boundary $\partial \Delta^2$ of Δ^2 is closed in Δ^2 and is a homeomorphic copy of the 1-sphere, while $\Delta^2 - \partial \Delta^2$ is open in Δ^2 and is a copy of Euclidean 2-space.

We use these copies of 2-space $\mathbb{R}^2$ and the 1-space S^1 together with the mathematics of the address map $\phi : N(\{0,1,2,3\}) \to \Delta^2$ induced by the IFS $\mathcal{F}_2^*$ to obtain a representation of 2-space and the 1-sphere.

78.1 Theorem *Let ϕ be the address map from $N(0,1,2,3)$ onto the 2-simplex Δ^2. Let F be the ϕ-inverse subspace of $N(0,1,2,3)$ that consists of (1) the constant sequences $\overline{0}$, $\overline{1}$, and $\overline{2}$; (2) the sequences in a doubleton set $K \subset \{0,1,2\}$ with no tail index; and (3) the matched tripleton sequences. Let*

$G = N(0,1,2,3) - F$. Then G is also a ϕ-inverse subspace of $N(0,1,2,3)$. Moreover,

$$\phi|_G : G \rightarrow (\Delta^2 - \partial\Delta^2) \qquad and \qquad \phi|_F : F \rightarrow \partial\Delta^2$$

are quotient maps, the former onto a copy of 2-space, and the latter onto a copy of the 1-sphere.

PROOF. The three $\phi^{-1}(\mathbf{x})$ sets where $\mathbf{x}$ is a vertex of Δ^2 are singleton sets whose members are the singleton sequences defined in (1). The $\phi^{-1}(\mathbf{x})$ sets where $\mathbf{x}$ is an irrational point in (the interior of) one of the edges of Δ^2 are singleton sets whose members are the singleton sequences defined in (2). And the $\phi^{-1}(\mathbf{x})$ sets where $\mathbf{x}$ is a rational point in one of the edges of Δ^2 are tripleton sets whose members are the tripleton sequences defined in (3). It follows that $F = \phi^{-1}(\partial\Delta^2)$ and that $G = \phi^{-1}(\Delta^2 - \partial\Delta^2)$. Since ϕ is continuous, $\phi|_F$ and $\phi|_G$ are continuous. In addition, since $\partial\Delta^2$ is closed in Δ^2 and $(\Delta^2 - \partial\Delta^2)$ is open in Δ^2, it follows that both $\phi|_F$ and $\phi|_G$ are quotient maps (see Dugundji [1966, page 122, Theorem 2.1]).   □

§79 Comments

As for the problem of creating a 3-simplex system that extends the 3-web system, the approach used in this chapter was not obviously extendable to the 3-simplex case. The 3-simplex case, however, was solved by Lipscomb [2007] who used another approach which is the topic of the following chapter. While the solution for the 3-simplex case is intuitive, the number of cardinalities of inverse sets of the address map is significant. The approach that led to the solution for the 3-simplex case does seem to be general enough to at least suggest an approach to the 4-simplex case. The problem appears to be the number of technicalities that one would encounter. That is, if the increase in technicalities that occurred in going from the 2-simplex case to the 3-simplex case is an indication, then it would take a significant effort to track all of the kinds of sequences that one encounters in the 4-simplex case.

Nevertheless, from the very beginning it was the 4-simplex case that was the goal of this author. In particular, it was the 3-sphere S^3 in an "adjacent-endpoint identification" IFS context that served as motivation — the one-dimensional edges of the 4-simplex is a level-1 4-web, i.e., a picture of a level-1 J_5. Moreover, topological studies of S^3 are both extensive and historically significant. For an introduction to such studies and models of S^3 see, for example, Bing [1988] and Wilder [1938].

CHAPTER 14

From 3-Web IFS to 3-Simplex IFS
3-Space and the 2-Sphere

The n-web fractal ω^n — the attractor of the IFS of $n+1$ contractions by $1/2$ toward the vertices of an n-simplex — emerges from a manifold (n-simplex). The classical example is Sierpiński's gasket (2-web) which emerges from a 2-simplex. It is therefore natural (inverse of moving from manifolds to fractals) to seek, for each $n \geq 2$, an extension of the n-web system to an n-simplex system. A solution for $n = 2$ is presented in Chapter 13 where an application yielded a representation of 2-space and the 1-sphere. In this chapter, we provide a solution for $n = 3$, which yields a representation of 3-space and the 2-sphere. The presentation follows Lipscomb [2007].

§80 Overview

The problem of extending the n-web ω^n IFS to an n-simplex Δ^n IFS was introduced in the previous chapter and then solved for the $n = 2$ case. In this chapter we present a solution for $n = 3$.

We continue to use the standard simplex Δ^3 as our model 3-simplex. That is, we use the four unit basis vectors

$$u_0 = (1, 0, 0, 0) \qquad u_1 = (0, 1, 0, 0) \qquad u_2 = (0, 0, 1, 0) \qquad u_3 = (0, 0, 0, 1)$$

in 4-space $\mathbb{R}^4$ as the vertices of Δ^3. For this Δ^3, the barycentric and Cartesian coordinates of any $x \in \Delta^3$ are equal.

We also have $\mathcal{F}_3 = \{w_0, w_1, w_2, w_3\}$ as the ω^3 IFS where for each k, $w_k(x) = 1/2(x+u_k)$ is a contraction by $1/2$ toward u_k. The *attractor* ω^3 of $\mathcal{F}_3$ is called the *3-web* and the *code space* of $\mathcal{F}_3$ is the Baire space $N(\{0, 1, 2, 3\})$.

Our goal here runs parallel to the goal of Chapter 13, namely to *emulate and extend the adjacent-endpoint approach*: Extend the $\mathcal{F}_3$ IFS to an IFS $\mathcal{F}_3^*$ whose address map yields Δ^3, and then recognize the basic forms of the sequences in $N(\{0, 1, 2, 3\})$ that when identified (via the address mapping) yield Δ^3.

The approach in the 2-web case required one additional affine transformation that served to iteratively fill the "holes" in ω^2. The approach in the 3-web case requires four additional transformations that combine to iteratively form octahedra that fill the "holes" in ω^3.

S.L. Lipscomb, *Fractals and Universal Spaces in Dimension Theory*,
DOI 10.1007/978-0-387-85494-6_14, © Springer Science+Business Media, LLC 2009

§81 Decomposing the 3-Simplex

Recall the scheme (complex) of triangles displayed in Figure 70.1 that motivated the 2-simplex extension $\mathcal{F}_2^*$ of $\mathcal{F}_2$. Unlike the 2-simplex, where the "hole" $\Delta^2 \setminus \cup_{k=0}^2 w_k(\Delta^2)$ is an (open) 2-simplex, the "hole" $\Delta^3 \setminus \cup_{k=0}^3 w_k(\Delta^3)$ in the 3-simplex is not an (open) 3-simplex. Rather, it is an octahedron minus its 1-skeleton. Since nonsingular affine transformations map (nondegenerate) tetrahedra to (nondegenerate) tetrahedra, we decompose this octahedron into a union of four tetrahedra.

As indicated in Figure 81.1, u_0, u_1, u_2, and u_3 denote the *vertices* of Δ^3, and m_{ij} denotes the *midpoint* of the edge (1-simplex) $[u_i, u_j]$. The line segment PQ, where $P = 1/2(u_0 + u_1) = m_{01}$ and $Q = 1/2(u_2 + u_3) = m_{23}$, is central to the construction — PQ is pictured as the dashed line interior to the "octahedral hole." With PQ, the hole decomposes into a union of four (closed) 3-simplexes $T_{0'}$, $T_{1'}$, $T_{2'}$, and $T_{3'}$, exposing Δ^3 as the union of eight (closed) simplexes T_q where $q \in A = \{0, 1, 2, 3, 0', 1', 2', 3'\}$:

$T_0 =$	$T_1 =$	$T_2 =$	$T_3 =$
$[u_0, P, m_{02}, m_{03}]$	$[u_1, P, m_{12}, m_{13}]$	$[u_2, Q, m_{02}, m_{12}]$	$[u_3, Q, m_{03}, m_{13}]$
$T_{0'} =$	$T_{1'} =$	$T_{2'} =$	$T_{3'} =$
$[P, Q, m_{12}, m_{13}]$	$[P, Q, m_{02}, m_{03}]$	$[T_{2'} = [P, Q, m_{03}, m_{13}]$	$[P, Q, m_{02}, m_{12}]$.

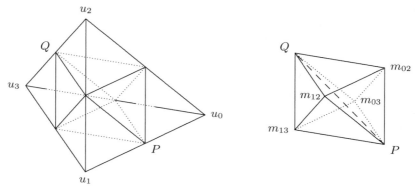

Fig. 81.1 A 3-simplex, octahedral-shaped hole, and segment PQ.

Because of our choice of PQ, the set $S = \{\{0, 1\}, \{2, 3\}\}$ plays a special role. For example, a primed subscript, say i', where $\{i, j\} \in S$, indicates that $T_{i'}$ has faces "opposite" the vertices u_i and u_j. To illustrate, the subscript $0'$ indicates that the "hole $[Q, m_{12}, m_{13}]$ opposite u_0" and "the face $[P, m_{12}, m_{13}]$ of T_1 opposite u_1" are faces of $T_{0'}$. In passing, note that PQ is an edge of each $T_{i'}$.

§82 A 3-Simplex IFS

We extend $\mathcal{F}_3$ to a 3-simplex system $\mathcal{F}_3^*$ according to the decomposition of Δ^3 provided in the previous section. Since we shall express the additional affine transformations as matrices, we shall write our u_k vectors as column vectors, i.e., we let $u_0 = [1,0,0,0]^T$, $u_1 = [0,1,0,0]^T$, $u_2 = [0,0,1,0]^T$, and $u_3 = [0,0,0,1]^T$ denote the vertices of Δ^3. In addition, we consider the barycenters of the 2-faces of Δ^3, namely $u_{0'} = (1/3)[0,1,1,1]^T$, $u_{1'} = (1/3)[1,0,1,1]^T$, $u_{2'} = (1/3)[1,1,0,1]^T$, and $u_{3'} = (1/3)[1,1,1,0]^T$. That is, $u_{k'}$ is the barycenter of the face opposite vertex u_k. We also define

$$
L_0 = \begin{bmatrix} 1 & 0 & 0 & 0 \\ 2/3 & -1/3 & 2/3 & 2/3 \\ -1/3 & 2/3 & -1/3 & 2/3 \\ -1/3 & 2/3 & 2/3 & -1/3 \end{bmatrix}
\qquad
L_1 = \begin{bmatrix} -1/3 & 2/3 & 2/3 & 2/3 \\ 0 & 1 & 0 & 0 \\ 2/3 & -1/3 & -1/3 & 2/3 \\ 2/3 & -1/3 & 2/3 & -1/3 \end{bmatrix}
$$

and

$$
L_2 = \begin{bmatrix} -1/3 & 2/3 & -1/3 & 2/3 \\ 2/3 & -1/3 & -1/3 & 2/3 \\ 0 & 0 & 1 & 0 \\ 2/3 & 2/3 & 2/3 & -1/3 \end{bmatrix}
\qquad
L_3 = \begin{bmatrix} -1/3 & 2/3 & 2/3 & -1/3 \\ 2/3 & -1/3 & 2/3 & -1/3 \\ 2/3 & 2/3 & -1/3 & 2/3 \\ 0 & 0 & 0 & 1 \end{bmatrix}.
$$

Correspondingly, we also define

$$
w_{k'}(x) = (1/2)L_k x + (1/2)u_{k'} \qquad (k = 0,1,2,3),
$$

and then let $\mathcal{F}_3^* = \mathcal{F}_3 \cup \{w_{0'}, w_{1'}, w_{2'}, w_{3'}\}$. Each $w_{k'} \in \mathcal{F}_3^*$ is nonsingular since each L_k has determinant "1", and except for notation, the subsystem $\{w_0, w_1, w_2, w_{3'}\} \subset \mathcal{F}_3^*$ restricted to the face $[u_0, u_1, u_2]$ is the 2-simplex system $\mathcal{F}_2^*$. To be sure that $\mathcal{F}_3^*$ has an attractor, however, we show that its contractivity factor is less than "1".

82.1 Theorem *The contractivity factor of $\mathcal{F}_3^*$ is $\leq \sqrt{(4+\sqrt{7})/12} \approx .744 < 1$.*

PROOF. Let I denote the 4×4 identity matrix, and let $k = 0,1,2,3$. Then each w_k has contractivity factor $1/2$: For $z = x - y$ in Euclidean 4-space and $B = (1/2)I$, we have

$$
|w_k(x) - w_k(y)|^2 = |(1/2)I(z)|^2 = |Bz|^2 = z^T B^T B z = \Sigma_i (1/4) z_i^2 = (1/4)|z|^2.
$$

Turning to the $w_{k'}$, we first consider $w_{0'}$: In this case, let $B = (1/2)L_0$. Then

$$
|w_{0'}(x) - w_{0'}(y)|^2 = |1/2 L_0(z)|^2 = |Bz|^2 = z^T B^T B z
$$

where $M = B^T B$ is a real symmetric matrix. The characteristic equation $\det(M - \lambda I) = 0$ of M is

$$
(1/4 - \lambda)^2 (\lambda^2 - (2/3)\lambda + (1/16)) =
$$
$$
(1/4 - \lambda)^2 (\lambda - (4 + \sqrt{7})/12)(\lambda - (4 - \sqrt{7})/12) = 0.
$$

Since M is real symmetric, there exist matrices

$$P=\begin{bmatrix} 0 & 0 & \left(\frac{7+\sqrt{7}}{14}\right)^{1/2} & -\left(\frac{7-\sqrt{7}}{14}\right)^{1/2} \\ 0 & \frac{1}{\sqrt{3}} & -\left(\frac{7-\sqrt{7}}{21}\right)^{1/2} & -\left(\frac{7+\sqrt{7}}{21}\right)^{1/2} \\ \frac{1}{\sqrt{2}} & \frac{1}{\sqrt{3}} & \left(\frac{7-\sqrt{7}}{84}\right)^{1/2} & \left(\frac{7+\sqrt{7}}{84}\right)^{1/2} \\ -\frac{1}{\sqrt{2}} & \frac{1}{\sqrt{3}} & \left(\frac{7-\sqrt{7}}{84}\right)^{1/2} & \left(\frac{7+\sqrt{7}}{84}\right)^{1/2} \end{bmatrix}; P^{-1}MP=\begin{bmatrix} \frac{1}{4} & 0 & 0 & 0 \\ 0 & \frac{1}{4} & 0 & 0 \\ 0 & 0 & \frac{4+\sqrt{7}}{12} & 0 \\ 0 & 0 & 0 & \frac{4-\sqrt{7}}{12} \end{bmatrix}$$

where the columns of P form an orthonormal basis for 4-space ($P^{-1} = P^T$), making $P^{-1}MP$ similar to M. It follows that

$$|w_{0'}(x) - w_{0'}(y)|^2 \;\; = z^T(B^TB)z \;\; = (P^{-1}z)^T(P^{-1}MP)(P^{-1}z)$$
$$\leq \left((4+\sqrt{7})/12\right)|z|^2.$$

For the remaining $w_{k'}$, let ρ denote an order-2 permutation of $\{1, 2, 3, 4\}$, and let Q_ρ denote the 4×4 permutation matrix obtained from I by interchanging row i with row $\rho(i)$. Then, using cycle notation, it is easily checked that

$$Q_{(12)}L_0Q_{(12)} = L_1, \quad Q_{(13)(24)}L_0Q_{(13)(24)} = L_2, \quad Q_{(14)(23)}L_0Q_{(14)(23)} = L_3.$$

Moreover, for each ρ, we have $Q_\rho^{-1} = Q_\rho = Q_\rho^T$ and $\det Q_\rho = \pm 1$. Thus, for each $k \in \{1, 2, 3\}$,

$$\det\left([(1/2)L_k]^T[(1/2)L_k] - \lambda I\right)$$
$$= \det\left([(1/2)Q_\rho L_0 Q_\rho]^T[(1/2)Q_\rho L_0 Q_\rho] - \lambda I\right)$$
$$= \det\left(Q_\rho(M - \lambda I)Q_\rho\right) = \det(M - \lambda I).$$

So each $w_{k'}$ is contractive with the same contraction factor as $w_{0'}$. □

Since the contractivity factor of $\mathcal{F}_3^*$ is less than 1, we know that it has an attractor. And since the attractor is characterized as the unique compact set K that satisfies $K = \cup_{a \in A} w_q(K)$, it follows from Figure 81.1 that the attractor of $\mathcal{F}_3^*$ is Δ^3.

§83 IFS-Induced Simplicial Complex K_n

Let $v_0, \ldots, v_n$ denote $n + 1$ linearly independent points in some Euclidean space. The *closed simplex* $\sigma = [v_0, \ldots, v_n]$ with vertices $v_0, \ldots, v_n$ is the set of points $x = \Sigma_{i=0}^n x_i v_i$ with each x_i non-negative and $\Sigma_i x_i = 1$. The *open simplex* $(v_0, \ldots, v_n)$ consists of those $x \in [v_0, \ldots, v_n]$ with each x_i positive. It follows that for $n = 0$, $(v_0) = [v_0] = \{v_0\}$ is both an open and closed simplex, and for $n \geq 1$, that $(v_0, \ldots, v_n)$ is the interior of $[v_0, \ldots, v_n]$. By convention, the empty set is also both an open and closed simplex. For $n \geq -1$, the *dimension* of a simplex with $n + 1$ vertices is n. A *face* (or *k-face* for $-1 \leq k \leq n$) of $[v_0, \ldots, v_n]$ is a simplex (of dimension k) whose vertices form a subset (of size $k + 1$) of $\{v_0, \ldots, v_n\}$.

A (simplicial) *complex* K is a finite collection of closed simplexes such that (1) K contains every face of every $\sigma \in K$; and (2) if $\sigma_1, \sigma_2 \in K$, then $\sigma_1 \cap \sigma_2$ is a common face of σ_1 and σ_2. The standard example is the collection of all faces of a given simplex. The *geometrical representation* (polyhedron) $|K|$ of K is the point set (with the Euclidean induced topology) that is the union of the members of K. A complex K^* is a *subdivision* of a complex K if $|K^*| = |K|$ and each simplex in K^* is a subset of some member of K. A subcollection L of K is a *subcomplex* of K whenever L is also a complex. The *m-skeleton* of a complex K is the subcomplex K^m of all simplexes in K whose dimension is $\leq m$.

83.1 Lemma *Let $\mathcal{T} = \{T_\alpha\}$ be a collection of simplexes with the property that each $T_\alpha \cap T_\beta$ is a face common to T_α and T_β. Then $K = \{\sigma : \sigma$ is a face of some $T \in \mathcal{T}\}$ is a simplicial complex.*

PROOF. Let f_α, f_β be faces of T_α, T_β respectively. We show $f_\alpha \cap f_\beta = \sigma \in K$. Clearly $\sigma \subset F = T_\alpha \cap T_\beta$, a face common to T_α and T_β. Since F and f_α are faces of the simplex T_α, their intersection $F \cap f_\alpha$ is a face of T_α and a face of F. Likewise, $F \cap f_\beta$ is a face of F. Since F is a simplex, $(F \cap f_\alpha) \cap (F \cap f_\beta) = f_\alpha \cap f_\beta = \sigma$ is a face of F, and consequently a face of T_α, i.e., $\sigma \in K$. $\qquad\square$

Let $\mathcal{T}_0 = \{\Delta^3\}$, and given $\mathcal{T}_{n-1}$, let $\mathcal{T}_n$ be the collection $\{w(T) \,|\, T \in \mathcal{T}_{n-1}$ and $w \in \mathcal{F}_3^*\}$. Members of $\mathcal{T}_n$ may be represented as $T_{\alpha_1 \cdots \alpha_n} = w_{\alpha_1} \circ \cdots \circ w_{\alpha_n}(\Delta^3)$.

83.2 Lemma *Let each $\mathcal{T}_n$ be defined as above. Then (a) $\Delta^3 = \cup_{q \in A} T_q$ and $T_{\alpha_1 \cdots \alpha_{n-1}} = \cup_{q \in A} T_{\alpha_1 \cdots \alpha_{n-1} q}$; and (b) $T_\alpha, T_\beta \in \mathcal{T}_n$ implies $T_\alpha \cap T_\beta$ is a face common to T_α and T_β.*

PROOF. We prove (a) first. Let w_α denote $w_{\alpha_1} \circ \cdots \circ w_{\alpha_{n-1}}$. Since $\Delta^3 = \cup_{q \in A} w_q(\Delta^3)$,

$$
\begin{aligned}
T_{\alpha_1 \cdots \alpha_{n-1}} = w_\alpha(\Delta^3) &= w_\alpha \left(\cup_{q \in A} w_q(\Delta^3) \right) \\
&= \cup_{q \in A} \left(w_\alpha \circ w_q(\Delta^3) \right) = \cup_{q \in A} T_{\alpha_1 \cdots \alpha_{n-1} q},
\end{aligned}
$$

which finishes the proof of (a). For the proof of (b), consider that (b) is clearly true for $n \leq 1$. So we assume $n \geq 2$ and continue by induction. For each $q \in A$, we to relate $T_{\gamma_1 \cdots \gamma_{n-1} q}$ in $\mathcal{T}_n$ to its supertetra $T_{\gamma_1 \cdots \gamma_{n-1}}$ in $\mathcal{T}_{n-1}$:

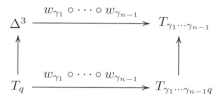

Fig. 83.3 Relating members of $\mathcal{T}_n$ to those in $\mathcal{T}_{n-1}$.

Indeed, viewing the vertical arrows in Figure 83.3 as inclusion mappings, we see that the eight tetra $T_{\gamma_1 \cdots \gamma_{n-1} q}$, $q \in A$, "combinatorially reside" in $T_{\gamma_1 \cdots \gamma_{n-1}}$ exactly (homeomorphically via the barycentric mapping $w_{\gamma_1} \circ \cdots \circ w_{\gamma_{n-1}}$) as the eight tetra T_q "reside" in Δ^3.

So if the ordered $(n-1)$-tuples $(\alpha_1, \ldots, \alpha_{n-1})$ and $(\beta_1, \ldots, \beta_{n-1})$ are equal, then T_α meets T_β in a common face since T_{α_n} meets T_{β_n} in a common face. In the other case, $T_{\alpha_1 \cdots \alpha_{n-1}}$ and $T_{\beta_1 \cdots \beta_{n-1}}$ are distinct, but meet (by the inductive assumption) in a common face F, which is a superset of $T_\alpha \cap T_\beta$. We suppose $T_\alpha \cap T_\beta$ is non-empty, and then consider the possible dimensions of F: If F is a 0-simplex, then T_α meets T_β in a common vertex. If F is a 1-simplex, then using Figure 83.3 with $(\gamma_1, \ldots, \gamma_{n-1}) = (\alpha_1, \ldots, \alpha_{n-1})$, we may first identify F with its inverse-image F^{-1}, an edge of Δ^3. Since each T_q meets F^{-1} in a simplex contained in the barycentric subdivision of F^{-1}, it follows that F meets T_α in some simplex in the barycentric subdivision of F. Similarly, using Figure 83.3 for $(\gamma_1, \ldots, \gamma_{n-1}) = (\beta_1, \ldots, \beta_{n-1})$, we find that T_β meets F in some simplex in the barycentric subdivision of F. Since this subdivision of F is itself a complex, T_α meets T_β in a common face when F is a 1-simplex. Finally, if F is a 2-simplex, then again we use Figure 83.3 with $(\gamma_1, \ldots, \gamma_{n-1}) = (\alpha_1, \ldots, \alpha_{n-1})$. In this case, F^{-1} is a 2-face of Δ^3. The subdivision of F^{-1} induced via each T_q is the one of "connecting midpoints" of its 1-faces. Thus, F meets T_α in some simplex in the "connecting midpoints" subdivision of F. Similarly, F meets T_β in some simplex in the (same) "connecting midpoints" subdivision of F. Since this subdivision is a complex, T_α meets T_β in a common face. □

Define, for each $n \geq 0$, $K_n = \{\sigma : \sigma \text{ is a closed face of some } T \text{ in } \mathcal{T}_n\}$.

83.4 Lemma *Let each K_n be defined as above. Then K_n is a simplicial complex, $|K_n| = \cup_{T \in \mathcal{T}_n} T = \Delta^3$, and K_n is a subdivision of K_{n-1}.*

PROOF. Since K_n is the collection of all faces of all members of $\mathcal{T}_n$, Lemmas 83.1 and 83.2(b) show that K_n is a complex. That $|K_n| = \cup_{T \in \mathcal{T}_n} T = \Delta^3$ follows by induction: For $n = 0$ the result is obvious. And if these equalities hold for $n - 1 \geq 0$, then $T_{\alpha_1 \cdots \alpha_{n-1}} = \cup_{q \in A} T_{\alpha_1 \cdots \alpha_{n-1} q}$ (Lemma 83.2(a)) shows that these equalities hold for n. To see that K_n is a subdivision of K_{n-1}, note that $|K_n| = |K_{n-1}|$, and $\sigma \in K_n$ implies $\sigma \subset T_{\alpha_1 \cdots \alpha_n} \subset T_{\alpha_1 \cdots \alpha_{n-1}}$. □

83.5 Lemma (induction for Δ^3) *For each $n \geq 2$ and each $m \in \{0, 1, 2\}$, let K_n^m denote the m-skeleton of K_n, i.e., all k-simplexes, $k \leq m$, in K_n. Then $K_n^m = \{w_{\alpha_1} \circ \cdots \circ w_{\alpha_{n-1}}(\sigma) : \sigma \in K_1^m\}$.*

PROOF. Consider $\tau \in K_n^m$ where $n \geq 2$. Then τ is a k-face $(k \leq m)$ of some $T_{\alpha_1 \cdots \alpha_n} = w_{\alpha_1} \circ \cdots \circ w_{\alpha_n}(\Delta^3) \in \mathcal{T}_n \subset K_n$. It follows that a k-face τ^* of Δ^3 exists such that $\tau = w_{\alpha_1} \circ \cdots \circ w_{\alpha_n}(\tau^*) = w_{\alpha_1} \circ \cdots \circ w_{\alpha_{n-1}}(w_{\alpha_n}(\tau^*))$. Thus, $\sigma = w_{\alpha_n}(\tau^*)$ is a k-simplex in K_1^m, yielding $K_n^m \subset \{w_{\alpha_1} \circ \cdots \circ w_{\alpha_{n-1}}(\sigma) | \sigma \in K_1^m\}$. The reverse inclusion is similarly straightforward. □

§84 The Subcomplex F_n

By restricting K_n to a 2-face of Δ^3, we obtain the subcomplex F_n: Let $\{\{i,j\},\{k,\ell\}\} = \{\{0,1\},\{2,3\}\}$, and let Δ^3_{ijk} be the 2-face of Δ^3 that is opposite u_ℓ. Define, for each $n \geq 0$, $F_n = \{\sigma \in K_n : \sigma \subset \Delta^3_{ijk}\}$.

84.1 Lemma Let $\Delta^3_{ijk} = [u_i, u_j, u_k]$ and $\Delta^3_{rst} = [u_r, u_s, u_t]$ be faces of Δ^3 of dimension 2, and let $q \in A$. Then $w_q(\Delta^3_{rst}) \cap \Delta^3_{ijk}$ contains a 2-simplex if and only if $\{r,s,t\} = \{i,j,k\}$ and $q \in \{i,j,k,\ell'\}$.

PROOF. The "if" part is clear. Conversely, let $w_q(\Delta^3_{rst}) \cap \Delta^3_{ijk}$ contain a 2-simplex. Then, since the 3-simplex $w_q(\Delta^3)$ meets Δ^3_{ijk} in a simplex of dimension < 2 whenever $q \notin \{i,j,k,\ell'\}$, we must have $q \in \{i,j,k,\ell'\}$. Keep this q fixed and observe that $\Delta^3_{ijk} = \cup\{w_m(\Delta^3_{ijk}) | m \in i,j,k,\ell'\}$ where each distinct pair of the four $w_m(\Delta^3_{ijk})$ meet in a simplex of dimension < 2. Thus, $w_q(\Delta^3_{rst})$ meeting Δ^3_{ijk} in a 2-simplex is equivalent to $w_q(\Delta^3_{rst})$ meeting $w_q(\Delta^3_{ijk})$ in a 2-simplex, which yields $\{r,s,t\} = \{i,j,k\}$. □

84.2 Lemma (representations of F_n and its members) For each $n \geq 0$,

(a) the collection F_n is a subcomplex of K_n,
(b) $|F_n| = \Delta^3_{ijk}$,
(c) $|F_n| = \cup\{\sigma \in F_n | \sigma$ is a 2-simplex$\}$, and
(d) $F_n = \{\tau : \tau$ is a face of some 2-simplex $\sigma \in F_n\}$.

And when $n \geq 1$,

(e) a 2-simplex $\sigma_n \in F_n \Leftrightarrow \sigma_n = w_{\alpha_1} \circ \cdots \circ w_{\alpha_n}(\Delta^3_{ijk})$ for some $\alpha_1, \ldots, \alpha_n \in \{i,j,k,\ell'\}$.

Moreover, the representation given in (e) is unique. That is, if $\sigma_n = w_{\beta_1} \circ \cdots \circ w_{\beta_n}(\sigma_0) \in F_n$ for some 2-simplex $\sigma_0 \in K_0$ and some list $\beta_1, \ldots, \beta_n \in A$, then $\sigma_0 = \Delta^3_{ijk}$ and each $\beta_m = \alpha_m$.

PROOF. First, we prove (e) for each $n \geq 1$. The $n = 1$ case is clear (Figure 70.1) because a 2-simplex $\sigma_1 \in F_1$ if and only if there is a unique $\alpha_1 \in \{i,j,k,\ell'\}$ such that $\sigma_1 = w_{\alpha_1}(\Delta^3_{ijk})$. So suppose $n \geq 2$ and that each 2-simplex in F_{n-1} has the indicated unique representation. Consider any 2-simplex $\sigma_n \in F_n$. Then $\sigma_n \in K_n$ implies σ_n is a 2-face of some $w_{\alpha_1} \circ \cdots \circ w_{\alpha_n}(\Delta^3) \in K_n$, i.e., for some 2-simplex $\sigma_0 \in K_0$, we have $\sigma_n = w_{\alpha_1} \circ \cdots \circ w_{\alpha_n}(\sigma_0)$. So $\sigma_n = w_{\alpha_1}(\sigma_{n-1})$ for some 2-simplex $\sigma_{n-1} = w_{\alpha_2} \circ \cdots \circ w_{\alpha_n}(\sigma_0) \in K_{n-1}$. We show that $\sigma_{n-1} \subset \Delta^3_{ijk}$: Since the 2-simplex $\sigma_{n-1} \in K_{n-1}$ and K_{n-1} is a subdivision of K_0, σ_{n-1} is either a subset of some 2-face Δ^3_{rst} of Δ^3 or meets the interior of Δ^3. The latter case is impossible because $w_{\alpha_1}(\text{int}(\Delta^3)) \subset \text{int}(\Delta^3)$ but we know that $w_{\alpha_1}(\sigma_{n-1}) = \sigma_n \subset \Delta^3_{ijk}$. So $\sigma_{n-1} \subset \Delta^3_{rst}$. It follows that $w_{\alpha_1}(\Delta^3_{rst}) \cap \Delta^3_{ijk}$ contains a 2-simplex. By Lemma 84.1, $\{r,s,t\} = \{i,j,k\}$ and $\alpha_1 \in \{i,j,k,\ell'\}$.

Thus, $\sigma_{n-1} = w_{\alpha_2} \circ \cdots \circ w_{\alpha_n}(\sigma_0) \in F_{n-1}$, and the inductive hypothesis shows that $\sigma_0 = \Delta^3_{ijk}$ and that each $\alpha_2, \ldots, \alpha_n \in \{i, j, k, \ell'\}$ is unique. This finishes the "representation half" of the proof of (e). For the converse, i.e., each such "form" yields a 2-simplex $\sigma_n \in F_n$, we note that $w_q(\Delta^3_{ijk}) \subset \Delta^3_{ijk}$ if and only if $q \in \{i, j, k, \ell'\}$. And so the proof of (e) is complete. Statement (a) is straightforward. To prove (b) and (c), first let "q" range over the members of $B = \{i, j, k, \ell'\}$ and let "α" range over the members $(\alpha_1, \ldots, \alpha_{n-1})$ of the product set B^{n-1}, making "$\cup_q = \cup_{q \in B}$" and "$\cup_\alpha = \cup_{\alpha \in B^{n-1}}$." And for each such α, let w_α denote $w_{\alpha_1} \circ \cdots \circ w_{\alpha_{n-1}}$. Second, observe that $\Delta^3_{ijk} = \cup_q w_q(\Delta^3_{ijk})$. Third, since $|F_0| = \Delta^3_{ijk}$ and $|F_0|$ is the union of the 2-simplexes that it contains, assume, for $n \geq 1$, that $|F_{n-1}| = \Delta^3_{ijk}$ and that $|F_{n-1}|$ is the union of the 2-simplexes that it contains ($|F_{n-1}| = \cup_\alpha w_\alpha(\Delta^3_{ijk})$). Then

$$
\begin{aligned}
\Delta^3_{ijk} &= |F_{n-1}| = \cup_\alpha w_\alpha(\Delta^3_{ijk}) \\
&= \cup_\alpha w_\alpha \left(\cup_q w_q(\Delta^3_{ijk}) \right) = \cup_\alpha \left(\cup_q \left(w_\alpha \circ w_q(\Delta^3_{ijk}) \right) \right) \subset |F_n| \subset \Delta^3_{ijk},
\end{aligned}
$$

where the next-to-last inclusion follows from (e). This finishes the proof for (b) and (c). Finally, to prove (d) let $\tau \in F_n$ have dimension ≤ 1. From (c), there is a 2-simplex σ that meets the interior of τ, and since F_n is a complex, $\tau = \tau \cap \sigma$ must be a face of σ. $\qquad \square$

84.3 Lemma (induction for Δ^3_{ijk}) *For each $n \geq 2$ and each $m \in \{0, 1\}$, let F^m_n denote the m-skeleton of F_n. Then $F^m_n = \{ w_{\alpha_1} \circ \cdots \circ w_{\alpha_{n-1}}(\sigma) | \sigma \in F^m_1$ and $\alpha_1, \ldots, \alpha_{n-1} \in \{i, j, k, \ell'\} \}$.*

PROOF. Consider $\tau \in F^m_n$. Then from Lemma 84.2 τ is a k-face ($k \leq m$) of some 2-simplex $\tau_n \in F_n$ that has a unique representation $\tau_n = w_{\alpha_1} \circ \cdots \circ w_{\alpha_n}(\Delta^3_{ijk})$. This representation provides $\sigma = w_{\alpha_n}(\Delta^3_{ijk}) \in F^m_1$ that satisfies the required condition. The reverse inclusion is obvious. $\qquad \square$

§85 Calculating Addresses

Unlike the 2-simplex case, where addresses are readily exposed via Figure 70.1, the analogous Figure 85.1 for the 3-simplex case is visually complicated. In Figure 85.1, the "dark edges" represent PQ and its eight images $w_q(PQ)$. Indeed, if P is pictured as the midpoint on the left-side edge of Δ^3 and Q the midpoint on the bottom-right edge, then PQ may be pictured as the "long dark segment" connecting those midpoints, and, then the eight "shorter dark segments" represent the eight $w_q(PQ)$ images of PQ. From that observation, we then see that "four short dark segments" meet PQ at its midpoint m_{PQ}, which, in turn, reveals that each $w_{k'}$-image of the octahedron-shaped hole meets the 1-simplex $[P, Q]$ at its midpoint m_{PQ}.

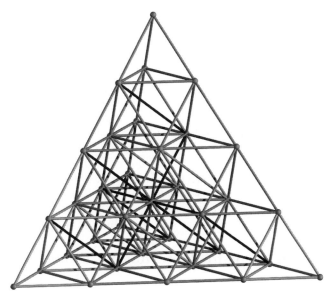

Fig. 85.1 The K_2 subdivision of K_1.

In Figure 85.2 we see the "level-1 decomposition" $\{w_{k'} \circ w_q(\Delta^3) : w_q \in \mathcal{F}_3^*\}$ of $w_{k'}(\Delta^3)$, which is part of the subdivision K_2 of K_1. Also note that the $w_{k'}$-image of the octahedron-shaped hole meets the 1-simplex $[P, Q]$ at its midpoint m_{PQ}.

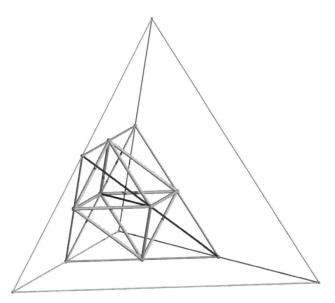

Fig. 85.2 "Level-1 decomposition" of $T_{k'}$.

85.3 Lemma (vertices) *Let* $w_q \in \mathcal{F}_3^*$, *let* ϕ *denote the address map, and let* $\alpha \in N(A)$. *Then* (a) $w_q(\phi(\alpha)) = \phi(q\alpha_1\alpha_2 \cdots)$; *and* (b) $w_q(\phi(\alpha)) = \phi(\alpha) = u_q$ *if and only if* $\alpha = \overline{q}$.

PROOF. We begin with (a): By definition, $\phi(\alpha)$ is the lone element in $\cap_{j=1}^{\infty} W_j$ where $W_j = w_{\alpha_1} \circ \cdots \circ w_{\alpha_j}(\Delta^3)$. So $\phi(\alpha) \in W_j$ for each j, showing that $w_q(\phi(\alpha)) \in w_q(W_j) = w_q \circ w_{\alpha_1} \circ \cdots \circ w_{\alpha_j}(\Delta^3)$ for each j. It follows that $w_q(\phi(\alpha))$ is the lone element in $\cap_{j=1}^{\infty} w_q(W_j)$, i.e., $w_q(\phi(\alpha)) = \phi(q\alpha_1\alpha_2 \cdots)$. Turning to (b), since a contraction has only one fixed point and each w_q is a contraction such that $w_q(u_q) = u_q$, we have $w_q(x) = x$ if and only if $x = u_q$. Now suppose $\alpha = \overline{q}$. Then $w_q(\phi(\alpha)) = \phi(q\alpha_1\alpha_2 \cdots) = \phi(\alpha)$, showing that $\phi(\alpha) = u_q$. Conversely, suppose $\phi(\alpha) = u_q$. For q fixed, the only index i such that $u_q \in w_i(\Delta^3)$ is $i = q$. It follows that $\alpha_1 = q$. If $\alpha_1 = \cdots = \alpha_{n-1} = q$, then $w_{\alpha_1} \circ \cdots \circ w_{\alpha_{n-1}}(u_q) = u_q$. It follows, since $u_q \in w_{\alpha_1} \circ \cdots \circ w_{\alpha_n}(\Delta^3)$, that $u_q \in w_{\alpha_n}(\Delta^3)$. That is, $\alpha_n = q$, and by induction $\alpha = \overline{q}$. $\qquad \square$

§86 Steps for Determining Fibers

In sequence, we shall calculate fibers of points in the following sets: (i) vertices and midpoints of edges (Table 87.3); (open) edges (Table 88.2); (open) edges of the hole and (P, Q) (Tables 88.3 and 88.4); (open) 2-faces (Table 89.2); and (open) 2-simplexes of K_1^2 (Tables 90.3 and 90.4). Then we shall finish the task by applying Theorem 86.2, which tells us that the fibers of points in $\Delta^3 \setminus |K_1^2|$ are either singleton fibers or "shifts" of those of points in $|K_1^2|$. All results are summarized in Parts I and II of Table 93.1.

86.1 Theorem (singleton fibers I) *A point* $x \in \Delta^3$ *satisfies* $|\phi^{-1}(x)| = 1$ *if and only if for each* $n \geq 1$, *x is contained in only one member of* $\mathcal{T}_n$.

86.2 Theorem (inductive step) *Let* $A = \{0, 1, 2, 3, 0', 1', 2', 3'\}$. *Let* x *be a point in* $\Delta^3 - |K_1^2|$, *and let* ϕ *denote the* $\mathcal{F}_3^*$ *address map. Then either* (a) *there is an* $(m \geq 1)$-*length list* $\varepsilon_1, \ldots, \varepsilon_m \in A$ *such that* $x \in w_{\varepsilon_1} \circ \cdots \circ w_{\varepsilon_m}(|K_1^2|)$, *in which case there is a* $y \in \mathrm{int}(\Delta^3) \cap |K_1^2|$ *and an* r, $1 \leq r \leq m$, *such that* $\phi^{-1}(x) = \{\varepsilon_1 \cdots \varepsilon_r \beta \mid \beta \in \phi^{-1}(y)\}$, *or* (b) $x \notin w_{\varepsilon_1} \circ \cdots \circ w_{\varepsilon_m}(|K_1^2|)$ *for every* $(m \geq 1)$-*list* $\varepsilon_1, \ldots, \varepsilon_m \in A$, *in which case* $|\phi^{-1}(x)| = 1$.

PROOF. It is obvious that the hypothesis of either (a) or (b) holds. In the (a) case, let $x = w_{\varepsilon_1} \circ \cdots \circ w_{\varepsilon_m}(y_m)$ where $y_m \in |K_1^2|$. If $y_m \in \partial\Delta^3$, then $y_{m-1} = w_{\varepsilon_m}(y_m) \in |K_1^2|$, and if $y_{m-1} \in \partial\Delta^3$, then $y_{m-2} = w_{\varepsilon_{m-1}}(y_{m-1}) \in |K_1^2|$, etc. In short, since $x \notin |K_1^2|$ there is an $r \geq 1$ such that $y = y_r \in \mathrm{int}(\Delta^3) \cap |K_1^2|$ and $x = w_{\varepsilon_1} \circ \cdots \circ w_{\varepsilon_r}(y)$. Moreover, if $r > 1$, then since $q \in A$ implies $w_q(\mathrm{int}(\Delta^3)) = \mathrm{int}(T_q) \subset \mathrm{int}(\Delta^3) - |K_1^2|$, each y_k $(1 \leq k < r)$ cannot be a member of $|K_1^2|$. From each such $y_k \notin |K_1^2|$, we may deduce that every ϕ-address γ of x must have its first r values given by $\gamma_1 = \varepsilon_1, \ldots, \gamma_r = \varepsilon_r$. (Otherwise, the minimum $k \leq r$ among these subscripts such that $\gamma_k \neq \varepsilon_k$

yields $x \in \sigma_3 = T_{\varepsilon_1 \cdots \varepsilon_k} \cap T_{\gamma_1 \cdots \gamma_k} \in K_k^2$, where $k > 1$ because $x \notin |K_1^2|$. So $k \geq 2$. But then $x \in \sigma_3 = w_{\varepsilon_1} \circ \cdots \circ w_{\varepsilon_{k-1}}(\sigma)$ where $1 \leq k - 1 < r$ and $\sigma \in K_1^2$. Thus $x = w_{\varepsilon_1} \circ \cdots \circ w_{\varepsilon_{k-1}}(y_{k-1})$ where $y_{k-1} \in \sigma \subset |K_1^2|$, which contradicts $y_{k-1} \notin |K_1^2|$.) It follows by Lemma 85.3(a) that each ϕ-address β of y determines a ϕ-address of x via the following formula:

$$x = w_{\varepsilon_1} \circ \cdots \circ w_{\varepsilon_r}(\phi(\beta)) = \phi(\varepsilon_1 \cdots \varepsilon_r \beta_1 \beta_2 \cdots).$$

Now suppose γ is any ϕ-address of x. Then, (from the argument above) $\gamma_1 = \varepsilon_1, \ldots, \gamma_r = \varepsilon_r$, and

$$x = \phi(\gamma_1 \gamma_2 \cdots) = \phi(\varepsilon_1 \cdots \varepsilon_r \gamma_{r+1} \gamma_{r+2} \cdots) = w_{\varepsilon_1} \circ \cdots \circ w_{\varepsilon_r}(\phi(\beta)),$$

where $\beta_1 = \gamma_{r+1}$, $\beta_2 = \gamma_{r+2}$, etc. And since $w_{\varepsilon_1} \circ \cdots \circ w_{\varepsilon_r}$ is one-to-one, $\phi(\beta) = y \in \operatorname{int}(\Delta^3) \cap |K_1^2|$, showing that when (a) holds, the addresses of x are determined as claimed. Next, suppose (b): Let ε be an address of x. Then, since $\partial \Delta^3 \subset |K_1^2|$, we have $x \notin w_{\varepsilon_1}(\partial \Delta^3) = \partial T_{\varepsilon_1}$, showing $x \in \operatorname{int}(T_{\varepsilon_1})$. And $x \notin w_{\varepsilon_1} \circ w_{\varepsilon_2}(\partial \Delta^3) = \partial T_{\varepsilon_1 \varepsilon_2}$, showing $x \in \operatorname{int}(T_{\varepsilon_1 \varepsilon_2})$, etc. Thus, Theorem 86.1 shows that $|\phi^{-1}(x)| = 1$, i.e., x has only one address. $\square$

86.3 Theorem (addresses for a 2-face) *Let x be a point in $\Delta_{ijk}^3 \setminus |F_1^1|$, and let ϕ denote the $\mathcal{F}_3^*$ address map. Then either (a) there is an $(m \geq 1)$-length list $\delta_1, \ldots, \delta_m \in \{i, j, k, \ell'\}$ such that $x \in w_{\delta_1} \circ \cdots \circ w_{\delta_m}(|F_1^1|)$, in which case there is a $y \in \operatorname{int}(\Delta_{ijk}^3) \cap |F_1^1|$ and there is an n, $1 \leq n \leq m$, such that $\phi^{-1}(x) = \{\delta_1 \cdots \delta_n \beta \mid \beta \in \phi^{-1}(y)\}$, or (b) $x \notin w_{\delta_1} \circ \cdots \circ w_{\delta_m}(|F_1^1|)$ for every $(m \geq 1)$-length list $\delta_1, \ldots, \delta_m \in \{i, j, k, \ell'\}$, in which case $|\phi^{-1}(x)| = 1$ and the lone address of x is a sequence in $\{i, j, k, \ell'\}$ with either a constant subsequence $\ell' \ell' \cdots$ or three constant subsequences.*

PROOF. It is obvious that the hypothesis of either (a) or (b) holds. The proof in the (a) case runs parallel to its counterpart in the proof of Theorem 86.2. So suppose the hypothesis of (b) holds: Since $x \in \Delta_{ijk}^3$ it has at least one address with each of its values in $\{i, j, k, \ell'\}$. (Recall that $\{w_i, w_j, w_k, w_{\ell'}\} \subset \mathcal{F}_3^*$ generates addresses for Δ_{ijk}^3 that correspond in an obvious manner to the addresses of Δ^2 induced from $\mathcal{F}_2^*$.) So we may select a sequence δ in $\{i, j, k, \ell'\}$ that is an address of x. Since $x \in \Delta_{ijk}^3 \subset \partial \Delta^3$, we have $x \in \partial T_{\delta_1} \cap \Delta_{ijk}^3$, i.e., x is in the 2-face $[w_{\delta_1}(u_i), w_{\delta_1}(u_j), w_{\delta_1}(u_k)]$. In addition, however, since $x \notin w_{\delta_1}(|F_1^1|)$ the point x is in the open 2-face $(w_{\delta_1}(u_i), w_{\delta_1}(u_j), w_{\delta_1}(u_k))$ of T_{δ_1}. Thus, the only simplex T_q that contains x is T_{δ_1}. Similarly, $x \notin w_{\delta_1} \circ w_{\Delta^2}(|F_1^1|)$ implies that the only 3-simplex $T_{\delta_1 q}$ that contains x is $T_{\delta_1 \delta_2}$. And so on. It follows from Theorem 86.1 that $|\phi^{-1}(x)| = 1$. The lone address δ has the required properties because the identification of the subsystem $\{w_i, w_j, w_k, w_{\ell'}\}$ with that of $\mathcal{F}_2^*$ matches δ with a singleton sequence in the $\operatorname{int}(\Delta^2)$. $\square$

To illustrate the idea common to Theorems 86.2 and 86.3, consider the $\mathcal{F}_2^*$ fibers of points in the $\operatorname{int}(\Delta^2)$ (§75 and Theorem 78.1). The nonsingleton fibers of points in $\mathrm{int}(\Delta^2)$ are simply *shifts of fibers* of points in $(m_{01}, m_{02}) \cup (m_{01}, m_{12}) \cup (m_{02}, m_{12})$. (In the display below, the addresses of points in $(m_{01}, m_{02}) \cup (m_{01}, m_{12}) \cup (m_{02}, m_{12})$ are underlined.)

$$\alpha = d_1 \cdots d_{r-1} c_r a_{r+1} a_{r+2} \cdots$$
$$\beta = d_1 \cdots d_{r-1} \underline{3\, b_{r+1} b_{r+2} \cdots}$$

, or,

$$\alpha = d_1 \cdots d_{r-1} c_r a_{r+1} \cdots a_{t-1} a_t \overline{a_{t+1}}$$
$$\beta = d_1 \cdots d_{r-1} c_r a_{r+1} \cdots a_{t-1} a_{t+1} \overline{a_t}$$
$$\gamma = d_1 \cdots d_{r-1} c_r a_{r+1} \cdots a_{t-1} 3 \overline{c_{t+1}}$$
$$\delta = d_1 \cdots d_{r-1} \underline{3\, b_{r+1} \cdots b_{t-1} a_t \overline{a_{t+1}}}$$
$$\epsilon = d_1 \cdots d_{r-1} \underline{3\, b_{r+1} \cdots b_{t-1} a_{t+1} \overline{a_t}}$$
$$\zeta = d_1 \cdots d_{r-1} \underline{3\, b_{r+1} \cdots b_{t-1} 3 \overline{c_{t+1}}}.$$

In short, the "$d_1 \cdots d_{r-1}$-shift" is analogous to the "ε" and "δ" shifts in Theorems 86.2 and 86.3.

§87 Fibers and the 0-Skeleton K_1^0

The 0-skeleton K_1^0 contains the empty set and ten 0-simplexes, namely the four vertices $[u_0]$, $[u_1]$, $[u_2]$, $[u_3]$ and the six midpoints $[P] = [m_{01}]$, $[m_{02}]$, $[m_{03}]$, $[m_{12}]$, $[m_{13}]$, $[Q] = [m_{23}]$ (Figure 81.1).

87.1 Theorem (addresses of vertices) *Let x be a vertex of some $T_{\alpha_1 \cdots \alpha_t} \in \mathcal{T}_t$. Then there is one and only one address of x whose first t terms are $\alpha_1, \ldots, \alpha_t$. Moreover, the unique address is $\alpha_1 \cdots \alpha_t \overline{r}$ where $r \in \{0, 1, 2, 3\}$ is the index of the unique vertex u_r that satisfies $w_{\alpha_1} \circ \cdots \circ w_{\alpha_t}(u_r) = x$.*

PROOF. Since each $w \in \mathcal{F}_3^*$ is nonsingular and affine, any finite composition of members of $\mathcal{F}_3^*$ is nonsingular and affine. It follows, since such a composition maps vertices to vertices, that there is a unique vertex u_r of Δ^3 such that $w_{\alpha_1} \circ \cdots \circ w_{\alpha_t}(u_r) = x$. Lemma 85.3(b) followed by t applications of Lemma 85.3(a) yield the desired result. $\qquad\square$

Theorem 87.1 along with Figure 87.2 provide information about the fibers of the vertices.

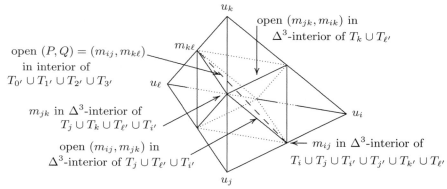

Fig. 87.2 Open simplexes in the Δ^3-interior of unions of 3-simplexes.

In more detail, Theorem 87.1 tells us that since there are eight $T_q \in \mathcal{T}_1$ sharing ten vertices (the points in $|K_1^0|$), there are $32 = 8 \cdot 4$ addresses "$q\bar{r}$" ($r = 0, 1, 2, 3$) that partition into ten fibers (four of size 1, four of size 4, and two of size 6). The distinct kinds of fibers may be derived from Figure 87.2 ("X in Δ^3-interior of Y" means $X \subset (Y \cap G)$ for some G in the topology of Δ^3.) The details are summarized in Table 87.3. For example, upon selecting $j, k \in \{0, 1, 2, 3\}$ such that $\{\{i, j\}, \{k, \ell\}\} = \{\{0, 1\}, \{2, 3\}\}$, one may find the fiber of m_{jk}. Table 87.3 is complete (it contains all addresses of all points in $|K_1^0|$) because any vertex x of any T_q is Δ^3-interior to the union of those $T \in \mathcal{T}_1$ that have x as a vertex.

NUMBER AND LOCATION	$\mathcal{F}_3^*$ ADDRESSES	$|\phi^{-1}(x)|$	TYPE
4 Vertices u_r ($r = 0, 1, 2, 3$)	$\bar{r}$	1	1.1
4 Midpoints m_{jk}	$j\bar{k}, k\bar{j}, i'\bar{\ell}, \ell'\bar{i}$	4	4.1
2 Midpoints m_{ij}	$i\bar{j}, j\bar{i}, i'\bar{i}, j'\bar{j}, k'\bar{\ell}, \ell'\bar{k}$	6	6.1

Table 87.3 Fibers of vertices and midpoints of edges.

In passing, we note that the "TYPE" column will serve to classify the $\mathcal{F}_3^*$ fibers and also aid in cross-referencing tables.

§88 Fibers and Open Edges of K_1^1

The 1-skeleton K_1^1 contains 25 1-simplexes, which partition into two subsets, one generating the edges of Δ^3 and the other the edges of the (octahedral) hole and $[P, Q]$. In both cases, we use the following theorem, whose (omitted) proof is similar to the proof of Theorem 87.1.

88.1 Theorem (transfer of addresses) *Let $t - 1 \geq 1$, and let x be a point in an open 1-face τ of some $T_{\alpha_1 \cdots \alpha_{t-1}} \in \mathcal{T}_{t-1}$. Then the number of $\mathcal{F}_3^*$-addresses of x whose first $t - 1$ terms are $\alpha_1, \ldots, \alpha_{t-1}$ is the number of $\mathcal{F}_3^*$-addresses of the unique y satisfying $x = w_{\alpha_1} \circ \cdots \circ w_{\alpha_{t-1}}(y)$. Moreover, each $\mathcal{F}_3^*$-address of x whose first $t - 1$ terms are $\alpha_1, \ldots, \alpha_{t-1}$ has the form $\alpha_1 \cdots \alpha_{t-1}\beta$, where β is an $\mathcal{F}_3^*$-address of y.*

Fibers of points in open edges of Δ^3. We summarize in Table 88.2 where distinct $r, s \in \{0, 1, 2, 3\}$; "a_{rs}" denotes any address of the midpoint m_{rs}; and "$\alpha^{(t-1)}$" denotes $\alpha_1, \ldots, \alpha_{t-1} \in \{r, s\}$ or the empty string.

$x \in (u_r, u_s)$	CONSTRAINTS	$\mathcal{F}_3^*$ ADDRESSES	$|\phi^{-1}(x)|$	TYPE
x $\mathcal{F}_3$-irrational	$\alpha \in N(r, s) - \{\bar{r}, \bar{s}\}$ with no tail index	α	1	1.2
x $\mathcal{F}_3$-rational	$\{r, s\} \notin \{\{0, 1\}, \{2, 3\}\}$	$\alpha^{(t-1)}a_{rs}$	4	4.1
	$\{r, s\} \in \{\{0, 1\}, \{2, 3\}\}$	$\alpha^{(t-1)}a_{rs}$	6	6.1

Table 88.2 Fibers of points in open edges of Δ^3.

Consider the "x is irrational" row of Table 88.2. Then x is a Δ^3-interior point of $T_{\alpha_1} \in \mathcal{T}_1$, a Δ^3-interior point of $T_{\alpha_1\alpha_2} \in \mathcal{T}_2$, etc. So Theorem 86.1 shows that x satisfies $|\phi^{-1}(x)| = 1$.

For the "x is rational" row, suppose $x = m_{jk}$ is the midpoint of (u_j, u_k). If $\alpha_1, \ldots, \alpha_{t-1}$ denotes the empty list, then the corresponding addresses a_{jk} in the third column match those of m_{jk} in Table 87.3. Next, suppose $x = (3/4)u_j + (1/4)u_k \in (u_j, u_k)$. Then x is the midpoint of the open 1-face (u_j, m_{jk}) of $T_j = T_{\alpha_1}$. (In this instance, $t - 1 = 1$.) Also, $y = m_{jk}$ is the unique point that satisfies $w_{\alpha_1}(y) = w_j(m_{jk}) = x$. Since the four $\mathcal{F}_3^*$-addresses of m_{jk} are known (Table 87.3), an application of (the transfer addresses) Theorem 88.1 yields the four (indicated) addresses of x.

x IRRATIONAL $\{\{i,j\},\{k,\ell\}\}=$ $\{\{0,1\},\{2,3\}\}$	CONSTRAINTS $\alpha \in N(r,s) - \{\overline{r},\overline{s}\}$ no tail index	$\mathcal{F}_3^*$ ADDRESSES	SIZE OF FIBER	TYPE
$x \in (m_{jk}, m_{ik})$	$\{r,s\} = \{j,i\}$ and $(\alpha_m, \beta_m) \in \{(j,i),(i,j)\}$	$k\alpha, \ell'\beta$	2	2.1
$x \in (m_{ij}, m_{jk})$	$\{r,s\} = \{i,k\}$; $(\alpha_m, \beta_m, \gamma_m) \in \{(i,k,i),(k,i,\ell)\}$	$j\alpha, \ell'\beta, i'\gamma$	3	3.1
$x \in (P,Q)$	$\{r,s\} = \{0,1\}$; $(\alpha_m, \beta_m, \gamma_m, \delta_m) \in \{(0,1,3,2), (1,0,2,3)\}$	$0'\alpha, 1'\beta, 2'\gamma, 3'\delta$	4	4.2

Table 88.3 Fibers of $\mathcal{F}_3$-irrational points in open edges of hole and (P,Q).

In Table 88.4 distinct $r, s \in \{0,1,2,3\}$; "a_{rs}" denotes any address of the midpoint m_{rs}; and "$\alpha^{(t-1)}$" is $\alpha_1, \ldots, \alpha_{t-1} \in \{r,s\}$ or the empty string.

x $\mathcal{F}_3$-RATIONAL $\{\{i,j\},\{k,\ell\}\}=$ $\{\{0,1\},\{2,3\}\}$	CONSTRAINTS	$\mathcal{F}_3^*$ ADDRESSES	SIZE OF FIBER	TYPE
$x \in (m_{jk}, m_{ik})$	$\{r,s\} = \{j,i\}$ and $(\alpha_m, \beta_m) \in \{(j,i),(i,j)\}$	$k\alpha^{(t-1)}a_{ji}$, $\ell'\beta^{(t-1)}a_{ij}$	$2 \cdot 6$ $= 12$	12.1
$x \in (m_{ij}, m_{jk})$	$\{r,s\} = \{i,k\}$ and $(\alpha_m, \beta_m, \gamma_m) \in \{(i,k,i),(k,i,\ell)\}$	$j\alpha^{(t-1)}a_{ik}$, $\ell'\beta^{(t-1)}a_{ki}$, $i'\gamma^{(t-1)}a_{i\ell}$	$3 \cdot 4$ $= 12$	12.2
$x \in (P,Q)$	$\{r,s\} = \{0,1\}$ and $(\alpha_m, \beta_m, \gamma_m, \delta_m) \in \{(0,1,3,2), (1,0,2,3)\}$	$0'\alpha^{(t-1)}a_{01}$, $1'\beta^{(t-1)}a_{10}$, $2'\gamma^{(t-1)}a_{32}$, $3'\delta^{(t-1)}a_{23}$	$4 \cdot 6$ $= 24$	24.1

Table 88.4 Fibers of $\mathcal{F}_3$-rational points in open edges of hole and in (P,Q).

Fibers of points in open edges of hole and in (P,Q). These are summarized in Tables 88.3 and 88.4. To begin, first note that a choice of an open edge of the hole or the lone (P,Q) forces a choice of (at least one) doubleton set

$\{i,j\} \in \{\{0,1\},\{2,3\}\}$, and since $(m_{ij}, m_{k\ell}) = (P,Q)$ is the only open edge in the interior of Δ^3, the three forms (m_{jk}, m_{ik}), (m_{ij}, m_{jk}), (P,Q) cover all possibilities. Second, $x \in (P,Q)$ implies $x \notin \omega^3$, bringing into question the meaning of "x is a rational point in (P,Q):"

88.5 Definition (rational and irrational points in (PQ)**)**   A point x in the open 1-simplex (P,Q) is a *rational point* if its unique barycentric representation $x = x_P P + x_Q Q$ is such that $x_P \in \{m/2^n | 1 \le m < 2^n\}$ for some $n = 1, 2, \ldots$. Otherwise, $x \in (P,Q)$ is an *irrational point*.

Tables 88.3 and 88.4 show that this definition is consistent with identification of adjacent endpoints.

Next, note that the edges of the hole and the 1-simplex $[P,Q]$ first appear in K_1, i.e., are not members of K_0. These "new" simplexes exist as a result of identifications of faces of various distinct $T_p, T_q \in \mathcal{T}_1$. Along with each such identification there corresponds an identification of addresses. In detail, consider the following development for the "$x \in (m_{jk}, m_{ik})$" of Table 88.3.

Fibers of points in (m_{jk}, m_{ik}). Figure 87.2 shows (m_{jk}, m_{ik}) as a subset of the Δ^3-interior of $T_k \cup T_{\ell'}$. Both barycentric mappings w_k and $w_{\ell'}$ map the open simplex (u_j, u_i) onto (m_{jk}, m_{ik}), the former preserving and the latter reversing the indicated orientation, i.e.,

$$
\begin{aligned}
m_{jk} &= w_k(u_j) = w_{\ell'}(u_i) \\
m_{ik} &= w_k(u_i) = w_{\ell'}(u_j).
\end{aligned}
$$

Let θ be the orientation-reversing barycentric map $(u_j, u_i) \to (u_i, u_j)$, i.e.,

$$\theta(a_j u_j + a_i u_i) = a_j u_i + a_i u_j \qquad (a_j + a_i = 1 \text{ and } a_j, a_i \ge 0).$$

Let $\psi = \rho \times \rho \times \cdots$ be the product map with ρ the transposition (ji), i.e.,

$$\psi(\alpha_1 \alpha_2 \cdots) = \rho(\alpha_1)\rho(\alpha_2)\cdots \qquad \alpha_1 \alpha_2 \cdots \in N(\{j,i\}) \setminus \{\overline{j}, \overline{i}\}.$$

And let ϕ denote the identification-of-adjacent-endpoints map, i.e., on $N(i,j)$, the mapping ϕ is the restriction of the address map (also denoted ϕ) induced by $\mathcal{F}_3^*$. Then, since $a_j = \Sigma_{m=1}^\infty \delta_{\alpha_m}^j / 2^m$ where $\delta_{\alpha_m}^j = 1$ when $\alpha_m = j$ and 0 otherwise, the diagram in Figure 88.6 is commutative:

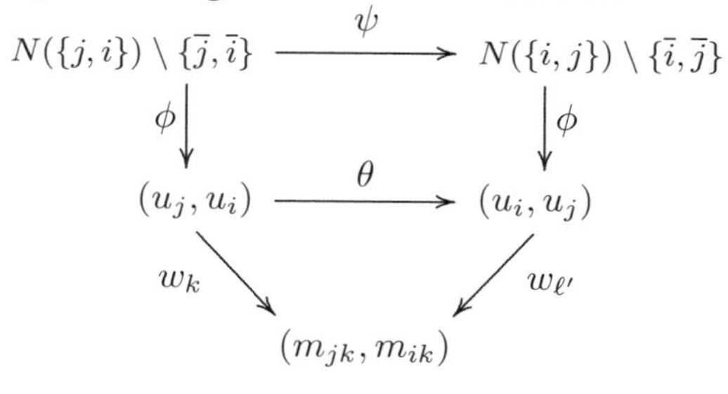

$$
\begin{array}{ccc}
N(\{j,i\}) \setminus \{\overline{j}, \overline{i}\} & \xrightarrow{\ \psi\ } & N(\{i,j\}) \setminus \{\overline{i}, \overline{j}\} \\
\phi \downarrow & & \downarrow \phi \\
(u_j, u_i) & \xrightarrow{\ \theta\ } & (u_i, u_j) \\
& \searrow^{w_k} \qquad \swarrow^{w_{\ell'}} & \\
& (m_{jk}, m_{ik}) &
\end{array}
$$

Fig. 88.6 Pasting diagram for (m_{jk}, m_{ik}).

For instance, if $\alpha \in N(\{j, i\}) \setminus \{\bar{j}, \bar{i}\}$ is an $\mathcal{F}_3^*$ address of the unique $a \in (u_j, u_i)$ such that $w_k(a) = x$, then $\beta = \psi(\alpha)$ is an $\mathcal{F}_3^*$ address of the unique $b = \theta(a) \in (u_i, u_j)$ such that $w_{\ell'}(b) = x$, i.e.,

$$
\begin{aligned}
w_k(a) &= w_k(a_j u_j + a_i u_i) = a_j w_k(u_j) + a_i w_k(u_i) \\
&= a_j w_{\ell'}(u_i) + a_i w_{\ell'}(u_j) = w_{\ell'}(a_j u_i + a_i u_j) = w_{\ell'}(b)
\end{aligned}
$$

where $b = \theta(a)$, showing that $w_k(a) = w_{\ell'} \circ \theta(a)$. Moreover,

$$
\begin{aligned}
\theta \circ \phi(\alpha) &= \theta \left((\Sigma_{m=1}^{\infty} \delta_{\alpha_m}^j / 2^m) u_j + (\Sigma_{m=1}^{\infty} \delta_{\alpha_m}^i / 2^m) u_i \right) \\
&= (\Sigma_{m=1}^{\infty} \delta_{\alpha_m}^j / 2^m) u_i + (\Sigma_{m=1}^{\infty} \delta_{\alpha_m}^i / 2^m) u_j \\
&= (\Sigma_{m=1}^{\infty} \delta_{\beta_m}^i / 2^m) u_i + (\Sigma_{m=1}^{\infty} \delta_{\beta_m}^j / 2^m) u_j = \phi(\beta) = \phi \circ \psi(\alpha).
\end{aligned}
$$

It follows that the pair α and $\beta = \psi(\alpha)$ provide $\mathcal{F}_3^*$ addresses $k\alpha_1 \alpha_2 \cdots$ and $\ell'\beta_1 \beta_2 \cdots$ of x where each ordered pair (α_n, β_n) is a member of $\{(j, i), (i, j)\}$. It remains, however, to compute the entire fiber $\phi^{-1}(x)$. So first suppose, as in the "$x \in (m_{jk}, m_{ik})$ row" of Table 88.3, that x is irrational. Then α and hence β have no tail index, and it follows from Table 88.2 that the "$x \in (m_{jk}, m_{ik})$ row" of Table 88.3 is correct. Second, suppose, as in the "$x \in (m_{jk}, m_{ik})$ row" of Table 88.4, that x is rational. Then α and hence β have a common tail index, and it follows from Table 88.2 that the "$x \in (m_{jk}, m_{ik})$ row" of Table 88.4 is correct.

To see that every address of $x \in (m_{jk}, m_{ik})$ has one of the forms indicated in either Table 88.3 or Table 88.4, let γ be any address of x. Then $\phi(\gamma) = x$ is Δ^3-interior to $T_k \cup T_{\ell'}$, which yields $\gamma_1 \in \{k, \ell'\}$. Suppose, for example, that $\gamma_1 = k$. Then $x = \phi(\gamma) = w_k(\phi(\gamma_2 \gamma_3 \cdots))$, showing that $\gamma_2 \gamma_3 \cdots$ is an address of $a \in (u_j, u_i)$. But then Table 88.2 shows that γ is among the forms listed in either the first row of Table 88.3 or the first row of Table 88.4. Since the $\gamma = \ell'$ case is similar, we conclude that the first rows of Tables 88.3 and 88.4 are complete. For the second rows of these tables, we have a similar development:

Fibers of points in (m_{ij}, m_{jk}). Observe that (m_{ij}, m_{jk}) is included in the Δ^3-interior of $T_j \cup T_{\ell'} \cup T_{i'}$; and that

$$
\begin{aligned}
m_{ij} &= w_j(u_i) = w_{\ell'}(u_k) = w_{i'}(u_i) \\
m_{jk} &= w_j(u_k) = w_{\ell'}(u_i) = w_{i'}(u_\ell).
\end{aligned}
$$

These equations yield the commutative diagram in Figure 88.7, where $\theta_{ji'} = \theta_{\ell'i'} \circ \theta_{j\ell'}$ and $\psi_{ji'} = \psi_{\ell'i'} \circ \psi_{j\ell'}$ have the obvious definitions. It follows that if $\alpha \in N(\{i, k\}) \setminus \{\bar{i}, \bar{k}\}$, then $\beta = \psi_{j\ell'}(\alpha)$, and $\gamma = \psi_{ji'}(\alpha)$ are such that $\beta \in N(\{k, i\}) \setminus \{\bar{k}, \bar{i}\}$ and $\gamma \in N(\{i, \ell\}) \setminus \{\bar{i}, \bar{\ell}\}$, and, $\phi(\alpha) = a \in (u_i, u_k)$, $\phi(\beta) = b \in (u_k, u_i)$, and $\phi(\gamma) = c \in (u_i, u_\ell)$ are such that $w_j(a) = w_{\ell'}(b) = w_{i'}(c) = x$. Thus, $j\alpha_1 \alpha_2 \cdots$, $\ell'\beta_1 \beta_2 \cdots$, and $i'\gamma_1 \gamma_2 \cdots$ are addresses of x. Moreover, because of the "ψ-mappings" we know that each ordered triple $(\alpha_n, \beta_n, \gamma_n) \in \{(i, k, i), (k, i, \ell)\}$. Finally, arguments similar to those following

Figure 88.6 show that the second rows of Tables 88.3 and 88.4 are both complete and correct.

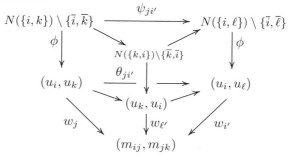

Fig. 88.7 Pasting diagram for (m_{ij}, m_{jk}).

Fibers of points in (P, Q). In this case, $(m_{01}, m_{23}) = (P, Q) \subset \text{int}(T_{0'} \cup T_{1'} \cup T_{2'} \cup T_{3'})$, and

$$
\begin{aligned}
m_{01} &= w_{0'}(u_0) = w_{1'}(u_1) = w_{2'}(u_3) = w_{3'}(u_2) \\
m_{23} &= w_{0'}(u_1) = w_{1'}(u_0) = w_{2'}(u_2) = w_{3'}(u_3).
\end{aligned}
$$

A "pasting diagram" similar to those above may be constructed where again, with orientations in mind, appropriate θ- and ψ-maps may be defined. Moreover, if $\alpha \in N(\{0,1\}) \setminus \{\overline{0}, \overline{1}\}$ and β, γ, and δ are given via $(\alpha_m, \beta_m, \gamma_m, \delta_m) \in \{(0,1,3,2), (1,0,2,3)\}$, then $0'\alpha$, $1'\beta$, $2'\gamma$, and $3'\delta$ are the addresses of $x = w_{0'}(\phi(\alpha))$ as listed in Table 88.3. Turning to Table 88.4, the constraints on (α_n, β_n), $(\alpha_n, \beta_n, \gamma_n)$, and $(\alpha_n, \beta_n, \gamma_n, \delta_n)$ are the same as those in Table 88.3 because the same pasting diagrams apply. (Each diagram is independent of the points being irrational or rational.) In the rational case, however, every address is eventually an address of a midpoint, which has a tail index.

§89 Fibers and Open 2-Faces of Δ^3

We begin with Table 89.1, (addresses for points in 2-faces via Theorem 86.3).

$x \in \Delta_{ijk}^3 \setminus \lvert F_1^1 \rvert$	CONSTRAINTS	ADDRESSES OF x ($\mathcal{F}_3^*$)	FIBER SIZE
for some $m \geq 1$ $x \in w_{\delta_1} \circ \cdots \circ w_{\delta_m}(\lvert F_1^1 \rvert)$ where $\delta_1, \ldots, \delta_m \in \{i, j, k, \ell'\}$	there is an $n \leq m$, $x = w_{\delta_1} \circ \cdots \circ w_{\delta_n}(y)$ for some $y \in (m_{ij}, m_{ik}) \cup (m_{ij}, m_{jk}) \cup (m_{ik}, m_{jk})$	$\delta_1 \cdots \delta_n a_y$ (a_y is any address of y)	$\lvert \phi^{-1}(y) \rvert$
$x \notin w_{\delta_1} \circ \cdots \circ w_{\delta_m}(\lvert F_1^1 \rvert)$ for every $m \geq 1$ where $\delta_1, \ldots, \delta_m \in \{i, j, k, \ell'\}$	one address $\alpha \in N(i,j,k,\ell')$ of x has either $\ell'\ell' \cdots$ or 3 constant subsequences	α	1

Table 89.1 Fibers of points in $\Delta_{ijk}^3 \setminus \lvert F_1^1 \rvert$.

An open 2-face of Δ^3 may be expressed as $(\Delta^3_{ijk} \setminus |F^1_1|) \cup (m_{jk}, m_{ik}) \cup$ $(m_{ij}, m_{jk}) \cup (m_{ij}, m_{ik})$. We calculate the fibers of points in this set using Tables 87.3, 88.2, 88.3, 88.4, and 89.1. First, fibers of points in $\Delta^3_{ijk} \setminus |F^1_1|$ are obtained by expanding the "a_y" term in Table 89.1. Second, fibers of points in the other sets follow directly from the other tables.

In Table 89.2 we continue to use the following Notation:

$$\{\{i,j\},\{k,\ell\}\} = \{\{0,1\},\{2,3\}\}, \quad w^{(n)} = \begin{cases} \text{identity} & \text{if } n = 0; \\ w_{\delta_1} \circ \cdots \circ w_{\delta_n} & \text{otherwise} \end{cases}$$

where $\delta_1, \ldots, \delta_n$ is a finite, possibly empty, list of members of $\{i, j, k, \ell'\}$. This list is denoted "$\delta^{(n)}$". Similarly, "$\alpha^{(t-1)}$" denotes a finite, possibly empty, list $\alpha_1, \ldots, \alpha_{t-1}$ of members of the doubleton set $\{r, s\} \subset \{0, 1, 2, 3\}$. And for the doubleton set $\{r, s\}$, recall that $\phi(a_{rs}) = m_{rs}$, i.e., "a_{rs}" denotes any address of the midpoint m_{rs}.

$a \in (u_i, u_j, u_k)$	Constraints	$\mathcal{F}^*_3$ Addresses of a	Size of Fiber	Type
$a \notin w^{(n)}(|F^1_1|)$ for every $n \geq 0$	$\alpha \in N(i,j,k,\ell')$ has $\overline{\ell'}$ or three constant subsequences	α	1	1.3
$a = w^{(n)}(y):$ $y \in (m_{jk}, m_{ik})$ is irrational	$\alpha \in N(\{j,i\}) \setminus \{\overline{j}, \overline{i}\}$ has no tail index; $(\alpha_m, \beta_m) \in \{(j,i),(i,j)\}$	$\delta^{(n)}\underline{k\alpha}$, $\delta^{(n)}\underline{\ell'\beta}$	2	2.1
$a = w^{(n)}(y):$ $y \in (m_{jk}, m_{ik})$ is rational	$\{r,s\} = \{j,i\}$ $(\alpha_m, \beta_m) \in \{(j,i),(i,j)\}$	$\delta^{(n)}\underline{k\alpha}^{(t-1)}a_{ji}$, $\delta^{(n)}\underline{\ell'\beta}^{(t-1)}a_{ij}$	12	12.1
$a = w^{(n)}(y):$ $y \in (m_{ij}, m_{jk})$ is irrational	$\alpha \in N(\{i,k\}) \setminus \{\overline{i}, \overline{k}\}$ has no tail index; $(\alpha_m, \beta_m, \gamma_m)$ $\in \{(i,k,i),(k,i,\ell)\}$	$\delta^{(n)}\underline{j\alpha}$, $\delta^{(n)}\underline{\ell'\beta}$, $\delta^{(n)}\underline{i'\gamma}$	3	3.1
$a = w^{(n)}(y):$ $y \in (m_{ij}, m_{jk})$ is rational	$\{r,s\} = \{i,k\}$; $(\alpha_m, \beta_m, \gamma_m)$ $\in \{(i,k,i),(k,i,\ell)\}$	$\delta^{(n)}\underline{j\alpha}^{(t-1)}a_{ik}$, $\delta^{(n)}\underline{\ell'\beta}^{(t-1)}a_{ki}$, $\delta^{(n)}\underline{i'\gamma}^{(t-1)}a_{i\ell}$	12	12.2

Table 89.2 Fibers of points in the open 2-face $(u_i, u_j, u_k) \subset \Delta^3_{ijk}$.

In Table 89.2, where the possible forms of fibers are listed, compare the "underlined strings" to the entries in Tables 88.3 and 88.4, and note that nonsingleton fibers are simply "shifts" of fibers of points in $(m_{jk}, m_{ik}) \cup$ $(m_{ij}, m_{jk}) \cup (m_{ij}, m_{ik})$, which is the pattern in the Δ^2 case (end of §86).

Fibers of points $a = w_{\delta_1} \circ \cdots \circ w_{\delta_n}(y)$ for an irrational point $y \in (m_{jk}, m_{ik})$. Table 88.3 provides $\alpha \in N(\{j, i\}) \setminus \{\overline{j}, \overline{i}\}$ with no tail index, each $(\alpha_m, \beta_m) \in$

$\{(j,i),(i,j)\}$, and $a_y = k\alpha$ or $a_y = \ell'\beta$. So $\phi^{-1}(a)$ contains ("*" indicates a Δ^2-address, i.e., an address in $N(\{i,j,k,\ell'\})$)

$$* \quad \delta_1 \cdots \delta_n k\alpha_1\alpha_2 \cdots$$
$$* \quad \delta_1 \cdots \delta_n \ell'\beta_1\beta_2 \cdots$$

Fibers of points $a = w_{\delta_1} \circ \cdots \circ w_{\delta_n}(y)$ *for an irrational point* $y \in (m_{ij}, m_{jk})$. Table 88.3 provides $\alpha \in N(\{i,k\}) \backslash \{\bar{i}, \bar{k}\}$ with no tail index, each $(\alpha_m, \beta_m, \gamma_m) \in \{(i,k,i),(k,i,\ell)\}$, and $a_y = j\alpha$, $\ell'\beta$, or $i'\gamma$. So $\phi^{-1}(a)$ contains ("*" indicates a Δ^2-address, i.e., an address in $N(\{i,j,k,\ell'\})$)

$$* \quad \delta_1 \cdots \delta_n j\alpha_1\alpha_2 \cdots$$
$$* \quad \delta_1 \cdots \delta_n \ell'\beta_1\beta_2 \cdots$$
$$\delta_1 \cdots \delta_n i'\gamma_1\gamma_2 \cdots$$

Fibers of points $a = w_{\delta_1} \circ \cdots \circ w_{\delta_n}(y)$ *for a rational point* $y \in (m_{jk}, m_{ik})$. Table 88.4 provides $\alpha_1, \ldots, \alpha_{t-1} \in \{j,i\}$ or the empty list; each $(\alpha_m, \beta_m) \in \{(j,i),(i,j)\}$; and $a_y = k\alpha^{(t-1)}a_{ji}$ or $\ell'\beta^{(t-1)}a_{ij}$. So $\phi^{-1}(a)$ contains ("*" indicates a Δ^2-address, i.e., an address in $N(\{i,j,k,\ell'\})$)

$* \quad \delta_1 \cdots \delta_n k\alpha_1 \cdots \alpha_{t-1}i\bar{j}$	$* \quad \delta_1 \cdots \delta_n \ell'\beta_1 \cdots \beta_{t-1}i\bar{j}$
$* \quad \delta_1 \cdots \delta_n k\alpha_1 \cdots \alpha_{t-1}j\bar{i}$	$* \quad \delta_1 \cdots \delta_n \ell'\beta_1 \cdots \beta_{t-1}j\bar{i}$
$\delta_1 \cdots \delta_n k\alpha_1 \cdots \alpha_{t-1}i'\bar{i}$	$\delta_1 \cdots \delta_n \ell'\beta_1 \cdots \beta_{t-1}i'\bar{i}$
$\delta_1 \cdots \delta_n k\alpha_1 \cdots \alpha_{t-1}j'\bar{j}$	$\delta_1 \cdots \delta_n \ell'\beta_1 \cdots \beta_{t-1}j'\bar{j}$
$\delta_1 \cdots \delta_n k\alpha_1 \cdots \alpha_{t-1}k'\bar{\ell}$	$\delta_1 \cdots \delta_n \ell'\beta_1 \cdots \beta_{t-1}k'\bar{\ell}$
$* \quad \delta_1 \cdots \delta_n k\alpha_1 \cdots \alpha_{t-1}\ell'\bar{k}$	$* \quad \delta_1 \cdots \delta_n \ell'\beta_1 \cdots \beta_{t-1}\ell'\bar{k}$

Fibers of points $a = w_{\delta_1} \circ \cdots \circ w_{\delta_n}(y)$ *for a rational point* $y \in (m_{ij}, m_{jk})$. Table 88.4 provides $\alpha_1, \ldots, \alpha_{t-1} \in \{i,k\}$ or the empty list; each $(\alpha_m, \beta_m, \gamma_m) \in \{(i,k,i),(k,i,\ell)\}$; and $a_y = j\alpha^{(t-1)}a_{ik}$, $\ell'\beta^{(t-1)}a_{ki}$, or $i'\gamma^{(t-1)}a_{i\ell}$. So $\phi^{-1}(a)$ contains ("*" indicates a Δ^2-address, i.e., an address in $N(\{i,j,k,\ell'\})$)

$* \, \delta_1 \cdots \delta_n j\alpha_1 \cdots \alpha_{t-1}i\bar{k}$	$* \, \delta_1 \cdots \delta_n \ell'\beta_1 \cdots \beta_{t-1}i\bar{k}$	$\delta_1 \cdots \delta_n i'\gamma_1 \cdots \gamma_{t-1}i\bar{\ell}$
$* \, \delta_1 \cdots \delta_n j\alpha_1 \cdots \alpha_{t-1}k\bar{i}$	$* \, \delta_1 \cdots \delta_n \ell'\beta_1 \cdots \beta_{t-1}k\bar{i}$	$\delta_1 \cdots \delta_n i'\gamma_1 \cdots \gamma_{t-1}\ell\bar{i}$
$\delta_1 \cdots \delta_n j\alpha_1 \cdots \alpha_{t-1}j'\bar{\ell}$	$\delta_1 \cdots \delta_n \ell'\beta_1 \cdots \beta_{t-1}j'\bar{\ell}$	$\delta_1 \cdots \delta_n i'\gamma_1 \cdots \gamma_{t-1}j'\bar{k}$
$* \, \delta_1 \cdots \delta_n j\alpha_1 \cdots \alpha_{t-1}\ell'\bar{j}$	$* \, \delta_1 \cdots \delta_n \ell'\beta_1 \cdots \beta_{t-1}\ell'\bar{j}$	$\delta_1 \cdots \delta_n i'\gamma_1 \cdots \gamma_{t-1}k'\bar{j}$

§90 Fibers and Open 2-Simplexes of K_1^2

The 2-skeleton K_1^2 contains 24 2-simplexes, which partition into three groups: The $4 \cdot 4 = 16$ in the boundary $\partial\Delta^3$, the four 2-faces $T_{i'} \cap T_{k'}$ that contain $[P,Q]$ as an edge, and the four 2-faces $T_\ell \cap T_{k'}$ where the tetrahedra T_ℓ $(\ell = 0,1,2,3)$ meet the octahedral hole (see Figure 90.1). Since the fibers of points in $\partial\Delta^3 = \cup\Delta_{ijk}^3$ were determined above, here we consider the latter two groups.

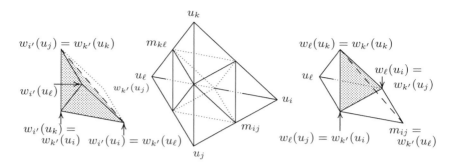

Fig. 90.1 Common face $T_{i'} \cap T_{k'}$ (left), and common face $T_\ell \cap T_{k'}$ (right).

Fibers of points in the open 2-simplex $(m_{ij}, m_{k\ell}, m_{j\ell})$. This 2-simplex, a subset of the interior of $T_{i'} \cup T_{k'}$, appears as the shaded area on the left side of Figure 90.1. Its vertices are given by

$$\left. \begin{array}{ccccc} m_{ij} & = & w_{i'}(u_i) & = & w_{k'}(u_\ell) \\ m_{k\ell} & = & w_{i'}(u_j) & = & w_{k'}(u_k) \\ m_{j\ell} & = & w_{i'}(u_k) & = & w_{k'}(u_i) \end{array} \right\} \text{ which yields Figure 90.2.}$$

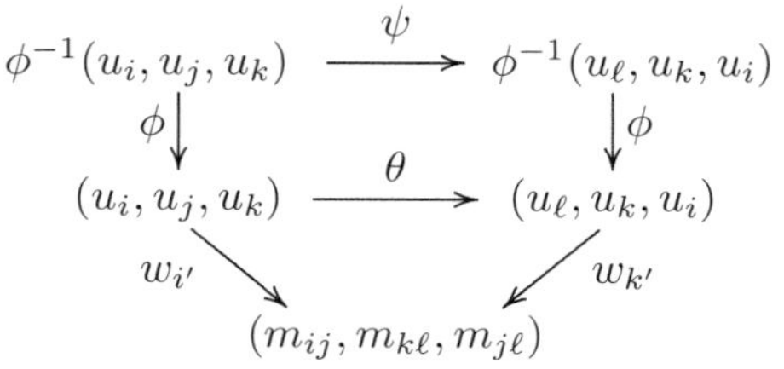

$$\begin{array}{ccc} \phi^{-1}(u_i, u_j, u_k) & \xrightarrow{\psi} & \phi^{-1}(u_\ell, u_k, u_i) \\ \phi \downarrow & & \downarrow \phi \\ (u_i, u_j, u_k) & \xrightarrow{\theta} & (u_\ell, u_k, u_i) \\ w_{i'} \searrow & & \swarrow w_{k'} \\ & (m_{ij}, m_{k\ell}, m_{j\ell}) & \end{array}$$

Fig. 90.2 Pasting diagram for $(m_{ij}, m_{k\ell}, m_{j\ell})$.

In Figure 90.2, $\psi = \rho \times \rho \times \cdots$ where ρ is the permutation $(i\ell jk)(i'\ell'j'k')$; and θ is the barycentric map that identifies Δ^3_{ijk} with $\Delta^3_{\ell ki}$ via $u_i \mapsto u_\ell$, $u_j \mapsto u_k$, and $u_k \mapsto u_i$. So

$$\phi \circ \psi = \theta \circ \phi.$$

(Recall that Table 89.2, as presented, provides the members of $\phi^{-1}(u_i, u_j, u_k)$ with input $a \in (u_i, u_j, u_k) \subset \Delta^3_{ijk}$. The substitutions ψ and θ appropriately applied to the data in Table 89.2 yield the corresponding table of members of $\phi^{-1}(u_\ell, u_k, u_i)$ with input $\theta(a) \in (u_\ell, u_k, u_i) \subset \Delta^3_{\ell ki}$.)

Moreover, if $a \in (u_i, u_j, u_k)$, then we may calculate that

$$\begin{aligned} w_{i'}(a) &= w_{i'}(a_i u_i + a_j u_j + a_k u_k) = a_i w_{i'}(u_i) + a_j w_{i'}(u_j) + a_k w_{i'}(u_k) \\ &= a_i w_{k'}(u_\ell) + a_j w_{k'}(u_k) + a_k w_{k'}(u_i) = w_{k'}(a_i u_\ell + a_j u_k + a_k u_i) \\ &= w_{k'} \circ \theta(a) = w_{k'}(b) \end{aligned}$$

where $\theta(a) = b$. It follows that if $\alpha \in \phi^{-1}(u_i, u_j, u_k)$ is an address of the unique $a \in (u_i, u_j, u_k)$ such that $w_{i'}(a) = x$, then $\psi(\alpha) = \beta$ is an address of the unique $b \in (u_\ell, u_k, u_i)$ such that $w_{k'}(b) = x$. So such an α provides addresses $i'\alpha_1\alpha_2\cdots$ and $k'\beta_1\beta_2\cdots$ of x. Details appear in Table 90.3.

Turning to Table 90.3, we again detail the meaning of the notation: In the "Type" column we use "$xx.y$" notation — the "xx" denotes the size of the fiber, while the "y" serves to index the fibers of size xx described in the adjacent "$\mathcal{F}_3^*$ Addresses" column. Also recall the assumption $\{\{i, j\}, \{k, \ell\}\} = \{\{0, 1\}, \{2, 3\}\}$ and the fact that "$w^{(n)}$" denotes either $w_{\delta_1} \circ \cdots \circ w_{\delta_n}$ or the identity when $n = 0$. And in the "$\mathcal{F}_3^*$ Addresses" column, "$\delta^{(n)}$" denotes either $\delta_1, \ldots, \delta_n \in \{i, j, k, \ell'\}$ or the empty list when $n = 0$, and, "$\alpha^{(t-1)}$" denotes either $\alpha_1, \ldots, \alpha_{t-1} \in \{r, s\} \subset \{0, 1, 2, 3\}$ or the empty list when $t = 1$. And recall that "a_{rs}" denotes any address of the midpoint m_{rs}.[1]

$x = w_{i'}(a)$ WHERE $a \in (u_i, u_j, u_k)$	CONSTRAINTS	$\mathcal{F}_3^*$ ADDRESSES OF x; $\psi = \rho \times \rho \times \cdots$ WHERE $\rho = (i\ell jk)(i'\ell'j'k')$	TYPE
$a \notin w^{(n)}(\lvert F_1^1 \rvert)$ for every $w^{(n)}$ $(n \geq 0)$	$\alpha \in N(\{i, j, k, \ell'\})$ has $\overline{\ell'}$ or 3 constant subsequences	$i'\alpha$ $k'\psi(\alpha)$	2.2
$a = w^{(n)}(y)$ $y \in (m_{jk}, m_{ik})$ irrational	$\alpha \in N(\{j, i\}) \setminus \{\overline{j}, \overline{i}\}$ no tail index (α_m, β_m) $\in \{(j, i), (i, j)\}$	$i'\delta^{(n)}k\alpha,$ $i'\delta^{(n)}\ell'\beta$ $k'\psi(\delta^{(n)}k\alpha), k'\psi(\delta^{(n)}\ell'\beta)$	4.3
$a = w^{(n)}(y):$ $y \in (m_{jk}, m_{ik})$ rational	$\{r, s\} = \{j, i\}$ (α_m, β_m) $\in \{(j, i), (i, j)\}$	$i'\delta^{(n)}k\alpha^{(t-1)}a_{ji}$ $i'\delta^{(n)}\ell'\beta^{(t-1)}a_{ij}$ $k'\psi(\delta^{(n)}k\alpha^{(t-1)}a_{ji})$ $k'\psi(\delta^{(n)}\ell'\beta^{(t-1)}a_{ij})$	24.2
$a = w^{(n)}(y):$ $y \in (m_{ij}, m_{jk})$ irrational	$\alpha \in N(\{i, k\}) \setminus \{\overline{i}, \overline{k}\}$ no tail index; $(\alpha_m, \beta_m,$ $\gamma_m) \in \{(i, k, i), (k, i, \ell)\}$	$i'\delta^{(n)}j\alpha, i'\delta^{(n)}\ell'\beta,$ $i'\delta^{(n)}i'\gamma, k'\psi(\delta^{(n)}j\alpha)$ $k'\psi(\delta^{(n)}\ell'\beta), k'\psi(\delta^{(n)}i'\gamma)$	6.2
$a = w^{(n)}(y):$ $y \in (m_{ij}, m_{jk})$ rational	$\{r, s\} = \{i, k\}$ $(\alpha_m, \beta_m, \gamma_m)$ $\in \{(i, k, i), (k, i, \ell)\}$	$i'\delta^{(n)}j\alpha^{(t-1)}a_{ik},$ $i'\delta^{(n)}\ell'\beta^{(t-1)}a_{ki}$ $i'\delta^{(n)}i'\gamma^{(t-1)}a_{i\ell}$ $k'\psi(\delta^{(n)}j\alpha^{(t-1)}a_{ik})$ $k'\psi(\delta^{(n)}\ell'\beta^{(t-1)}a_{ki})$ $k'\psi(\delta^{(n)}i'\gamma^{(t-1)}a_{i\ell})$	24.3

Table 90.3 Fibers of points in $(m_{ij}, m_{k\ell}, m_{j\ell})$.

Fibers of points in the open 2-simplex $(m_{\ell i}, m_{\ell j}, m_{\ell k})$. This open 2-simplex, which is interior to $T_\ell \cup T_{k'}$, appears as the shaded area on the right-side of Figure 90.1. Its vertices are given by

$$
\begin{aligned}
m_{\ell i} &= w_\ell(u_i) &= w_{k'}(u_j) \\
m_{\ell j} &= w_\ell(u_j) &= w_{k'}(u_i) \\
m_{\ell k} &= w_\ell(u_k) &= w_{k'}(u_k).
\end{aligned}
$$

---

[1] For more details on the notation, see the paragraph preceding Table 89.2.

This case is similar to the previous "2-simplex $(m_{ij}, m_{k\ell}, m_{j\ell})$" case: In Figure 90.2, replace $\phi^{-1}(u_\ell, u_k, u_i)$ with $\phi^{-1}(u_j, u_i, u_k)$, replace ψ with $\zeta = \rho \times \rho \times \cdots$ where ρ is the permutation $(ij)(k)(\ell)(i'j')(k')(\ell')$, make θ compatible to ζ, i.e., make θ the barycentric map that identifies Δ^3_{ijk} with Δ^3_{jik} via $u_i \mapsto u_j$, $u_j \mapsto u_i$, and $u_k \mapsto u_k$. Then replace (u_ℓ, u_k, u_i) with (u_j, u_i, u_k), replace $w_{i'}$ with w_ℓ, and finally replace $(m_{ij}, m_{k\ell}, m_{j\ell})$ with $(m_{\ell i}, m_{\ell j}, m_{\ell k})$. (It may help to notice the $w_{k'}$ and w_ℓ labels on the right-side graphic in Figure 90.1.)

The resulting diagram is commutative, implying that if $\alpha \in \phi^{-1}(u_i, u_j, u_k)$ is an address of a where $w_\ell(a) = x$, then $\beta = \zeta(\alpha)$ is an address of $\theta(a) = b$ and $w_\ell(a) = x = w_{k'}(b)$. So each such α produces two addresses $\ell\alpha_1\alpha_2 \cdots$ and $k'\beta_1\beta_2 \cdots$ of x. Parallel to the previous case, we know the members of $\phi^{-1}(a)$ from Table 89.2, and we calculate those of $\phi^{-1}(b)$ using ζ instead of ψ and using our newly defined θ. The results are summarized in Table 90.4.

$x = w_{i'}(a)$ where $a \in (u_i, u_j, u_k)$	CONSTRAINTS	$\mathcal{F}_3^*$ ADDRESSES OF x; $\zeta = \rho \times \rho \times \cdots$ where $\rho = (ij)(i'j')$	TYPE
$a \notin w^{(n)}(\lvert F_1^1 \rvert)$ for every $w^{(n)}$ $(n \geq 0)$	$\alpha \in N(\{i,j,k,\ell'\})$ has $\overline{\ell'}$ or 3 constant subsequences	$\ell\alpha$, $k'\zeta(\alpha)$	2.3
$a = w^{(n)}(y)$ $y \in (m_{jk}, m_{ik})$ irrational	$\alpha \in N(\{j,i\}) \setminus \{\overline{j}, \overline{i}\}$ no tail index; $(\alpha_m, \beta_m) \in \{(j,i),(i,j)\}$	$\ell\delta^{(n)}k\alpha$, $\ell\delta^{(n)}\ell'\beta$ $k'\zeta(\delta^{(n)}k\alpha)$, $k'\zeta(\delta^{(n)}\ell'\beta)$	4.4
$a = w^{(n)}(y)$: $y \in (m_{jk}, m_{ik})$ rational	$\{r,s\} = \{j,i\}$ (α_m, β_m) $\in \{(j,i),(i,j)\}$	$\ell\delta^{(n)}k\alpha^{(t-1)}a_{ji}$ $\ell\delta^{(n)}\ell'\beta^{(t-1)}a_{ij}$ $k'\zeta(\delta^{(n)}k\alpha^{(t-1)}a_{ji})$ $k'\zeta(\delta^{(n)}\ell'\beta^{(t-1)}a_{ij})$	24.4
$a = w^{(n)}(y)$: $y \in (m_{ij}, m_{jk})$ irrational	$\alpha \in N(\{i,k\}) \setminus \{\overline{i}, \overline{k}\}$ no tail index; $(\alpha_m, \beta_m, \gamma_m) \in \{(i,k,i),(k,i,\ell)\}$	$\ell\delta^{(n)}j\alpha$, $\ell\delta^{(n)}\ell'\beta$, $\ell\delta^{(n)}i'\gamma$, $k'\zeta(\delta^{(n)}j\alpha)$, $k'\zeta(\delta^{(n)}\ell'\beta)$, $k'\zeta(\delta^{(n)}i'\gamma)$	6.3
$a = w^{(n)}(y)$: $y \in (m_{ij}, m_{jk})$ rational	$\{r,s\} = \{i,k\}$ $(\alpha_m, \beta_m, \gamma_m)$ $\in \{(i,k,i),(k,i,\ell)\}$	$\ell\delta^{(n)}j\alpha^{(t-1)}a_{ik}$ $\ell\delta^{(n)}\ell'\beta^{(t-1)}a_{ki}$ $\ell(\delta^{(n)}i'\gamma^{(t-1)}a_{i\ell})$ $k'\zeta(\delta^{(n)}j\alpha^{(t-1)}a_{ik})$ $k'\zeta(\delta^{(n)}\ell'\beta^{(t-1)}a_{ki})$ $k'\zeta(\delta^{(n)}i'\gamma^{(t-1)}a_{i\ell})$	24.5

Table 90.4 Fibers of points in $(m_{\ell i}, m_{\ell j}, m_{\ell k})$.

§91 Singleton Fibers

Our first theorem characterizes singleton fibers of points in $\partial\Delta^3$.

91.1 Theorem (singleton fibers and $\partial\Delta^3$) *Let ϕ be the $\mathcal{F}_3^*$ address map, and let $\lvert \phi^{-1}(x) \rvert = 1$. Then $\phi(\alpha) = x \in \partial\Delta^3$ if and only if α is either $(i)_\partial$*

constant, $(ii)_\partial$ in a doubleton $K \subset \{0, 1, 2, 3\}$ and has no tail index, $(iii)_\partial$ in a quadruple $\{i, j, k, \ell'\}$ and has subsequence $\overline{\ell'}$, or, $(iv)_\partial$ in a quadruple $\{i, j, k, \ell'\}$ and has subsequences $\overline{i}$, $\overline{j}$, and $\overline{k}$.

PROOF. Since $|\phi^{-1}(x)| = 1$, Tables 87.3 and 88.2 show that $\phi(\alpha) = x$ is in an edge $[u_r, u_s]$ of Δ^3 if and only if α satisfies either $(i)_\partial$ with $\alpha \in \{\overline{0}, \overline{1}, \overline{2}, \overline{3}\}$ or $(ii)_\partial$. Similarly, since $|\phi^{-1}(x)| = 1$, Table 89.2 shows that $\phi(\alpha) = x$ is in an open 2-face (u_i, u_j, u_k) of Δ^3 if and only if α satisfies either $(i)_\partial$ with $\alpha = \overline{\ell'}$, or $(iii)_\partial$, or $(iv)_\partial$. $\square$

For the results below, we shall use the following notation: For any $\alpha \in N(A)$ and any $t \geq 0$, we let $\alpha^t = \alpha_{t+1}\alpha_{t+2}\cdots$ and refer to α^t as a *tail of* α. Our next two lemmas combine to characterize those sequences α that occupy singleton fibers of points x in the interior $\mathrm{int}(\Delta^3)$ of Δ^3.

91.2 Lemma (singleton fibers, $\mathrm{int}(\Delta^3)$, a necessary condition) *Let ϕ be the $\mathcal{F}_3^*$ address map, and let $|\phi^{-1}(x)| = 1$. Then $\phi(\alpha) = x \in \mathrm{int}(\Delta^3)$ implies α has either $(i)_{\mathrm{int}}$ two (primed) subsequences $\overline{a'}$ and $\overline{b'}$ ($a, b \in \{0, 1, 2, 3\}$), or, $(ii)_{\mathrm{int}}$ subsequences $\overline{\ell}$ and $\overline{\ell'}$, or, $(iii)_{\mathrm{int}}$ subsequences $\overline{0}$, $\overline{1}$, $\overline{2}$, $\overline{3}$.*

PROOF. Since $x \in \mathrm{int}(\Delta^3)$, either $x \in |K_1^2| \cap \mathrm{int}(\Delta^3)$ or $x \in \mathrm{int}(\Delta^3) \setminus |K_1^2|$. But the former case is not possible because $|\phi^{-1}(x)| = 1$ together with Tables 88.3 and 88.4 show that $x \notin (P, Q)$, and $|\phi^{-1}(x)| = 1$ together with Tables 90.3 and 90.4 show that x is not in a 2-simplex of K_1^2. So $x \in \mathrm{int}(\Delta^3) \setminus |K_1^2|$. Next, we note that

$$x \notin w_{\delta_1} \circ \cdots \circ w_{\delta_n}(|K_1^2|) \qquad \text{for each } (n \geq 1)\text{-list } \delta_1, \ldots, \delta_n \in A.$$

(Otherwise, $x = w_{\delta_1} \circ \cdots \circ w_{\delta_n}(y)$ for some $y \in |K_1^2|$, and Theorem 86.2 tells us that y has only one address, which implies by Tables 87.3 through 93.1 that $y \in \partial\Delta^3$. Thus, x is in the boundary of the 3-simplex $w_{\delta_1} \circ \cdots \circ w_{\delta_n}(\Delta^3) \in K_n$. So x, being in the interior of Δ^3, must have more than one address, which contradicts $|\phi^{-1}(x)| = 1$.) It follows, since $x = \phi(\alpha) = w_{\alpha_1} \circ \cdots \circ w_{\alpha_t}(\phi(\alpha^t))$, that no tail $\alpha^t = \alpha_{t+1}\alpha_{t+2}\cdots$ ($t \geq 1$) of α is an address of any $y \in |K_1^2|$. And since $x \notin |K_1^2|$, $\alpha = \alpha_1\alpha_2\cdots$ itself is not an address of any $y \in |K_1^2|$. More concisely, for each $t \geq 0$, $\phi(\alpha^t) \notin |K_1^2|$, which implies (since $\partial\Delta^3 \subset |K_1^2|$) that for each $t \geq 0$, $\phi(\alpha^t) \notin \partial\Delta^3$. So Theorem 91.1 tells us that each α^t cannot satisfy any of $(i)_\partial$, $(ii)_\partial$, $(iii)_\partial$, or $(iv)_\partial$. That is, each α^t is

- (P1) not constant;
- (P2) not in a doubleton $K \subset \{0, 1, 2, 3\}$, or, is in such a K and has a tail index;
- (P3) not in an $F = \{i, j, k, \ell'\}$, or, is in such an F and has no subsequence $\overline{\ell'}$; and
- (P4) not in an $F = \{i, j, k, \ell'\}$, or, is in such an F and has at most two subsequences $\overline{i}$, $\overline{j}$.

In particular, since each α^t satisfies (P1), α must have at least two constant subsequences. For the first case, suppose two subsequences are $\overline{i'}$ and $\overline{\ell'}$: Then α satisfies $(i)_{\text{int}}$. For the second case, suppose two subsequences are $\overline{\ell}$ and $\overline{\ell'}$: Then α satisfies $(ii)_{\text{int}}$. For the third case, suppose α satisfies neither $(i)_{\text{int}}$ nor $(ii)_{\text{int}}$, but has two subsequences $\overline{i}$ and $\overline{\ell'}$. So there is a tail α^t with each of its values in $A \setminus \{\ell, i', j', k'\} = \{i, j, k, \ell'\}$. Since α^t must satisfy (P3), however, we have a contradiction. So our third case is impossible. Since the first three cases exhaust the possibilities of having an $\overline{\ell'}$ as a subsequence of α, we consider the fourth case, where α has no $\overline{\ell'}$ subsequence, but has two "unprimed" subsequences $\overline{i}$ and $\overline{j}$: In this case, there is an $m \geq 0$ such that α^m is a sequence in $\{0, 1, 2, 3\}$. Since each tail of α^m is a sequence in $\{0, 1, 2, 3\}$ that satisfies (P2), however, α^m (and hence α and each tail α^t of α) must have yet another constant subsequence $\overline{k}$ in $\{0, 1, 2, 3\}$. Then, since each α^t has three subsequences $\overline{i}$, $\overline{j}$, $\overline{k}$ and also satisfies (P4), each α^t cannot be a sequence in any set of the form $\{i, j, k, \ell'\}$. Thus, each α^t must contain the four subsequences $\overline{0}$, $\overline{1}$, $\overline{2}$, $\overline{3}$. We conclude that α satisfies $(iii)_{\text{int}}$. Since all possibilities of pairs of subsequences have been exhausted, we conclude that α satisfies one of $(i)_{\text{int}}$, $(ii)_{\text{int}}$, or $(iii)_{\text{int}}$.   □

91.3 Lemma (singleton fibers, int(Δ^3), a sufficiency condition) *Let ϕ be the $\mathcal{F}_3^*$ address map, and let α have either $(i)_{\text{int}}$ two (primed) subsequences $\overline{a'}$ and $\overline{b'}$ ($a, b \in \{0, 1, 2, 3\}$), $(ii)_{\text{int}}$ subsequences $\overline{\ell}$ and $\overline{\ell'}$, or $(iii)_{\text{int}}$ subsequences $\overline{0}$, $\overline{1}$, $\overline{2}$, $\overline{3}$. Then $\phi(\alpha) = x \in \text{int}(\Delta^3)$ and $|\phi^{-1}(x)| = 1$.*

PROOF. Tables 87.3, 88.2, 88.3, 88.4, and 89.2 show that each tail of any address of a boundary point contains neither two constant "primed" subsequences, nor, a pair ℓ and ℓ', nor, the four sequences $\overline{0}$, $\overline{1}$, $\overline{2}$, $\overline{3}$ as subsequences. It follows that $\phi(\alpha) = x \in \text{int}(\Delta^3)$. Likewise, $\phi(\alpha) = x \notin (P, Q)$ (and similarly, each $\phi(\alpha^t) \notin (P, Q)$), and $\phi(\alpha) = x \notin |K_1^2|$ (and similarly, each $\phi(\alpha^t) \notin |K_1^2|$). We also claim that

$$\phi(\alpha) \notin w_{\delta_1} \circ \cdots \circ w_{\delta_n}(|K_1^2|)$$

for every $(n \geq 1)$-list $\delta_1, \ldots, \delta_n \in A$. Suppose otherwise. Then there is a minimum $(n \geq 1)$-list $\delta_1, \ldots, \delta_n$ such that $\phi(\alpha) = w_{\delta_1} \circ \cdots \circ w_{\delta_n}(\phi(\beta))$ where $\phi(\beta) \in |K_1^2|$. Since each $\delta_m = \alpha_m$ (see the proof of Theorem 86.2) and since $\phi(\alpha) = w_{\alpha_1} \circ \cdots \circ w_{\alpha_n}(\phi(\alpha^n))$, we find that $\phi(\alpha^n) = \phi(\beta) \in |K_1^2|$. But this containment contradicts $\phi(\alpha^n) \notin |K_1^2|$. So the displayed equation holds under the stated conditions, and, then an application of Theorem 86.2 shows that $|\phi^{-1}(x)| = 1$.   □

From Lemmas 91.2 and 91.3 we have the following theorem, which characterizes those sequences that occupy singleton fibers of points in int(Δ^3).

91.4 Theorem *Let ϕ be the $\mathcal{F}_3^*$ address map, and let $\phi(\alpha) = x \in \text{int}(\Delta^3)$. Then $|\phi^{-1}(x)| = 1$ if and only if α has either $(i)_{\text{int}}$ two (primed) subsequences $\overline{a'}$ and $\overline{b'}$ ($a, b \in \{0, 1, 2, 3\}$), or, $(ii)_{\text{int}}$ subsequences $\overline{\ell}$ and $\overline{\ell'}$, or, $(iii)_{\text{int}}$ subsequences $\overline{0}$, $\overline{1}$, $\overline{2}$, $\overline{3}$.*

So Theorems 91.1 and 91.4 yield the following theorem.

91.5 Theorem (singleton fibers II) *Let ϕ be the $\mathcal{F}_3^*$ address map, and let $\phi(\alpha) = x$. Then $|\phi^{-1}(x)| = 1$ if and only if α is either constant, or, in a doubleton $K \subset \{0,1,2,3\}$ with no tail index, or, in a quadruple $\{i,j,k,\ell'\}$ with either $\overline{\ell'}$ or each of $\overline{i}, \overline{j}, \overline{k}$ as subsequences, or, is in A and, for some representation $\{i,j,k,\ell,i',j',k',\ell'\}$ of A, has either $(\overline{i'}$ and $\overline{\ell'})$ or $(\overline{\ell}$ and $\overline{\ell'})$ or each of $\overline{0}, \overline{1}, \overline{2}, \overline{3}$ as subsequences.*

Each condition listed in Theorem 91.5 concerns constant subsequences: Call a sequence α "(m,k)" whenever its range has size m and it has exactly k constant subsequences. Then α is $(1,1)$ if and only if it is constant. And if α is in a doubleton $K \subset \{0,1,2\}$ with no tail index, then it is $(2,2)$. It is easy to show that in the Δ^2-case, a sequence α is a member of a singleton fiber $\{\alpha\}$ if and only if it contains the subsequence $\overline{3} = 333\cdots$ or is either $(1,1)$, $(2,2)$, $(3,3)$, or $(4,3)$.

§92 Fibers of Points in $\Delta^3 \setminus |K_1^2|$

Knowing all singleton fibers of all points in Δ^3 as well as all fibers of all points in $|K_1^2|$, we turn to fibers of points in $\Delta^3 \setminus |K_1^2|$. The approach is that of merging the (singleton fibers II) Theorem 91.5 and the (inductive step) Theorem 86.2.

$x \in \Delta^3 \setminus |K_1^2|$	CONSTRAINTS	$\mathcal{F}_3^*$ ADDRESSES	$|\phi^{-1}(x)|$
for some $m \geq 1$, $x \in$ $w_{\varepsilon_1} \circ \cdots \circ w_{\varepsilon_m}(|K_1^2|)$ where $\varepsilon_1, \ldots, \varepsilon_m \in A$	there is an r, $1 \leq r$ $\leq m$, $x = w_{\varepsilon_1} \circ \cdots$ $\cdots \circ w_{\varepsilon_r}(y)$ for some $y \in \text{int}(\Delta^3) \cap |K_1^2|$	$\varepsilon_1 \cdots \varepsilon_r a_y$ where a_y is any address of y	$|\phi^{-1}(y)|$
$x \notin w_{\varepsilon_1} \circ \cdots$ $\cdots \circ w_{\varepsilon_m}(|K_1^2|)$ for every $(m \geq 1)$-length string $\varepsilon_1, \ldots, \varepsilon_m \in A$	lone address α of x satisfies one of the conditions in Theorem 91.5	α	1

Table 92.1 Fibers of points in $\Delta^3 \setminus |K_1^2|$.

So Table 92.1 implies that nonsingleton fibers of points in $\Delta^3 \setminus |K_1^2|$ are "shifts" of those fibers of points in

$$\text{int}(\Delta^3) \cap |K_1^2| = \bigcup_U \{(m_{ij}, m_{k\ell}, m_{j\ell}) \cup (m_{\ell i}, m_{\ell j}, m_{\ell k}) \cup (m_{ij}, m_{k\ell})\}$$

where

$$U = \{\{i,j\}, \{k,\ell\}\} = \{\{0,1\}, \{2,3\}\}.$$

In other words, the "a_y" term in Table 92.1 may be calculated via Tables 90.3 and 90.4 for points in the open 2-faces, and via Tables 88.3 and 88.4 for points in $(m_{ij}, m_{k\ell})$.

§93 Summary Table, Octic Group, Choice of Letters

Cardinalities of fibers of the address map ϕ of $\mathcal{F}_3^*$ are 1, 2, 3, 4, 6, 12, and 24. The details appear in the (summary) Table 93.1 (Parts I, II, and III), where the data are encoded using the letters i, j, k, ℓ such that $\{\{i,j\},\{k,\ell\}\} = \{\{0,1\},\{2,3\}\}$.

Using this encoding, we classified *eight* 2-simplexes in K_1 (those that meet the interior of Δ^3) into two kinds, namely, $[m_{ij}, m_{k\ell}, m_{j\ell}]$ and $[m_{\ell i}, m_{\ell j}, m_{\ell k}]$ (Figure 90.1). In more detail, let us calculate the *four* (A, B, C, D) of type $[m_{ij}, m_{k\ell}, m_{j\ell}]$ — using the constraint $\{\{i,j\},\{k,\ell\}\} = \{\{0,1\},\{2,3\}\}$ we calculate the following:

$$\begin{smallmatrix}i=0,j=1\\k=2,\ell=3\end{smallmatrix} \to [m_{01}, m_{23}, m_{13}] = A \qquad \begin{smallmatrix}i=1,j=0\\k=2,\ell=3\end{smallmatrix} \to [m_{10}, m_{23}, m_{03}] = B$$

$$\begin{smallmatrix}i=0,j=1\\k=3,\ell=2\end{smallmatrix} \to [m_{01}, m_{32}, m_{12}] = C \qquad \begin{smallmatrix}i=1,j=0\\k=3,\ell=2\end{smallmatrix} \to [m_{10}, m_{32}, m_{02}] = D$$

$$\begin{smallmatrix}i=2,j=3\\k=0,\ell=1\end{smallmatrix} \to [m_{23}, m_{01}, m_{31}] = A \qquad \begin{smallmatrix}i=3,j=2\\k=0,\ell=1\end{smallmatrix} \to [m_{32}, m_{01}, m_{21}] = C$$

$$\begin{smallmatrix}i=2,j=3\\k=1,\ell=0\end{smallmatrix} \to [m_{23}, m_{10}, m_{30}] = B \qquad \begin{smallmatrix}i=3,j=2\\k=1,\ell=0\end{smallmatrix} \to [m_{32}, m_{10}, m_{20}] = D.$$

In turn, these two representations yielded, via corresponding "pastings," two permutations $\rho_\zeta = (ij)(k)(\ell)$ and $\rho_\psi = (i\ell jk)$ which, respectively, fix and transpose the members of $\{\{i,j\},\{k,\ell\}\}$.

In general, these two permutations form the *octic group*

$$G = \langle \rho_\zeta, \rho_\psi \rangle = \{(1, (ij), (k\ell), (ij)(k\ell), (i\ell jk), (kj\ell i), (ik)(j\ell), (i\ell)(jk)\},$$

a subgroup of symmetric group $S_{\{i,j,k,\ell\}}$ generated by ρ_ζ and ρ_ψ.

And for the set of letters

$$A^l = \{i, j, k, \ell, i', j', k', \ell'\},$$

the group G may also be viewed as the subgroup of the symmetric group S_{A^l} with generators

$$(ij)(k)(\ell)(i'j')(k')(\ell') \text{ and } (i\ell jk)(i'\ell'j'k').$$

The role of G relative to Table 93.1 may be summarized as follows: Call a bijection $x : \{i, j, k, \ell\} \to \{0, 1, 2, 3\}$ such that $\{\{x(i), x(j)\}, \{x(k), x(\ell)\}\} = \{\{0,1\},\{2,3\}\}$ a *choice* of i, j, k, ℓ. (Or view $x : A^l \to A^l$ as the obvious extension.) For each choice x, each $\kappa \in N(A)$ has the representation (string of letters)

$$\kappa^x = x^{-1} \circ \kappa = x^{-1}(\kappa_1) x^{-1}(\kappa_2) \cdots \in N(A^l).$$

NOTATION: $A = \{0, 1, 2, 3, 0', 1', 2', 3'\}$, $\{\{i, j\}, \{k, \ell\}\} = \{\{0, 1\}, \{2, 3\}\}$, int denotes 3-space interior, int_{Δ^3} the interior rel Δ^3, $\overline{\ell'}$ Type 1.1 & Type 1.3, $\delta^{(n)} = \delta_1, \cdots, \delta_n \in \{i, j, k, \ell'\}$ or empty list, $\alpha^{(t-1)} = \alpha_1, \cdots, \alpha_{t-1} \in \{r, s\}$ or empty list, $\varepsilon^{(r)} = \varepsilon_1, \cdots, \varepsilon_r \in A$ or empty list, $F = \{i, j, k, \ell'\}, K \subset \{0, 1, 2, 3\}$ denotes a doubleton subset, a_{rs} an address of m_{rs}, $m_{rs} = (1/2)(u_r + u_s)$, $\psi = \rho \times \rho \times \cdots$ where $\rho = (i\ell jk)(i'\ell'j'k')$, $\zeta = \rho \times \rho \times \cdots$ where $\rho = (ij)(i'j')$, and $(P, Q) = (m_{ij}, m_{k\ell})$.

TYPE	ADDRESSES	MNEMONIC	IN FIBER OF	REF.
1.1	$\overline{a}$	$a \in \{0, 1, 2, 3\} \cup$ $\{0', 1', 2', 3'\}$	vertex or 2-face barycenter	87.3 89.2
1.2	$\alpha \in N(K)$	non-constant with no tail index	irrational in an open edge	88.2 89.1
1.3	$\alpha \in N(F)$	$\overline{\ell'}/(\overline{i}\&\overline{j}\&\overline{k})$ subsequences	point in 2-face Δ^3_{ijk}	89.2
1.4	$\alpha \in N(A)$	subsequences $(\overline{i'}\&\overline{\ell'})/$ $(\overline{\ell}\&\overline{\ell'})/(\overline{0}\&\overline{1}\&\overline{2}\&\overline{3})$	point in $\mathrm{int}(\Delta^3)$	91.4
2.1	$\delta^{(n)}k\alpha$ $\delta^{(n)}\ell'\beta$	$\alpha, \beta \in N(i, j)$ of Type 1.2, each $\{\alpha_m, \beta_m\} = \{i, j\}$	shift of irrational in $(m_{jk}, m_{ik}) \subset$ $\mathrm{int}_{\Delta^3}(T_k \cup T_{\ell'})$	88.3/.6 89.2 87.2
2.2	$\varepsilon^{(r)}i'\alpha$ $\varepsilon^{(r)}k'\beta$	α Type 1.3, (α_m, β_m) $\in \{(i, \ell), (j, k),$ $(k, i), (\ell', j')\}$	shift of point in $(m_{ij}, m_{k\ell}, m_{j\ell}) \subset$ $\mathrm{int}(T_{i'} \cup T_{k'})$	90.3/.2 90.1 92.1
2.3	$\varepsilon^{(r)}\ell\alpha$ $\varepsilon^{(r)}k'\beta$	α Type 1.3, (α_m, β_m) $\in \{(i, j), (j, i),$ $(k, k), (\ell', \ell')\}$	shift of point in $(m_{\ell i}, m_{\ell j}, m_{\ell k}) \subset$ $\mathrm{int}(T_\ell \cup T_{k'})$	90.4 92.1 90.1
3.1	$\delta^{(n)}j\alpha$ $\delta^{(n)}\ell'\beta$ $\delta^{(n)}i'\gamma$	α Type 1.2, $(\alpha_m, \beta_m, \gamma_m) \in$ $\{(i, k, i), (k, i, \ell)\}$	shift of point in $(m_{ij}, m_{jk}) \subset \mathrm{int}_{\Delta^3}$ $(T_j \cup T_\ell \cup T_{i'})$	88.3/.7 89.2 87.2
4.1	$\alpha^{(t-1)}a_{jk}$	$a_{jk} \in \{j\overline{k}, k\overline{j}, i'\overline{\ell}, \ell'\overline{i}\}$	rational in (u_j, u_k)	87.3 88.2
4.2	$\varepsilon^{(r)}0'\alpha$ $\varepsilon^{(r)}1'\beta$ $\varepsilon^{(r)}2'\gamma$ $\varepsilon^{(r)}3'\delta$	α Type 1.2, $(\alpha_m, \beta_m,$ $\gamma_m, \delta_m) \in \{(0, 1, 3, 2),$ $(1, 0, 2, 3)\}$	shift of irrational in $(P, Q) \subset \mathrm{int}(T_{0'} \cup$ $T_{1'} \cup T_{2'} \cup T_{3'})$	88.3 92.1 87.2
4.3	$\varepsilon^{(r)}i'\lambda$ $\varepsilon^{(r)}i'\mu$ $\varepsilon^{(r)}k'\nu$ $\varepsilon^{(r)}k'\xi$	$\{\lambda = \lambda_\alpha, \mu = \mu_\beta\}$ Type 2.1, $\nu = \psi(\lambda_\alpha),$ $\xi = \psi(\mu_\beta)$	$\varepsilon^{(r)}$-shift of point in $(m_{ij}, m_{k\ell}, m_{j\ell}) \subset$ $\mathrm{int}(T_{i'} \cup T_{k'})$	90.3 92.1 90.2
4.4	$\varepsilon^{(r)}\ell\lambda$ $\varepsilon^{(r)}\ell\mu$ $\varepsilon^{(r)}k'\nu$ $\varepsilon^{(r)}k'\xi$	$\{\lambda = \lambda_\alpha, \mu = \mu_\beta\}$ Type 2.1, $\nu = \zeta(\lambda_\alpha),$ $\xi = \zeta(\mu_\beta)$	$\varepsilon^{(r)}$-shift of point in $(m_{\ell i}, m_{\ell j}, m_{\ell k}) \subset$ $\mathrm{int}(T_\ell \cup T_{k'})$	90.4 92.1

Table 93.1 (Part I) Summary table for fibers of sizes 1, 2, 3, and 4.

And for any choices x and y, we define the *change of choice* $y^{-1} \circ x = \pi \in G$. The upshot is that if κ^x is not in Table 93.1, then for some $\pi = y^{-1} \circ x$, the representation $\kappa^y = \pi \circ \kappa^x$ of κ explicitly appears in Table 93.1.

NOTATION: $A = \{0, 1, 2, 3, 0', 1', 2', 3'\}$, $\{\{i, j\}, \{k, \ell\}\} = \{\{0, 1\}, \{2, 3\}\}$, int denotes 3-space interior, int_{Δ^3} the interior rel Δ^3, $\overline{\ell'}$ Type 1.1 & Type 1.3, $\delta^{(n)} = \delta_1, \cdots, \delta_n \in \{i, j, k, \ell\}$ or empty list, $\alpha^{(t-1)} = \alpha_1, \cdots, \alpha_{t-1} \in \{r, s\}$ or empty list, $\varepsilon^{(r)} = \varepsilon_1, \cdots, \varepsilon_r \in A$ or empty list, $F = \{i, j, k, \ell'\}$, $K \subset \{0, 1, 2, 3\}$ denotes a doubleton subset, a_{rs} an address of m_{rs}, $m_{rs} = (1/2)(u_r + u_s)$, $\psi = \rho \times \rho \times \cdots$ where $\rho = (i\ell jk)(i'\ell' j'k')$, $\zeta = \rho \times \rho \times \cdots$ where $\rho = (ij)(i'j')$, and $(P, Q) = (m_{ij}, m_{k\ell})$.

TYPE	ADDRESSES	MNEMONIC	IN FIBER OF	REF.
6.1	$\alpha^{(t-1)} b_{nm}$	$b_{nm} = n\overline{m} \in \{i\overline{j}, j\overline{i},$ $i'\overline{i}, j'\overline{j}, k'\overline{\ell}, \ell'\overline{k}\}$	rational point in edge (u_i, u_j)	87.3 88.2
6.2	$\varepsilon^{(r)} i' \lambda$ $\varepsilon^{(r)} i' \mu$ $\varepsilon^{(r)} i' \nu$ $\varepsilon^{(r)} k' \xi$ $\varepsilon^{(r)} k' \pi$ $\varepsilon^{(r)} k' \sigma$	$\{\lambda = \lambda_\alpha, \mu = \mu_\beta,$ $\nu = \nu_\gamma\}$ Type 3.1, $\xi = \psi(\lambda_\alpha),$ $\pi = \psi(\mu_\beta),$ $\sigma = \psi(\nu_\gamma)$	$\varepsilon^{(r)}$-shift of point in $(m_{ij}, m_{k\ell}, m_{j\ell}) \subset$ $\mathrm{int}(T_{i'} \cup T_{k'})$	90.3 92.1 90.2
6.3	$\varepsilon^{(r)} \ell \lambda$ $\varepsilon^{(r)} \ell \mu$ $\varepsilon^{(r)} \ell \nu$ $\varepsilon^{(r)} k' \xi$ $\varepsilon^{(r)} k' \pi$ $\varepsilon^{(r)} k' \sigma$	$\{\lambda = \lambda_\alpha, \mu = \mu_\beta,$ $\nu = \nu_\gamma\}$ Type 3.1, $\xi = \zeta(\lambda_\alpha),$ $\pi = \zeta(\mu_\beta),$ $\sigma = \zeta(\nu_\gamma)$	$\varepsilon^{(r)}$-shift of point in $(m_{\ell i}, m_{\ell j}, m_{\ell k}) \subset$ $\mathrm{int}(T_\ell \cup T_{k'})$	90.4 92.1
12.1	$\delta^{(n)} k \lambda_\alpha$ $\delta^{(n)} \ell' \mu_\beta$	$\lambda_\alpha, \mu_\beta$ Type 6.1, $\{\alpha_m, \beta_m\} = \{i, j\}$	shift of rational in $(m_{jk}, m_{ik}) \subset$ $\mathrm{int}_{\Delta^3}(T_k \cup T_{\ell'})$	88.2 89.2 88.6
12.2	$\delta^{(n)} j \lambda_\alpha$ $\delta^{(n)} \ell' \mu_\beta$ $\delta^{(n)} i' \nu_\gamma$	$\lambda_\alpha, \mu_\beta, \nu_\gamma$ Type 4.1, $(\alpha_m, \beta_m, \gamma_m) \in$ $\{(i, k, i), (k, i, \ell)\}$	shift of rational in $(m_{ij}, m_{jk}) \subset$ $\mathrm{int}_{\Delta^3}(T_j \cup T_{\ell'} \cup T_{i'})$	88.2 89.1 88.7

Table 93.1 (Part II) Summary table for fibers of sizes 6 and 12.

To illustrate "change of choice" in the context of these tables, consider a string of letters $\kappa^x = \varepsilon_1 \cdots \varepsilon_r j' \beta_1 \beta_2 \cdots$ where $\beta_1 \beta_2 \cdots$ is Type 1.3. Then κ^x, except for the "j'" letter, matches the first entry of Type 2.2, i.e., κ^x "almost" matches an entry in Table 93.1 (Part I), but in fact does not match any entry in either part of Table 93.1. But using the choice y, i.e., the solution to $\rho_\zeta = (ij)(k)(\ell)(i'j')(k')(\ell') = y^{-1} \circ x$, we have $\kappa^y = y^{-1} \circ \kappa = y^{-1} \circ (x \circ x^{-1}) \circ \kappa = \rho_\zeta \circ \kappa^x$, which does match the first entry of Type 2.2 in Table 93.1 (Part I):

$$\kappa^y = \varepsilon_1 \cdots \varepsilon_r i' \alpha_1 \alpha_2 \cdots \qquad \alpha_1 \alpha_2 \cdots = \rho_\zeta(\beta_1 \beta_2 \cdots) \text{ is of Type 1.3.}$$

To dovetail these observations with the assumption that $\{\{i,j\},\{k,\ell\}\} = \{\{0,1\},\{2,3\}\}$, consider the (obviously commutative) diagram in Figure 93.2 which shows the relation between two choices x and y of letters i,j,k,ℓ'.

> NOTATION: $A = \{0,1,2,3,0',1',2',3'\}$, $\{\{i,j\},\{k,\ell\}\} = \{\{0,1\},\{2,3\}\}$, int denotes 3-space interior, int_{Δ^3} the interior rel Δ^3, $\overline{\ell'}$ Type 1.1 & Type 1.3, $\delta^{(n)} = \delta_1, \cdots, \delta_n \in \{i,j,k,\ell'\}$ or empty list, $\alpha^{(t-1)} = \alpha_1, \cdots, \alpha_{t-1} \in \{r,s\}$ or empty list, $\varepsilon^{(r)} = \varepsilon_1, \cdots, \varepsilon_r \in A$ or empty list, $F = \{i,j,k,\ell'\}, K \subset \{0,1,2,3\}$ denotes a doubleton subset, a_{rs} an address of m_{rs}, $m_{rs} = (1/2)(u_r + u_s)$, $\psi = \rho \times \rho \times \cdots$ where $\rho = (i\ell jk)(i'\ell'j'k')$, $\zeta = \rho \times \rho \times \cdots$ where $\rho = (ij)(i'j')$, and $(P,Q) = (m_{ij}, m_{k\ell})$.

TYPE	ADDRESSES	MNEMONIC	IN FIBER OF	REF.
24.1	$\varepsilon^{(r)}0'\lambda_\alpha$ $\varepsilon^{(r)}1'\mu_\beta$ $\varepsilon^{(r)}2'\nu_\gamma$ $\varepsilon^{(r)}3'\xi_\delta$	λ_α Type 6.1, $(\alpha_m, \beta_m, \gamma_m, \delta_m) \in$ $\{(0,1,3,2),(1,0,2,3)\}$	shift of rational in $(P,Q) \subset \text{int}(T_{0'} \cup$ $T_{1'} \cup T_{2'} \cup T_{3'})$	88.4 92.1 87.2
24.2	$\varepsilon^{(r)}i'\eta$ $\varepsilon^{(r)}i'\sigma$ $\varepsilon^{(r)}k'\xi$ $\varepsilon^{(r)}k'\omega$	$\{\eta = \eta_\alpha, \sigma = \sigma_\beta\}$ Type 12.1, $\xi = \psi(\eta)$, $\omega = \psi(\sigma)$	shift of point in $(m_{ij}, m_{k\ell}, m_{j\ell}) \subset$ $\text{int}(T_{i'} \cup T_{k'})$	90.3 92.1 90.2
24.3	$\varepsilon^{(r)}i'\eta$ $\varepsilon^{(r)}i'\sigma$ $\varepsilon^{(r)}i'\tau$ $\varepsilon^{(r)}k'\nu$ $\varepsilon^{(r)}k'\xi$ $\varepsilon^{(r)}k'\omega$	fiber $\{\eta = \eta_\alpha, \sigma = \sigma_\beta,$ $\tau = \tau_\gamma\}$ Type 12.2, $\nu = \psi(\eta)$, $\xi = \psi(\sigma)$, $\omega = \psi(\tau)$	shift of point in $(m_{ij}, m_{k\ell}, m_{j\ell}) \subset$ $\text{int}(T_{i'} \cup T_{k'})$	90.3 92.1 90.2
24.4	$\varepsilon^{(r)}\ell\eta$ $\varepsilon^{(r)}\ell\sigma$ $\varepsilon^{(r)}k'\xi$ $\varepsilon^{(r)}k'\omega$	fiber $\{\eta = \eta_\alpha, \sigma = \sigma_\beta\}$ Type 12.1, $\xi = \zeta(\eta)$, $\omega = \zeta(\sigma)$	shift of point in $(m_{\ell i}, m_{\ell j}, m_{\ell k}) \subset$ $\text{int}(T_\ell \cup T_{k'})$	90.4 92.1
24.5	$\varepsilon^{(r)}\ell\eta$ $\varepsilon^{(r)}\ell\sigma$ $\varepsilon^{(r)}\ell\tau$ $\varepsilon^{(r)}k'\nu$ $\varepsilon^{(r)}k'\xi$ $\varepsilon^{(r)}k'\omega$	fiber $\{\eta = \eta_\alpha, \sigma = \sigma_\beta$ $\tau = \tau_\gamma\}$ Type 12.2, $\nu = \zeta(\eta)$, $\xi = \zeta(\sigma)$, $\omega = \zeta(\tau)$	shift of point in $(m_{\ell i}, m_{\ell j}, m_{\ell k}) \subset$ $\text{int}(T_\ell \cup T_{k'})$	90.4 92.1

Table 93.1 (Part III) Summary table for fibers of size 24.

The homeomorphism $x^{-1} : N(A^l) \to N(A^l)$ (or $x^{-1} : N(A) \to N(A)$ if no "primes" are involved) matches $N(A^l)$ with $N(A^l)$, i.e., equates "strings of letters" $\kappa^x \in N(A^l)$ with sequences $\kappa \in N(A^l)$ by identifying $i \in A^l$ with $x(i) \in A^l$, $j \in A^l$ with $x(j) \in A^l$, etc. Under these identifications, the connection with Table 93.1 is clear since $\{\{x(i), x(j)\}, \{x(k), x(\ell)\}\} =$

$\{\{0,1\},\{2,3\}\}$ is equivalent to $\{\{i,j\},\{k,\ell\}\} = \{\{0,1\},\{2,3\}\}$. Moreover, these identifications allow us to use κ^x as κ and $\phi(\kappa^x)$ as $\phi(\kappa)$.

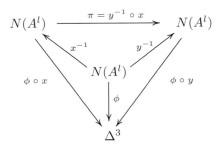

Fig. 93.2 Change of choices for Table 93.1.

§94 Octic Group Action and Induced Barycentric Maps

Throughout this section we shall assume, without loss of generality, that $u_0 = [1,0,0,0]^T, \ldots, u_3 = [0,0,0,1]^T$ are the standard basis vectors in 4-space. So the barycentric coordinates of Δ^3 are Cartesian coordinates. We shall also assume that the octic group is represented by

$$G = \{1, (01), (23), (01)(23), (0312), (2130), (02)(13), (03)(12)\}.$$

Then for each $\pi \in G$, we let the barycentric mapping $\theta_\pi : \Delta^3 \to \Delta^3$ be given by $\Sigma_{m=0}^3 x_m u_m \mapsto \Sigma_{m=0}^3 x_m u_{\pi(m)}$. The homeomorphism θ_π may be viewed as a restriction of a linear transformation represented by the 4×4 permutation matrix P_π whose $(m+1)$st column is $u_{\pi(m)} = P_\pi u_m$. That is, for $\Sigma x_m u_m \in \Delta^3$,

$$\theta_\pi(\Sigma x_m u_m) = \Sigma x_m u_{\pi(m)} = \Sigma x_m(P_\pi u_m) = P_\pi(\Sigma x_m u_m).$$

With these givens we shall show, for each $q \in A^l$, that the diagram In Figure 94.1 is commutative:

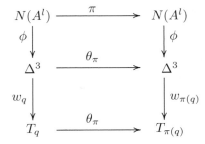

Fig. 94.1 Transformation of addresses induced by the octic group.

94.2 Lemma (lower square of Figure 94.1) *Let ϕ be the address map of $\mathcal{F}_3^* = \{w_q | q \in A^l\}$, let $\pi \in G$, and let θ_π denote the induced barycentric map. Then for each $q \in A^l$, $w_{\pi(q)} \circ \theta_\pi = \theta_\pi \circ w_q$.*

PROOF. For $q \in \{0, 1, 2, 3\}$, we have

$$
\begin{aligned}
w_{\pi(q)}(\theta_\pi(\Sigma x_m u_m)) &= w_{\pi(q)}(\Sigma x_m u_{\pi(m)}) = (1/2)\Sigma x_m u_{\pi(m)} + (1/2)u_{\pi(q)} \\
&= \theta_\pi\left((1/2)\Sigma x_m u_m + (1/2)u_q\right) = \theta_\pi(w_q(\Sigma x_m u_m)).
\end{aligned}
$$

And for $q' \in \{0', 1', 2', 3'\}$,

$$
\begin{aligned}
w_{\pi(q')}(\theta_\pi(\Sigma x_m u_m)) &= w_{\pi(q')}(\Sigma x_m u_{\pi(m)}) = w_{p'}(\Sigma x_m u_{\pi(m)}) \\
&= (1/2)L_p \Sigma x_m u_{\pi(m)} + (1/2)u_{p'} \\
&= (1/2)P_\pi L_q P_\pi^{-1}(\Sigma x_m u_{\pi(m)}) + (1/2)u_{\pi(q')} \\
&= \theta_\pi\left((1/2)L_q \Sigma x_m u_m + (1/2)u_{q'}\right) = \theta_\pi(w_{q'}(\Sigma x_m u_m)).
\end{aligned}
$$

So the lower square of the Figure 94.1 diagram is commutative. □

94.3 Lemma (upper square of Figure 94.1) *Let ϕ be the address map of $\mathcal{F}_3^*$, let $\pi \in G$, and let θ_π denote the induced barycentric map. Then $\phi \circ \pi = \theta_\pi \circ \phi$.*

PROOF. Let $\alpha \in N(A^l)$. Then $\phi(\alpha) = p$ is the lone member of $\cap_{j \geq 1} T_{\alpha_1 \cdots \alpha_j}$. It also follows from Lemma 94.2 that $p \in T_{\alpha_1}$ implies that $\theta_\pi(p) \in T_{\pi(\alpha_1)}$. So, using induction, we suppose $p \in T_{\alpha_1 \cdots \alpha_n}$ implies $\theta_\pi(p) \in T_{\pi(\alpha_1) \cdots \pi(\alpha_n)}$ is true and let $p \in T_{\alpha_1 \cdots \alpha_{n+1}}$. Then

$$
\begin{aligned}
\theta_\pi(p) \in \theta_\pi(T_{\alpha_1 \cdots \alpha_{n+1}}) &= \theta_\pi \circ w_{\alpha_1}(T_{\alpha_2 \cdots \alpha_{n+1}}) = w_{\pi(\alpha_1)} \circ \theta_\pi(T_{\alpha_2 \cdots \alpha_{n+1}}) \\
&\subset w_{\pi(\alpha_1)}(T_{\pi(\alpha_2) \cdots \pi(\alpha_{n+1})}) = T_{\pi(\alpha_1) \cdots \pi(\alpha_{n+1})}.
\end{aligned}
$$

So $\theta_\pi(\phi(\alpha))$ is the lone member of $\cap_{j \geq 1} T_{\pi(\alpha_1) \cdots \pi(\alpha_j)}$, i.e., $\phi(\pi(\alpha)) = \theta_\pi(\phi(\alpha))$. So the upper square of the diagram in Figure 94.1 is commutative. □

We note that θ_π permutes not only the vertices, but also midpoints, e.g., $p \in (m_{ij}, m_{k\ell}, m_{j\ell})$ implies $\theta_\pi(p) \in (m_{\pi(i)\pi(j)}, m_{\pi(k)\pi(\ell)}, m_{\pi(j)\pi(\ell)})$. Our next theorem concerns the action of G on the fibers of ϕ and its (omitted) proof rests on the fact that the top square in the diagram in Figure 94.1 is commutative.

94.4 Theorem *Let ϕ be the address map of $\mathcal{F}_3^*$, let π be a member of the octic group G, and let θ_π be the induced barycentric map. Then the fiber $\phi^{-1}(\theta_\pi(p)) = \pi(\phi^{-1}(p))$.*

§95 Completeness of Table 93.1

95.1 Theorem *Let $\kappa \in N(A^l)$. Then for some choice x of i, j, k, ℓ, the string of letters $\kappa^x = x^{-1} \circ \kappa = x^{-1}(\kappa_1)x^{-1}(\kappa_2) \cdots = \kappa_1^x \kappa_2^x \cdots$ is one of the types listed in Table 93.1.*

PROOF. If x exists such that κ^x is Type 1.1, 1.2, 1.3, or 1.4, we are finished. So suppose otherwise. Then "not Type 1.4" implies κ has at most one "primed" constant subsequence.

First, suppose κ has *exactly one "primed" constant subsequence*. Fix x such that $\overline{\ell'}$ is the "primed" constant subsequence of κ^x. Then κ^x not Type 1.4 implies $\overline{\ell}$ is not a subsequence of κ^x. And κ^x not Type 1.3 implies a minimum $r + 1 \geq 1$ exists such that for some $\varepsilon_{r+1}^x \in \{i', j', k', \ell\}$

$$(1) \qquad \kappa^x = \varepsilon_1^x \cdots \varepsilon_r^x \varepsilon_{r+1}^x \alpha^x \qquad \alpha^x \in N(i, j, k, \ell') \ \text{ is Type 1.3.}$$

If $\varepsilon_{r+1}^x = i'$, then κ^x is Type 2.2. If $\varepsilon_{r+1}^x = \ell$, then κ^x is Type 2.3. If $\varepsilon_{r+1}^x = k'$, then $\kappa^x = \varepsilon^{(r)} k' \alpha^x = \varepsilon^{(r)} k' \beta$ where $\beta = \alpha^x$ is Type 1.3, making κ^x Type 2.3. And finally, if $\varepsilon_{r+1}^x = j'$, the change of choice $\zeta = y^{-1} \circ x$ yields

$$\kappa^y = y^{-1} \circ \kappa = \zeta \circ x^{-1} \circ \kappa = \zeta(\kappa^x) = \varepsilon^{(r)} i' \zeta(\alpha^x), \quad \zeta(\alpha^x) \in N(i, j, k, \ell')$$

where $\zeta(\alpha^x)$ is Type 1.3. So κ^y is Type 2.2.

Second, if κ has *no "primed" constant subsequence*, then "not Type 1.4" implies

$$(2) \qquad \kappa^x = \varepsilon_1^x \cdots \varepsilon_r^x \varepsilon_{r+1}^x \alpha^x \qquad\qquad \alpha^x \in N(i, j, k)$$

for some choice x where either "(2*i*)," the prefix $\varepsilon_1^x \cdots \varepsilon_r^x \varepsilon_{r+1}^x$ is empty, or "(2*ii*)," the prefix is non-empty with $\varepsilon_{r+1}^x \in \{\ell, i', j', k', \ell'\}$.

Suppose 2(i) is true. Then $\kappa^x = \alpha^x$ "not Type 1.3" implies

$$(3) \qquad\qquad \kappa^x = \alpha^x = \delta^{(n)} \delta_{n+1} \alpha \qquad\qquad \alpha \in N(i, j)$$

where either the prefix $\delta^{(n)} \delta_{n+1}$ is empty or $\delta_{n+1} = k$; or

$$(4) \qquad\qquad \kappa^x = \alpha^x = \delta^{(n)} \delta_{n+1} \alpha \qquad\qquad \alpha \in N(j, k)$$

where either the prefix $\delta^{(n)} \delta_{n+1}$ is empty or $\delta_{n+1} = i$. If (3) holds with an empty prefix, then κ^x "not 1.2" implies κ^x is Type 6.1. And if (3) holds with a non-empty prefix, then κ^x is Type 12.1 when α is Type 6.1, and κ^x is Type 2.1 when α is Type 1.2. The subcase (4) runs parallel to the subcase (3), as illustrated in Figure 95.2.

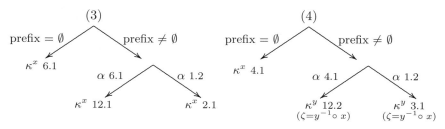

Fig. 95.2 Subcases (3) and (4) of 2(i), empty prefix $\varepsilon_1^x \cdots \varepsilon_{r+1}^x$ case.

Finally, suppose $2(ii)$ is true, i.e., $\varepsilon_{r+1}^x \in \{\ell, i', j', k', \ell'\}$: Using the comments surrounding Figure 93.2, we have $\varepsilon_{r+1} = \varepsilon_{r+1}^x \in \{\ell, i', j', k', \ell'\}$ and $\phi(\alpha) = \phi(\alpha^x)$ is in the 2-face $[u_i, u_j, u_k]$. So

$$w_{\varepsilon_{r+1}}(\phi(\alpha)) = \phi(\varepsilon_{r+1}\alpha) \in \begin{cases} [m_{\ell i}, m_{\ell j}, m_{\ell k}] & \text{if } \varepsilon_{r+1} = \ell, \\ [m_{ij}, m_{k\ell}, m_{j\ell}] & \text{if } \varepsilon_{r+1} = i', \\ [m_{k\ell}, m_{ij}, m_{i\ell}] & \text{if } \varepsilon_{r+1} = j', \\ [m_{j\ell}, m_{k\ell}, m_{i\ell}] & \text{if } \varepsilon_{r+1} = k', \\ [m_{jk}, m_{ik}, m_{ij}] & \text{if } \varepsilon_{r+1} = \ell'. \end{cases}$$

The two cases $\varepsilon_{r+1} \in \{\ell, k'\}$ yield $\phi(\varepsilon_{r+1}\alpha) \in [m_{\ell i}, m_{\ell j}, m_{\ell k}]$. If $\phi(\varepsilon_{r+1}^x \alpha^x)$ is in the interior of this 2-face it must match one of the entries in Table 90.4 (and consequently κ^x matches an entry in Table 93.1). Otherwise, $\phi(\varepsilon_{r+1}^x \alpha^x)$ is in an edge of $[m_{\ell i}, m_{\ell j}, m_{\ell k}]$, i.e., in

$$(m_{\ell i}, m_{\ell j}) \cup (m_{\ell j}, m_{\ell k}) \cup (m_{\ell i}, m_{\ell k}) \cup \{m_{\ell i}, m_{\ell j}, m_{\ell k}\}.$$

Consider $(m_{\ell i}, m_{\ell j})$ and midpoint $m_{\ell j}$. Then the change of choice $\pi = y^{-1} \circ x = (i)(j)(k\ell)(i')(j')(k'\ell')$ provides a match of $\varepsilon_{r+1}^y \alpha^y$ with an entry in Table 87.3, 88.3, or 88.4. For $(m_{\ell i}, m_{\ell k})$ and the other two midpoints, $\pi = (ik)(\ell j)(i'k')(\ell'j') \in G$ yields a match in one of those same tables. For $(m_{\ell j}, m_{\ell k})$, $\pi = (i\ell)(kj)(i'\ell')(k'j')$ provides a match of $\varepsilon_{r+1}^y \alpha^y$ with an entry in Table 88.3 or 88.4. Consequently, each corresponding κ^y matches an entry in Table 93.1.

In the two cases $\varepsilon_{r+1} \in \{i', j'\}$, we see that the "$j'$" case is "equivalent" to the other via a change of choice $(ij)(k)(\ell)(i'j')(k')(\ell')$. So it suffices to consider $\phi(i'\alpha^x) \in [m_{ij}, m_{k\ell}, m_{j\ell}]$. Then $i'\alpha^x$ must match either an entry in Table 90.3, or an entry in Table 87.3, 88.2, 88.3, or 88.4 (and consequently κ^x matches an entry in Table 93.1). For the final case, namely $\varepsilon_{r+1} = \ell'$, we have $\phi(\varepsilon_{r+1}^x \alpha^x)$ in a 2-simplex of K_1 that is a subset of the face $[u_i, u_j, u_k]$. Thus, $\varepsilon_{r+1}^x \alpha^x$ must match an entry in Table 87.3 or 89.2. So again there is a choice x such that κ^x matches an entry in Table 93.1. $\qquad\square$

§96 Representations of 3-Space and the 2-Sphere

Recall that in the previous chapter we considered the boundary $\partial\Delta^2$ of Δ^2, which is homeomorphic to the 1-sphere, and $\Delta^2 \setminus \partial\Delta^2$, which is homeomorphic to Euclidean 2-space. By grouping the fibers according to those that map to points in the former and then according to those that map to points in the latter, we obtained representations of the 1-sphere and 2-space, repectively. In this section, we provide a similar representation of 3-space and the 2-sphere.

The boundary $\partial\Delta^3$ of Δ^3 is homeomorphic to the 2-sphere, while $\Delta^3 \setminus \partial\Delta^3$ is homeomorphic to Euclidean 3-space.

96.1 Theorem *Let ϕ be the address map of $\mathcal{F}_3^* = \{w_q : q \in A^l\}$ from $N(A^l)$ onto the 3-simplex Δ^3. Let F be the ϕ-inverse subspace of $N(A^l)$ that* *contains all of the following fibers:* (1) *size* 1, *except those of Type* 1.4; (2) *size* 2 *and Type* 2.1; (3) *size either* 3 *or* 12; (4) *size* 4 *and Type* 4.1; *and* (5) *size* 6 *and Type* 6.1. *Let* $G = N(A) \setminus F$. *Then* G *is also a ϕ-inverse subspace of* $N(A)$. *Moreover,*

$$\phi|_G : G \to (\Delta^3 \setminus \partial\Delta^3) \quad and \quad \phi|_F : F \to \partial\Delta^3$$

are quotient maps, the former onto a copy of 3-space and the latter onto a copy of the 2-sphere.[2]

PROOF. From Table 93.1 it follows that $F = \phi^{-1}(\partial\Delta^3)$ and that $G = \phi^{-1}(\Delta^3 \setminus \partial\Delta^3)$. Since ϕ is continuous, $\phi|_F$ and $\phi|_G$ are continuous. In addition, since $\partial\Delta^3$ is closed in Δ^3 and $(\Delta^3 \setminus \partial\Delta^3)$ is open in Δ^3, it follows (see, e.g., Dugundji [1966, page 122, Theorem 2.1]) that both $\phi|_F$ and $\phi|_G$ are quotient maps.   □

§97 Comments

As Barnsley and Sloan [1988] indicate, an IFS may be used to compress a picture that requires 130 megabytes of memory by a factor of 10,000, requiring only a manageable 13,000 bytes. Here we used an IFS $\mathcal{F}_3^*$ of size eight to compress an *infinite number of pastings* and an *infinite number of cuttings* to "picture" a 3-simplex.

The pastings correspond to the affine transformations $w_{0'}$, $w_{1'}$, $w_{2'}$, and $w_{3'}$, while the cuttings correspond to the members w_0, w_1, w_2, and w_3 of the IFS $\mathcal{F}_3$ whose attractor is the 3-web ω^3 (the Sierpiński cheese). In other words, we used an IFS to view the (3-dimensional manifold) 3-simplex in the context of the (1-dimensional fractal) 3-web ω^3.

In the context of dimension theory of separable metric spaces at least, an IFS has some interesting, but yet to be explored features. For example, consider Morita's Theorem (Theorem 1.6): *A metric-space X has dimension $\leq n$ if and only if there exists a subspace S of $N(A)$ for suitable A and a closed continuous mapping f of P onto X such that for each point $q \in X$, $f^{-1}(q)$ consists of at most $n + 1$ points.* Morita's Theorem places a nice upper bound "$n + 1$" on the sizes of the fibers at the cost of having a rather nebulous domain (some subspace P of $N(\Omega)$). In contrast, an IFS $(\{w_a : a \in A\})$ provides a well-defined domain $N(A)$ (of the closed continuous address map ϕ onto its attractor X) at the cost of having rather nebulous sizes of fibers. An IFS also provides a "built-in" and "uniformly indexed" sequence $\mathcal{C}_k$ ($k = 1, 2, \ldots$) of ever-finer closed coverings of X, namely $\mathcal{C}_k = \{w_{a_1} \circ \cdots \circ w_{a_k}(X) : a_1, \ldots, a_k \in A\}$ that is indexed on the k-fold finite-product set $A^k = A \times \cdots \times A$.

Moreover, in this chapter we needed some general formulas (formulas that apply to any IFS and its address map), e.g., Lemma 85.3 part (a) that concern

---

[2]Recall (from §93) that $A^l = \{i, j, k, \ell, i', j', k', \ell'\}$ where $\{\{i, j\}, \{k, \ell\}\} = \{\{0, 1\}, \{2, 3\}\}$.

the interaction between the address map and the members of the IFS. Such formulas allowed us to navigate between fractals and manifolds, and should prove key to any solution of the extension problem for $n \geq 4$.

As to the open problem of extending the 4-web ω^4 IFS to a 4-simplex Δ^4 IFS, the approach developed in this chapter (for the ω^3 IFS extension) could serve as an outline or model for the 4-simplex case. The key problem, however, is that of understanding the "4-hole" $\Delta^4 \setminus \cup_0^4 w_k(\Delta^4)$ induced by the $\mathcal{F}_4$ IFS in Δ^4.

The fundamental requirement is that of obtaining a well defined description of the "4-hole." That is, obtain a representation of the "hole" induced by $\mathcal{F}_4$ in $\Delta^4 \subset \mathbb{R}^5$ that is analogous to the 3-hole generated by $\mathcal{F}_3$ in $\Delta^3 \subset \mathbb{R}^4$. Recall that the closure of the 3-hole may be viewed as an octahedron, which in turn was viewed as a realization of a 3-complex consisting of four tetrahedra. It is also worth noting that the 2-skeleton of an octahedron is homeomorphic to a 2-sphere.

To begin the search for the corresponding complex whose realization is the closure of the "4-hole" in the 4-simplex, this author believes that the 3-space representation of a level-1 J_5 should provide an intuitive background for "picturing the hole."

For example, the midpoints m_{ij}, for distinct $i, j \in \{0, 1, 2, 3, 4\}$, of edges $[u_i, u_j]$ of Δ^4 are easily pictured at the level-1 approximation of J_5. By considering all possible edges $[m_{ij}, m_{k\ell}]$ whose endpoints are these midpoints, one may see, at least combinatorially, five tetrahedra, each just touching the other four. In addition, one may also see five octahedra with interesting combinatorial properties. It is easy to conjecture that this subcomplex whose vertices are the midpoints m_{ij} may serve as a model for decomposing the closure of the 4-hole into 4-simplexes that could then serve to define the desired $\mathcal{F}_4^*$.

Another easy conjecture is that the boundary of the 4-hole is homeomorphic to a 3-sphere.

APPENDIX 1

Background Basics

We recall (with gateways to references) the most basic of relevant concepts
— notations, covers, and Cartesian poducts (§A1), topological spaces (§A2),
metric spaces (§A3), mappings (§A4), product, biquotient and perfect mappings (§A5), and topological dimension theory (§A6).

§A1 Notations, Covers, and Cartesian Products

A1.1 NOTATIONS. The set-theoretic notation used in this book is standard.
Nevertheless, we note that $f : X \to Y$ does not necessarily imply that f is
surjective; and, as is standard, we may call the inverse image "$f^{-1}(y)$" of the
point $y \in Y$ either a *point-inverse set* or a *fiber of f*.

A1.2 COVERS. A family $\mathcal{C}$ of subsets of $X \neq \emptyset$ *covers* X (or is a *covering of*
X) if each point in X is contained in at least one member of $\mathcal{C}$. When each
$C \in \mathcal{C}$ is open in (the topological space) X, then $\mathcal{C}$ is an *open cover*.

Any $\mathcal{C}' \subset \mathcal{C}$ that also covers X is a *subcover* of $\mathcal{C}$. The cover $\mathcal{C}$ itself is
irreducible when it has no *proper* subcover; it is *point finite* if each $x \in X$ is
contained in only finitely many members of $\mathcal{C}$; and it is *locally finite* (*discrete*)
if for each $x \in X$ there is an open set G_x such that $G_x \cap C \neq \emptyset$ is valid for
only finitely many (for at most one) members C of $\mathcal{C}$.[1]

A collection $\mathcal{B}$ is *σ-locally finite* (*σ-discrete*) if $\mathcal{B} = \cup_i \mathcal{B}_i$ is a countable
union where each $\mathcal{B}_i$ is locally finite (discrete).

Let $\mathcal{U}$ and $\mathcal{V}$ be coverings of X. Then $\mathcal{U}$ *refines* or *is a refinement of* $\mathcal{V}$
if for each $U \in \mathcal{U}$ there is some $V \in \mathcal{V}$ such that $U \subset V$. In such a case we
may write $\mathcal{U} \prec \mathcal{V}$. If $\mathcal{U} = \{U_a : a \in A\}$, $\mathcal{V} = \{V_a : a \in A\}$, and $U_a \subset V_a$ for
each $a \in A$, then $\mathcal{U}$ *precisely refines* (*is a precise refinement of*) $\mathcal{V}$.

For a covering $\mathcal{C}$ of X, the *star* $S(x, \mathcal{C})$ *of a point* $x \in X$ is given by
$S(x, \mathcal{C}) = \cup\{C : x \in C \in \mathcal{C}\}$. Similarly, the *star* $S(R, \mathcal{C})$ of $R \subset X$ is given
by $S(R, \mathcal{C}) = \cup\{C : C \cap R \neq \emptyset; C \in \mathcal{C}\}$. The induced *star-covering* $\mathcal{C}^*$ is
given by $\mathcal{C}^* = \{S(C, \mathcal{C}) : C \in \mathcal{C}\}$.

A1.3 CARTESIAN PRODUCTS. The times notation "$\times$" is used for (Cartesian)
products, e.g., $A \times B$, $A \times B \times C$, and $\times_i A_i$. The last example is only used

---

[1]One must be careful about the choice of indexing. For example, if we simply write
$\mathcal{C} = \{C_a : a \in A\}$, then it may be that for a fixed $a' \in A$ we have $a' \neq a$ but $C_{a'} = C_a$ for
an infinite number of $a \in A$. To avoid any confusion, when we say "$\{C_a : a \in A\}$ *is locally
finite*" it shall be understood that there exists an open set G_x such that $G_x \cap C_a \neq \emptyset$ is
valid for only finitely many $a \in A$.

when the index set and each *factor* A_i are known. For $A \neq \emptyset$, we write

$$N(A) = A \times A \times \cdots = \times_i A_i \qquad (i \in N = \{1, 2, \ldots\} \text{ and each } A_i = A),$$

which is the set of all sequences in A. The set of all mappings $X \to Y$ is denoted by either "$\times_{x \in X} Y_x$" (where each $Y_x = Y$) or "Y^X" (which is the *exponential* notation). For $X = \{1, 2, \ldots, 2n+1\}$ and $Y = I$, we may write

$$I^{2n+1} = I_1 \times I_2 \times \cdots \times I_{2n+1} = \{(x_1, \ldots, x_{2n+1}) : \text{ each } x_i \in I\},$$

and when $X = \{1, 2, \ldots\}$ we write $I^\infty = \times_i I_i$ where each factor $I_i = I$.

§A2 Topological Spaces

A *topological space* $X = (X, \mathcal{T})$ consists of a non-empty set X and a family (the *topology*) $\mathcal{T}$ of subsets of X such that (i) $\{G_\gamma | \gamma \in \Gamma\} \subset \mathcal{T}$ implies $\cup_\gamma G_\gamma \in \mathcal{T}$ and (ii) $\{G_\gamma | \gamma \in \Gamma\} \subset \mathcal{T}$ and Γ finite imply $\cap_\gamma G_\gamma \in \mathcal{T}$. Each $G \in \mathcal{T}$ is an *open set*, and its complement $F = X \setminus G$ is a *closed set*.

In this monograph, a *neighborhood* N_x of the point x is any subset N_x of X such that $x \in G \subset N_x$ for some open set G.

Let X be a topological space and let $R \subset X$. With the understanding that $x \in G \in \mathcal{T}$ is expressed as "G_x", recall the most basic concepts: The *interior* "int(R)" *of* R is the set of those x such that some G_x satisfies $G_x \subset R$; the *closure* "$\overline{R}$" *of* R consists of those x such that every G_x satisfies $G_x \cap R \neq \emptyset$; and the *boundary* "$B(R)$" *of* R consists of those x such that every G_x satisfies both $G_x \cap R \neq \emptyset$ and $G_x \cap (X \setminus R) \neq \emptyset$.

It follows that int(R) is open and that $\overline{R} = X - \cup\{G_x : G_x \cap R = \emptyset\}$ is closed. In turn, since $B(R) = \overline{R} \setminus \text{int}(R)$, we see that $B(R)$ is also closed. When $R = G$ itself is open, then int(R) $= R$ and $B(R) = \overline{R} \setminus R$, which yields $B(G) \cap G = \emptyset$ and $\overline{G} = G \cup B(G)$.

A2.1 SEPARATION AXIOMS. Again, "G_x" and "G_A" denote, respectively, "$x \in G \in \mathcal{T}$" and "$A \subset G \in \mathcal{T}$". A topological space X is a T_1-*space* if and only if each singleton set $\{x\}$ is a closed set; it is T_2 or *Hausdorff* if $x, y \in X$ and $x \neq y$, then there exist disjoint G_x and G_y; it is T_3 or *regular* if it is a T_1-space and if for each $x \in X$ and closed $F \subset X$ with $x \notin F$ there exist disjoint G_x and G_F; and it is T_4 or *normal* if it is a T_1-space, and, F and H disjoint and closed implies disjoint G_F and G_H exist.

Theorem (covering characterization of normality) *A topological space X is normal if and only if for each point-finite open covering $\mathcal{U} = \{U_a : a \in A\}$ of X there exists an open covering $\mathcal{V} = \{V_a : a \in A\}$ of X such that $\overline{V}_a \subset U_a$ for each $a \in A$ and $V_a \neq \emptyset$ when $U_a \neq \emptyset$.*[2]

In general, a closed precise refinement $\overline{\mathcal{V}} = \{\overline{V}_a : a \in A\}$ of $\{U_a : a \in A\} = \mathcal{U}$ is called a *shrinking of* $\mathcal{U}$.

---

[2]For a detailed proof see Dugundji [1966, §6, page 152].

A2.2 COMPACT AND PARACOMPACT. A topological space X is *compact* if each open covering $\mathcal{C}$ has a finite subcover $\mathcal{C}' \subset \mathcal{C}$; and X is *paracompact* if each open covering $\mathcal{C}$ has a locally finite subcover $\mathcal{C}' \subset \mathcal{C}$.

Theorem (sufficient condition for normality) *Let X be a topological space that is both Hausdorff and paracompact. Then X is normal.*[3]

A2.3 BASIS, WEIGHT, AND SUBBASIS. A *basis* $\mathcal{B}$ for a topological space $X = (X, \mathcal{T})$ is a subcollection of $\mathcal{T}$ such that each member of $\mathcal{T}$ is a union of members of $\mathcal{B}$. It follows, since X itself is a member of $\mathcal{T}$ that each basis is an open cover of X. Clearly, $\mathcal{T}$ itself is a basis for X.

If X has a countable basis $\mathcal{B}$, i.e., $\aleph(\mathcal{B}) \leq \aleph_0$, then X is *separable*. Otherwise, X is *nonseparable*. The *weight* of X is the cardinality of a minimum-size basis for X.

A basis $\mathcal{B}$ that is σ-locally finite (σ-discrete) is called a *σ-locally finite (σ-discrete) basis*.

Any non-empty collection $\mathcal{S}$ of subsets of a non-empty set X *generates* or *is a subbasis of* a topology for X: Since $\mathcal{S}$ is a subcollection of the set 2^X of all subsets of X and since 2^X is a topology for X, the topology $\mathcal{T}(\mathcal{S}) = \cap \{ \mathcal{T} : \mathcal{S} \subset \mathcal{T}$ and $\mathcal{T}$ is a topology on $X \}$ has $\mathcal{S}$ as a subbasis. Since $\mathcal{T}(\mathcal{S}) = \{ G : G$ is a union of finite intersections of members of $\mathcal{S} \}$, $\mathcal{B}(\mathcal{S}) = \{ B : B$ is a finite intersection of members of $\mathcal{S} \}$ is a basis for $\mathcal{T}(\mathcal{S})$.

§A3 Metric Spaces

Let X be a non-empty set and let $\rho : X \times X \to [0, \infty)$ be such that (i) $\rho(x, y) = 0$ if and only if $x = y$, (ii) $\rho(x, y) = \rho(y, x)$ for every $x, y \in X$, and (iii) $\rho(x, z) \leq \rho(x, y) + \rho(y, z)$ for every $x, y, z \in X$. Then $X = (X, \rho)$ is a *metric space* and ρ is a *a metric* or *distance function*.

For $x \in X$ and $\delta > 0$, the *open δ-ball $B_\delta(x)$ centered at x with radius δ* is given by $B_\delta(x) = \{ y \in X : \rho(x, y) < \delta \}$, while the corresponding *closed ball* is given by $\overline{B}_\delta(x) = \{ y \in X : \rho(x, y) \leq \delta \}$.

On the one hand, the collection $\mathcal{S} = \{ B_\delta(x) : x \in X$ and $\delta > 0 \}$ of all open balls is a subbasis for *the topology $\mathcal{T}_\rho$ induced by the metric ρ*. If metrics ρ and ρ' on X induce identical topologies $\mathcal{T}_\rho = \mathcal{T}_{\rho'}$, then they are *equivalent metrics*. And when a metric space (X, ρ) is called a topological space, it is understood that $X = (X, \mathcal{T}_\rho)$.

On the other hand, a topological space $(X, \mathcal{T})$ is *metrizable* if there exists a metric ρ on X such that $\mathcal{T}_\rho = \mathcal{T}$.

Characterizations of those regular spaces that are metrizable were created independently by Bing [1951], Nagata [1950], and Smirnov [1951].[4]

---

[3]For a detailed proof see Dugundji [1966, §2, page 162].

[4]For metrizable spaces and related topics see Nagata [1968, Chapter VI], and for imbeddings and metrization see Kelly [1955]. For characterizations of those T_0 and T_1 spaces that are metrizable, namely the Morita, Stone, and Arhangel'skii theorems, see Dugundji [1966, page 196, Theorem 9.5].

Theorem (Bing's Metrization Theorem) *A regular topological space X is metrizable if and only if it has a σ-discrete basis.*

Theorem (Nagata-Smirnov Metrization Theorem) *A regular space X is metrizable if and only if it has a σ-locally finite basis.*

The *distance of a point $x \in X$ from a set $R \subset X$* is given by $\rho(x, R) = \inf\{\rho(x, y) : y \in R\}$. And the *diameter of a set R* is given by $|R| = \sup\{\rho(x, y) : x, y \in R\}$. Some basic properties are listed below.

Theorem (properties of metric spaces) *Let $X = (X, \rho)$ be a metric space. Then*
  (i)   *The space X is paracompact and Hausdorff.*
  (ii)   *The space X satisfies the first axiom of countability.*
  (iii)   *The closure of $R \subset X$ is $\overline{R} = \{x \in X : \rho(x, R) = 0\}$.*
  (iv)   *A sequence $x_k \to x$ if and only if $\rho(x_k, x) \to 0$.*
  (v)   *The metric ρ is continuous, i.e., if a sequence $(x_n, y_n) \to (x, y)$ relative to the product topology, then $\rho(x_n, y_n) \to \rho(x, y)$.*

Finally, let $X = (X, \rho)$ be a metric space. Then a *ρ-Cauchy sequence* $x_1, x_2, \ldots$ in X is a sequence with the property that for each $\varepsilon > 0$ a positive integer n exists such that $k, m > n$ implies $\rho(x_k, x_m) < \varepsilon$. The metric ρ is *complete metric for X* if every ρ-Cauchy sequence in X converges to a point in X. And X is *topologically complete* whenever X has a complete metric. For a proof of the following theorem see Dugundji [1966, Chapter XIV, Theorem 2.5].

Theorem (completeness of countable product spaces) *A countable product space $\times_i X_i$ is topologically complete if and only if each factor X_i is topologically complete.*

§A4 Mappings

A function $f : Y \to X$ from a topological space Y to a topological space X is *continuous* if H open in X implies (*the inverse image*) $f^{-1}(H)$ is open in Y. A *homeomorphism* $Y \to X$ is a continuous bijection whose inverse $X \to Y$ is also continuous. When a homeomorphism $Y \to X$ exists we may say that Y and X are *homeomorphic* or that Y and X are *topologically equivalent*. Topological equivalence is denoted $Y =_t X$.

Let $f : Y \to X$ be a continuous surjection. Then f is a *quotient mapping* or *identification* when f satisfies "the converse of continuity" — $f^{-1}(H)$ open in Y implies H open in X.

The construction of quotient mappings is fundamental. The standard method is to begin with a topological space Y, an arbitrary set X, and a surjective function $f : Y \to X$. Then use f to induce the largest topology on X that makes f continuous. That is, the *identification* or *quotient* topology $\mathcal{T}(f)$ is given by $\mathcal{T}(f) = \{G \subset X : f^{-1}(G) \text{ is open in } Y\}$.

Typically, however, one obtains the set X via an equivalence relation $\sim$ on Y, i.e., one defines $X = Y/\sim$. In this case the points of X are the parts of the partition of Y induced by $\sim$. The mapping $f : Y \to X$ is then specified as the *natural map* — the point $f(y) = [y]$ is the part $[y]$ of the partition X that contains y.

A4.1 Theorem   *Let Y be a topological space, $\sim$ an equivalence relation on Y, and $X = Y/\sim$ the induced partition of Y with the quotient topology. Then the natural mapping $f : Y \to X$ is a quotient mapping.*

A surjection $f : Y \to X$ of topological spaces Y and X is a *closed mapping* if F closed in Y implies its image $f(F)$ is closed in X. Similarly, f is an *open mapping* if G open in Y implies its image $f(G)$ is open in X.

A4.2 Theorem   *Let $f : Y \to X$ be a continuous open (or closed) mapping. Then f is a quotient mapping.*

A4.3 Theorem   *Let $f : Y \to X$ be surjective and quotient, $g : Y \to J$ continuous, and $gf^{-1} : X \to J$ single valued, i.e., g is constant on each fiber $f^{-1}(x)$. Then gf^{-1} is continuous. Moreover, gf^{-1} is closed if and only if $g(F)$ is closed whenever F is a closed f-inverse set $(F = f^{-1}f(F))$.*

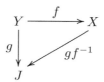

Fig. A4.4 Diagram used in proof of J_A^{n+1} Imbedding Theorem (see §41).

Proofs of A4.2 and A4.3 appear in Dugundji [1966, pages 121–123].

§A5 Product, Biquotient, and Perfect Mappings

The product $\times_\gamma X_\gamma$ provides, for each $\gamma' \in \Gamma$, a *projection map*

$$p_{\gamma'} : \times_\gamma X_\gamma \to X_{\gamma'}$$

given by $(x_\gamma) \mapsto x'_\gamma \in X_{\gamma'}$. And the mappings $f_\gamma : X_\gamma \to Y_\gamma$ (for each $\gamma \in \Gamma$) yield the *product map*

$$f = \times_\gamma f_\gamma : \times_\gamma X_\gamma \to \times_\gamma Y_\gamma$$

given by the formula $(x_\gamma) \mapsto (y_\gamma)$ where each $y_\gamma = f_\gamma(x_\gamma)$.

When each factor set X_γ is a topological space, $\times_\gamma X_\gamma$ inherits *the product topology* — the "smallest topology" such that each projection p_γ is continuous. The product topology $\mathcal{T}(\mathcal{S})$ is generated by the subbasis $\mathcal{S}$ of all sets of the form

$$\langle G \rangle = p_{\gamma'}^{-1}(G) = \times_\gamma U_\gamma \quad \text{where} \quad U_\gamma = \left\{ \begin{array}{ll} X_\gamma & \text{if } \gamma \neq \gamma'; \\ G \text{ open in } X_{\gamma'} & \text{if } \gamma = \gamma'. \end{array} \right.$$

It follows that the collection $\mathcal{B}(\mathcal{S})$ of sets

$$\langle G_1, G_2, \ldots, G_k \rangle = \langle G_1 \rangle \cap \langle G_2 \rangle \cap \cdots \cap \langle G_k \rangle \qquad \text{each } G_i \text{ open in } X_{\gamma_i}$$

is a basis for the product topology. A Cartesian product $\times_\gamma X_\gamma$ with the product topology is often referred to as a *product space*.

If $\times_\gamma X_\gamma$ and $\times_\gamma Y_\gamma$ are product spaces and each component f_γ of a product map $f = \times_\gamma f_\gamma$ is continuous, then f itself is continuous. Unlike continuous mappings, a product of quotient mappings is not necessarily quotient.[5]

A5.1 Definition A continuous surjection $f : X \to Y$ is *biquotient* if for each $y \in Y$ and each open covering $\mathcal{U} = \{U_a : a \in A\}$ of $f^{-1}(y)$, finitely many $f(U_a)$ cover some neighborhood of $y \in Y$.[6]

It is straightforward to show that each biquotient map is necessarily a quotient map. The nice behavior of biquotient maps — as opposed to the behavior of quotient maps in general — with respect to taking products is stated in the following result which is due to Michael [1968].

A5.2 Theorem *Any product of biquotient maps is a biquotient map.*

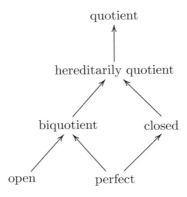

Fig. A5.3 Mappings and how they relate (open implies biquotient, etc.)

---

[5] For an example of a nonquotient product map $f \times g$ where both f and g are quotient see Brown [1968, Example 4, page 102].

[6] The concept of biquotient was introduced by Ernest Michael [1968] who states that these mappings are equivalent to "limit lifting maps" as defined in Hájek [1966].

A5.4 Definition   A *perfect* of *proper* mapping $p : X \to Y$ is a continuous closed surjection such that each fiber $p^{-1}(y)$ is compact.

A5.5 Definition   A continuous surjection $h : X \to Y$ is *hereditarily quotient* if for each non-empty $S \subset Y$, the restriction of h to $h^{-1}(S)$ is a quotient mapping $h^{-1}(S) \to S$.

A5.6 Theorem (Arhangel'skii [1963])   *A continuous surjection $h : X \to Y$ is hereditarily quotient if and only if $h(U)$ is a neighborhood of $y \in Y$ whenever U is a neighborhood of the fiber $h^{-1}(y) \subset X$.*

A5.7 Corollary   *Every biquotient map is hereditarily quotient, and every closed map is hereditarily quotient.*

The equivalence in Theorem A5.6 also yields the fact that any hereditarily quotient map $f : X \to Y$ with compact fibers $f^{-1}(y)$ is biquotient. It follows, since perfect maps are closed maps, that perfect implies biquotient.

To summarize, within the class of continuous surjections, the inclusions among the subclasses discussed above may be diagrammed as in Figure A5.3.[7]

For a proof of the following theorem see Bourbaki [1966, Proposition 4, page 98] [1961, Chapters 1 and 2]; the former concerns finite products.

A5.8 Theorem   *Any product of perfect maps is a perfect map.*

A5.9 Theorem (Morita and Hanai [1956] and Stone [1956])   *Let the map $p : X \to Y$ be a perfect map. Then X metrizable implies Y is metrizable.*

§A6 Topological Dimension Theory

For the prehistory of (topological) dimension theory see Crilly [1999]; for the separable-metric-space theory see Hurewicz and Wallman [1948]; for the status of dimension theory circa 1955 see Alexandroff [1955]; for the general (not necessarily separable) metric-space theory see Nagata [1965]; and for combinations or parts of these two theories along with evolving theories in general (not necessarily metric) spaces see Nagami (with an appendix by Kodama) [1970], Pears [1975], and Engelking [1978].

And to this list we add a rather concise and appropriate quotation of Kuratowski [1966, page 273] on its progress up to the 1950s:[8]

---

[7]For details and closely related references see Lašnev [1966], and, Michael [1972][1974].

[8]The original statements appear in a footnote that contains additional references. Also, the formatting of references has been adjusted to conform with those of this book.

The idea of a definition of dimension was originally due to Henri Poincaré [1912]. The definition was made precise by Brouwer [1913]. The dimension theory based on a definition rather close to that of Poincaré-Brouwer was created and developed independently by K. Menger and P. Uryshon in a number of papers beginning from 1922. See particularly Menger [1928] and Urysohn [1925b] [1926]. For a more modern exposition of dimension theory see Hurewicz and Wallman [1948]. For a dimension theory based on the notion of homology, see Alexandroff [1932].

A6.1 DEFINITIONS. We recall the definitions of the covering, the strong inductive, and the weak inductive dimension functions. In each case, X is any topological space and n is any non-negative integer.

Definition (ord$_x$ and ord $\mathcal{U}$) Let $\mathcal{U}$ a family of subsets of X, and let $x \in X$. Then "ord$_x$ $\mathcal{U}$" denotes *the order of $\mathcal{U}$ at x*, i.e., the number of members of $\mathcal{U}$ that contain x. If $x \in U \in \mathcal{U}$ for infinitely many U, then ord$_x$ $\mathcal{U} = \infty$. Moreover, the *order of $\mathcal{U}$* is given by ord $\mathcal{U} = \sup \{\text{ord}_x \mathcal{U} | x \in X\}$.[9]

The concept of ord $\mathcal{U}$ yields the covering/Lebesgue dimension.

Definition (covering dimension) The space X has *covering dimension* $\leq n$ if for each finite open covering $\{V_1, \ldots, V_k\}$ of X there is an open covering $\mathcal{U} = \{U_1, \ldots, U_k\}$ such that each $U_i \subset V_i$ and ord $\mathcal{U} \leq n + 1$. When X has covering dimension $\leq n$, we may write "dim $X \leq n$", and when dim $X \leq n$ and it is not true that dim $X \leq (n - 1)$, then X has covering dimension n and we may write "dim $X = n$". When no such n exists, then by definition dim $X = +\infty$. By convention, we define dim $\emptyset = -1$.[10]

The definition of the strong inductive dimension "Ind" involves the boundary "$B(G)$" of G.

Definition (strong inductive dimension) The definition is inductive, initiated with $X = \emptyset$ whose *strong inductive dimension* is defined as -1. And when the strong inductive dimension of X equals -1, we may write "Ind $X = -1$". In general, X has *strong inductive dimension* $\leq n$ if for each pair of disjoint closed subsets K and F of X there exists an open set G such that

$$F \subset G \subset X - K \qquad \text{and} \qquad \text{Ind } B(G) \leq n - 1.$$

When X has strong inductive dimension $\leq n$, then we may write "Ind $X \leq n$", and when Ind $X \leq n$ and it is not true that Ind $X \leq (n-1)$, then X *has*

---

[9]The definition of "order of a cover" varies from author to author: For example, the definition here agrees with Nagata [1965, page 9] but differs from Engelking [1978, page 54] and Pears [1975, page 111].

[10]A proof that this definition is equivalent to the one used by Nagata [1965, page 9] appears in Pears [1975, Proposition 1.2].

strong inductive dimension n and we may write "Ind $X = n$". If no such n exists, then by definition Ind $X = +\infty$.

Definition (weak inductive dimension) The definition is inductive, initiated with $X = \emptyset$ whose *weak inductive dimension* is defined as -1. And when the weak inductive dimension of X equals -1, we may write "ind $X = -1$". In general, let X be a topological space and let n be a non-negative integer. Then X has *ind inductive dimension* $\leq n$ if for each $x \in X$ and each open set G_x containing x, there exists an open set U such that

$$x \in U \subset G_x \qquad \text{and} \qquad \text{ind } B(U) \leq n - 1.$$

When X has weak inductive dimension $\leq n$, then we may write "ind $X \leq n$", and when ind $X \leq n$ and it is not true that ind $X \leq (n-1)$, then X *has weak inductive dimension* n and we may write "ind $X = n$". If no such n exists, then by definition ind $X = +\infty$.

When X is a separable metric space, then the covering, strong inductive, and weak inductive dimensions are equivalent, i.e., dim $X = $ Ind $X = $ ind X. And when X is a general (not necessarily separable) metric space dim $X = $ Ind X. (See Nagata [1965, Theorem II.7, page 27]. The first proofs were due to Katětov [1952] and, independently, Morita [1954].) However, Prabir Roy [1962, 1968] provided an example of a (nonseparable) metric space P such that ind $P = 0$ while Ind $P = $ dim $P = 1$.

A6.2 BASIC THEOREMS. Within the context of metric spaces, the most basic theorems of dimension theory are the Equivalence, Subspace, Sum, Decomposition, and Product theorems. Detailed proofs of these theorems appear in either Nagata [1965, Chapter II] or Engelking [1978, Chapter 4, Section 4.1]. Precise statements of these theorems are provided below, where it is assumed that X **is always a metric space**, and the page number and theorem number are from Engelking [1978].

Equivalence Theorem [page 254, Theorem 4.1.3] *For each X, Ind $X = $ dim X.*

Subspace Theorem [page 257, Theorem 4.1.7] *Let $S \subset X$. Then Ind $S \leq$ Ind X.*

Sum Theorem [page 257, Theorem 4.1.11] *Let $\{F_\gamma\}_{\gamma \in \Gamma}$ be a locally countable closed covering of X such that Ind $F_\gamma \leq n$ for each γ. Then Ind $X \leq n$.*

Decomposition Theorem [page 259, Theorem 4.1.17] *Let $n \geq 0$. Then Ind $X \leq n$ if and only if $X = \cup_{i=1}^{n+1} X_i$ where each $X_i \subset X$ satisfies Ind $X_i \leq 0$.*

Product Theorem [page 260, Theorem 4.1.21] *If Y is either a metric space or the empty set, then Ind $X \times Y \leq$ Ind $X + $ Ind Y.*

For extensions or analogues of these theorems in the context of general (not necessarily metric) spaces one may begin by reviewing Pears [1975] and the references listed therein.

The following theorem is applied in Chapter 7. And as presented below, the proof follows that of Proposition C) in Nagata [1965, page 12].

Theorem (open sets in the context of a zero-dimensional set) *Let* X *be a metric space, let* $X_0 \subset X$ *where* $\dim X_0 = 0$, *and, let* F *and* F' *be disjoint closed subsets of* X. *Then there exists an open set* M *such that* $F \subset M \subset \overline{M} \subset X \setminus F'$ *and* $B(M) \cap X_0 = \emptyset$.

PROOF. Since X is normal there are open sets V and V' such that

$$F \subset V, \quad F' \subset V', \quad \text{and} \quad \overline{V} \cap \overline{V'} = \emptyset.$$

Since Ind $X_0 = 0$, we can find an open and closed set U of the subspace X_0 such that
$$\overline{V} \cap X_0 \subset U \subset X_0 \setminus (\overline{V'} \cap X_0).$$

Since $C = F \cup U$ and $D = F' \cup (X_0 \setminus U)$ are separated, i.e., $(\overline{C} \cap D) \cup (C \cap \overline{D}) = \emptyset$, there exist open sets M and N such that

$$F \cup U \subset M, \quad F' \cup (X_0 \setminus U) \subset N, \quad \text{and} \quad M \cap N = \emptyset.$$

Now $B(M) \cap X_0 = \emptyset$ because N is open and U is both open and closed in X_0. Furthermore, $\overline{M} \subset X \setminus N$ because $M \cap N = \emptyset$ and N is open. So $X \setminus N \subset X \setminus F'$ yields $\overline{M} \subset X \setminus F'$. In short, $F \subset M \subset \overline{M} \subset X \setminus F'$. □

The following lemma and the next theorem are applied in Chapter 7. They are essentially "locally finite" versions of their "finite counterparts" Remark 2 and Remark 3 in Ostrand [1971] — the constructions used to prove the theorem parallel those of Ostrand in his proof of his Remark 3.

Lemma (covers at points in dim $\leq n$ closed subspaces) *Let* X *be a normal Hausdorff space, and let* F *be a closed subspace of* X *with* $0 \leq \dim F \leq n$. *Let* $\mathcal{U} = \{U_b : b \in B\}$ *be a locally finite open family of subsets of* X *that cover* F. *Then there is an open precise refinement* $\mathcal{V}$ *of* $\mathcal{U}$ *that covers* F *and satisfies* $\mathrm{ord}_x \mathcal{V} \leq n + 1$ *for each* $x \in F$.

PROOF. Consider the family $\{U_b \cap F\}_{b \in B}$. This family is a locally finite open (in F) cover of F. Since $\dim F \leq n$, there is a locally finite open (in F) precise refinement $\mathcal{U}'$ of $\{U_b \cap F\}_{b \in B}$ such that $\mathrm{ord}\,\mathcal{U}' \leq n + 1$. So for each $b \in B$ there is an open in X subset V'_b such that $V'_b \cap F = U'_b$. Then $(V'_b \cap U_b) \cap F = U'_b$. So let $V_b = V'_b \cap U_b$. Then each $V_b \subset U_b$ and each V_b is open in X. So $\mathcal{V} = \{V_b\}_{b \in B}$ is a precise refinement of $\mathcal{U}$. And if $x \in F$, then $\mathrm{ord}_x \mathcal{U}' \leq n + 1$ implies there are at most $n + 1$ distinct $U' \in \mathcal{U}'$ that contain x, which in turn implies there are at most $n + 1$ distinct $V \cap F$ that contain x, i.e., $\mathrm{ord}_x \mathcal{V} \leq n + 1$ for each $x \in F$. □

Theorem (refining covers of dim $\leq n$ closed subspaces) *Let X be a normal Hausdorff space, $F \subset X$ be closed with $0 \leq \dim F \leq n$, and $\mathcal{U} = \{U_b : b \in B\}$ be a locally finite open family that covers F. Then there is an open precise refinement $\mathcal{V}$ of $\mathcal{U}$ that covers F and satisfies $\operatorname{ord} \mathcal{V} \leq n+1$.*

PROOF. By the previous lemma, there exists a locally finite open family $\mathcal{U}'$ that shrinks $\mathcal{U}$ and covers F and satisfies $\operatorname{ord}_x \mathcal{U}' \leq n + 1$ for each $x \in F$. Consider $\mathcal{U}' \cup \{X \setminus F\}$, which is a locally finite open cover of X. By normality, we may shrink this cover to an open cover $\mathcal{U}'' = \{U_b'' : b \in B\} \cup \{G\}$ where each $\overline{U}_b'' \subset U_b'$ and $\overline{G} \subset X \setminus F$. Then $\mathcal{U}''$ covers F. Now let

$$N_2 = \left\{ C = \{b_1, \ldots, b_{n+2}\} \subset B : U_{b_1}'', \ldots, U_{b_{n+2}}'' \text{ are distinct} \right\}.$$

For $C \in N_2$, let $Y_C = \cap\{\overline{U}_b'' : b \in C\}$ and $Y = \cup_{C \in N_2} Y_C$. Then Y is closed (the family $\{Y_C\}_{C \in N_2}$ is locally finite) and $Y \cap F = \emptyset$. So for each $b \in B$, let $V_b = U_b'' \setminus Y$, and observe that $\mathcal{V} = \{V_b\}_{b \in B}$ is the desired family. $\square$

A6.3 CLASSICAL IMBEDDING THEOREM. Unless stated otherwise, each space in this book is a general (not necessarily separable) metric space. In this section we consider only separable metric spaces and the Classical Imbedding Theorem. Its statement is provided below, and an extensive discussion and development may be found in Hurewicz and Wallman [1948].

Any study of the Classical Imbedding Theorem shows that the mathematics used to develop the Classical Theorem is distinct from that used to in the development of the General J_A Imbedding Theorem. Indeed, the mathematics behind the Classical Theorem is extensively documented in several texts, while the mathematics for the J_A Theorem had, until the publication of this monograph, appeared only in the research literature.

Classical Imbedding Theorem (separable metric spaces) *Let $n \geq 0$, let I denote the unit interval, let I^{2n+1} denote the Cartesian product space of $2n + 1$ copies of I, and let*

$$I^{2n+1}(n) = \{x \in I^{2n+1} : x \text{ has at most } n \text{ rational coordinates}\}.$$

Then $\operatorname{Ind} I^{2n+1}(n) = n$, *and, if X is a separable metric space with $\operatorname{Ind} X \leq n$, then there exists an imbedding $f : X \to I^{2n+1}(n)$.*

APPENDIX 2

The Standard Simplex Δ^A in $l^2(A)$

We provide background material (with gateways to references) for the standard simplex $\Delta^A = \{(x_a) \in l^2(A) : 0 \le \Sigma_a x_a \le 1; 0 \le \text{each } x_a \le 1\}$ and show that Δ^A, like its counterpart $\Delta^n = \{(x_i) \in \mathbb{R}^{n+1} : \Sigma_i x_i = 1; 0 \le \text{each } x_i \le 1\}$, is the closed convex hull of the standard orthonormal basis.

§A7 Real Hilbert Spaces

We consider only real linear spaces. Let $\mathbb{R}$ denote the field of real numbers. Then a *real linear space or real vector space* is a set V together with *vector addition* $+ : V \times V \to V$ denoted $(\mathbf{v}, \mathbf{w}) \mapsto \mathbf{v} + \mathbf{w}$, and *scalar multiplication* $\mathbb{R} \times V \to V$ denoted $(\lambda, \mathbf{v}) \mapsto \lambda \mathbf{v}$, such that (i) $(V, +)$ is an Abelian group with identity "$\mathbf{0}$"; and (ii) $\lambda, \mu \in \mathbb{R}$ and $\mathbf{v}, \mathbf{w} \in V$ implies $\lambda(\mathbf{v} + \mathbf{w}) = \lambda \mathbf{v} + \lambda \mathbf{w}$, and, $(\lambda + \mu)\mathbf{v} = \lambda \mathbf{v} + \mu \mathbf{v}$, and, $\lambda(\mu \mathbf{v}) = (\lambda \mu)\mathbf{v}$, and, $1\mathbf{v} = \mathbf{v}$.

For any non-empty set A, the set $V = \mathbb{R}^A$ of tuples $(x_a)_{a \in A}$ (each $x_a \in \mathbb{R}$) with operations $(x_a) + (y_a) = (x_a + y_a)$ and $\lambda(x_a) = (\lambda x_a)$ is a linear space. In particular, we have *Hilbert's $l^2(A)$ space* given by[1],[2]

$$\{ \mathbf{x} = (x_a) \in \mathbb{R}^A : x_a \ne 0 \text{ for only countably many } a \in A \text{ and } \Sigma_a[x_a]^2 < \infty \}.$$

A7.1 LINEAR COMBINATIONS AND SUBSPACES. For $\mathbf{v}_1, \ldots, \mathbf{v}_k$ in the linear space V and $\lambda_1, \ldots, \lambda_k \in \mathbb{R}$, the finite sum $\lambda_1 \mathbf{v}_1 + \cdots + \lambda_k \mathbf{v}_k$ is called a *linear combination* of $\mathbf{v}_1, \ldots, \mathbf{v}_k$. Any non-empty subset S of a linear space V that contains all linear combinations of its members is a *linear subspace of V*. Given $n \ge 1$ vectors $\mathbf{v}_1, \ldots, \mathbf{v}_n$, the intersection of all linear subspaces $S \supset \{\mathbf{v}_1, \ldots, \mathbf{v}_n\}$ of V is *the linear subspace spanned by* $\mathbf{v}_1, \ldots, \mathbf{v}_n$ and may be specified as

$$\{\Sigma_1^n \lambda_i \mathbf{v}_i : \text{ each } \lambda_i \in \mathbb{R}\}.$$

A7.2 INDEPENDENCE, BASIS, AND DIMENSION. A non-empty *finite set* of vectors $\mathbf{v}_1, \ldots, \mathbf{v}_n$ in V is *linearly independent* if $\Sigma_1^n \lambda_i \mathbf{v}_i = \mathbf{0}$ implies $\lambda_1 = \cdots = \lambda_n = 0$; and a non-empty *arbitrary set* $\{\mathbf{v}_a\} \subset V$ is *linearly independent* if each of its finite non-empty subsets is linearly independent.

---

[1] In 1906 David Hilbert introduced $l^2(A)$ (for countably infinite A) in his research on the theory of integral equations as the natural infinite-dimensional analogue of Euclidean n-space (see Riesz and Sz-Nagy [1955, page 195] and Taylor [1965, page 155]). When A is finite, "$l^2(A)$" is often called *Euclidean space*. When A is uncountable, "$l^2(A)$" may be called *generalized Hilbert space with index set A* (Nagata [1968, page 95]).

[2] The fact that $x + y \in l^2(A)$ whenever $x, y \in l^2(A)$ follows from Minkowski's inequality $[\Sigma_a[x_a + y_a]^2]^{1/2} \le [\Sigma_a[x_a]^2]^{1/2} + [\Sigma_a[y_a]^2]^{1/2}$ (see Rudin [1966, Theorems 3.5 and 3.9] and use $p = 2$ and μ as the counting measure on A).

A maximal linearly independent subset $\{\mathbf{v}_a\}$ of V is a *basis* for V. And when $\{\mathbf{v}_a\}$ is a basis for V, then for each $\mathbf{v} \in V$ there is a unique finite subset $F = \{\mathbf{v}_i\}$ of $\{\mathbf{v}_a\}$ and a unique subset $\{\lambda_i\}$ of $\mathbb{R}$ such that $\mathbf{v} = \Sigma_F \lambda_i \mathbf{v}_i$.

Except for the *trivial space* ($V = \{\mathbf{0}\}$), every vector space V has a basis; and all bases of V have the same cardinality "dim V" (the *dimension of* V). If V is the trivial space, then by definition dim $V = \mathbf{0}$. Otherwise, either V is *finite-dimensional*, i.e., dim $V = n \geq 1$ is an integer, or, V is *infinite-dimensional*. The linear subspace spanned by $\mathbf{v}_1, \ldots, \mathbf{v}_n$ is *n-dimensional* if and only if the set $\{\mathbf{v}_1, \ldots, \mathbf{v}_n\}$ is linearly independent.

A non-empty *finite set* of vectors $\mathbf{v}_0, \mathbf{v}_1, \ldots, \mathbf{v}_n$ is *geometrically independent* if $\Sigma_0^n \lambda_i = 0$ and $\Sigma_0^n \lambda_i \mathbf{v}_i = \mathbf{0}$ imply that $\lambda_0 = \cdots = \lambda_n = 0$. One may show that $\{\mathbf{v}_0, \ldots, \mathbf{v}_n\}$ is geometrically independent if and only if $\{\mathbf{v}_1 - \mathbf{v}_0, \ldots, \mathbf{v}_n - \mathbf{v}_0\}$ is linearly independent.[3]

A *flat* (*plane, hyperplane,* or *linear manifold*) is a translation $\mathbf{v} + S = \{\mathbf{v} + \mathbf{w} : \mathbf{w} \in S\}$ of a subspace S of V. If dim $S = n$, then $\mathbf{v} + S$ may be called an *n-flat*. (A 2-flat in $\mathbb{R}^3$ corresponds to a plane not necessarily through the origin.) If $\{\mathbf{v}_0, \ldots, \mathbf{v}_n\}$ is a set of $n + 1$ points in V, then the smallest flat that contains $\mathbf{v}_0, \ldots, \mathbf{v}_n$ may be specified as[4]

$$(1) \qquad \left\{ \sum_0^n \lambda_i \mathbf{v}_i \; : \; \sum_0^n \lambda_i = 1 \; ; \; \text{each } \lambda_i \in \mathbb{R} \right\}$$

and is an *n-flat* if and only if $\{\mathbf{v}_0, \ldots, \mathbf{v}_n\}$ is geometrically independent.

A7.3 NORMED, METRIC, BANACH, AND LINEAR TOPOLOGICAL SPACES. A *normed linear space* is a linear space V with a *norm* $V \to \mathbb{R}$ denoted $\mathbf{v} \mapsto \|\mathbf{v}\|$ such that all $\mathbf{v}, \mathbf{w} \in V$ and all $\lambda \in \mathbb{R}$ satisfy $\|\mathbf{v}\| \geq 0$, and, ($\|\mathbf{v}\| = 0 \Leftrightarrow \mathbf{v} = 0$), and, $\|\mathbf{v} + \mathbf{w}\| \leq \|\mathbf{v}\| + \|\mathbf{w}\|$, and, $\|\lambda \mathbf{v}\| = |\lambda| \, \|\mathbf{v}\|$.

Any normed linear space V is a metric (hence topological) space with *norm-induced metric* $d(\mathbf{x}, \mathbf{y}) = \|\mathbf{x} - \mathbf{y}\|$, and its linear subspaces S are closed when dim $V < \infty$. Otherwise, its linear subspaces may not be closed.[5]

Normed spaces that are complete metric spaces relative to the norm-induced metric are called *Banach spaces*.[6]

Each normed space V is also a *linear topological space*, i.e., a linear space with a Hausdorff topology such that vector addition and scalar multiplication are continuous.[7]

---

[3]For a detailed proof, see Pontryagin [1952, page 3].

[4]Consider the subspace $S = \{\Sigma_1^n \mu_i (\mathbf{v}_i - \mathbf{v}_0) : \text{each } \mu_i \in \mathbb{R}\}$; and then consider $\mathbf{v}_0 + S = \{(1 - \Sigma_1^n \mu_i)\mathbf{v}_0 + \Sigma_1^n \mu_i \mathbf{v}_i : \text{each } \mu_i \in \mathbb{R}\} = \{\Sigma_0^n \lambda_i \mathbf{v}_i : \Sigma_0^n \lambda_i = 1; \text{each } \lambda_i \in \mathbb{R}\}$.

[5]Kolmogorov and Fomin [1957, Remark 1, page 73].

[6]For a proof that $\mathbb{R}^n$ is a Banach space see Simmons [1963, Theorem A, page 89]; for a proof that $l^2(A)$ is complete (and hence a Banach space) see Rudin [1966, Theorem 3.11] and use $p = 2$ and μ as the counting measure on A.

[7]For a concise proof see Simmons [1963, Section 46, page 212]. Both $\mathbb{R}^n$ and $l^2(A)$ are normed, metric, Banach, and linear topological spaces. For an extensive development of normed spaces see Kolmogorov and Fomin [1957, Chapter III]. For sufficient conditions for a linear topological space to be normable see Kelly and Namioka [1963, page 43].

For $\mathbf{v} \in V$, the continuous and mutually inverse *translations* $\mathbf{x} \mapsto \mathbf{x} + \mathbf{v}$ and $\mathbf{x} \mapsto \mathbf{x} - \mathbf{v}$ are homeomorphisms $V \to V$. Similarly, each *scaling* $\mathbf{x} \mapsto \lambda\mathbf{x}$ ($\lambda \neq 0$) is a homeomorphism. So $G + \mathbf{v}$ open $\Leftrightarrow G$ open $\Leftrightarrow \lambda G$ open.

A7.4 ABSTRACT REAL HILBERT SPACES. Following Kolmogorov and Fomin[8] we say that H is a *real Hilbert space* when the following axioms are satisfied: I. H is a linear space; II. an *inner product* $H \times H \to \mathbb{R}$ is defined on H, i.e., for $\mathbf{x}, \mathbf{y}, \mathbf{z} \in H$ and $\lambda \in \mathbb{R}$, the *inner product* $(\mathbf{x}, \mathbf{y}) \in \mathbb{R}$ satisfies (*i*) $(\mathbf{x}, \mathbf{y}) = (\mathbf{y}, \mathbf{x})$, (*ii*) $(\lambda\mathbf{x}, \mathbf{y}) = \lambda(\mathbf{x}, \mathbf{y})$, (*iii*) $(\mathbf{x} + \mathbf{y}, \mathbf{z}) = (\mathbf{x}, \mathbf{z}) + (\mathbf{y}, \mathbf{z})$, and (*iv*) $(\mathbf{x}, \mathbf{x}) > 0$ if $\mathbf{x} \neq \mathbf{0}$; III. H is complete in the *norm-induced* metric $d(\mathbf{x}, \mathbf{y}) = \|\mathbf{x} - \mathbf{y}\|$ where $\|\mathbf{x}\| = (\mathbf{x}, \mathbf{x})^{\frac{1}{2}}$ is the *norm* of $\mathbf{x}$; and IV. H is infinite-dimensional.

Hilbert spaces are normed, metric, Banach, and linear topological spaces. Moreover, the inner product function is also continuous, i.e., $(\mathbf{x}_n, \mathbf{y}_n) \to (\mathbf{x}, \mathbf{y})$ when $\|\mathbf{x}_n - \mathbf{x}\| \to 0$ and $\|\mathbf{y}_n - \mathbf{y}\| \to 0$. In particular, $(\mathbf{x}_n, \mathbf{x}_n) \to (\mathbf{x}, \mathbf{x})$ whenever $\|\mathbf{x}_n - \mathbf{x}\| \to 0$, showing that the norm $\mathbf{x} \mapsto \|\mathbf{x}\| = (\mathbf{x}, \mathbf{x})^{\frac{1}{2}}$ is also continuous.

For any infinite set A, the space $l^2(A)$ is a Hilbert space whose operations may be summarized as follows: For $\mathbf{x} = (x_a)$ and $\mathbf{y} = (y_a)$,

$$
\begin{aligned}
&(x_a) + (y_a) = (x_a + y_a), && \lambda(x_a) = (\lambda x_a), \\
&(\mathbf{x}, \mathbf{y}) = \Sigma_a x_a y_a, && \|\mathbf{x}\| = (\mathbf{x}, \mathbf{x})^{1/2} = (\Sigma_a (x_a)^2)^{1/2}, \\
&d((x_a), (y_a)) = (\Sigma_a (x_a - y_a)^2)^{\frac{1}{2}}.
\end{aligned}
$$

A7.5 ORTHONORMAL BASES. Let H be a Hilbert space. Then $\{\mathbf{u}_a : a \in A\} \subset H$ is an *orthonormal set* if $a, b \in A$ implies $(\mathbf{u}_a, \mathbf{u}_a) = 1$ and $(\mathbf{u}_a, \mathbf{u}_b) = 0$ whenever $a \neq b$. A *maximal orthonormal set* $\{\mathbf{u}_a\}$ in H is frequently called a *complete orthonormal set* or an *orthonormal basis*.[9]

Each $l^2(A)$ has its *standard orthonormal basis* $\{\mathbf{u}_a : a \in A\}$ given by

$$
\mathbf{u}_a = (u_b^a) \in l^2(A) \quad \text{where } u_b^a = \begin{cases} 1 & \text{if } b = a, \\ 0 & \text{otherwise.} \end{cases}
$$

Every orthonormal set is also a linearly independent set, but a *maximal* orthonormal set may not be a *maximal* linearly independent set. (So an orthonormal basis need not be a basis.)[10]

In passing, recall that the standard orthonormal basis $\{\mathbf{u}_a : a \in A\}$ of $l^2(A)$ provides an "inner-product representation" of values "x_a" of each function $\mathbf{x} = (x_a) \in l^2(A)$, namely, $x_a = (\mathbf{x}, \mathbf{u}_a)$ for each $a \in A$.

---

[8]Kolmogorov and Fomin [1961, Chapter IX]. For complex Hilbert spaces see Rudin [1966, Chapter 4].

[9]For characterizations of an orthonormal basis see Rudin [1966, Theorem 4.18] and, for a concise list of various aspects of Hilbert spaces, see problems H, I, J, K, and L in Kelly and Namioka [1963, pages 65–67].

[10]For infinite A, the orthonormal basis $\{\mathbf{u}_a : a \in A\}$ of $l^2(A)$ is not a basis of $l^2(A)$: If $\mathbf{x} = (x_a) \in l^2(A)$ where infinitely many $x_a \neq 0$, then for each finite $F \subset \{\mathbf{u}_a\}$ and all λ_a, we have $\mathbf{x} \neq \Sigma_F \lambda_a \mathbf{u}_a$.

A7.6 HILBERT SPACE ISOMORPHISMS. For linear spaces V and W, a *linear isomorphism* $\Lambda : V \to W$ is a bijective linear transformation, i.e., a bijection such that $\Lambda(\lambda\mathbf{x} + \mu\mathbf{y}) = \lambda\Lambda\mathbf{x} + \mu\Lambda\mathbf{y}$ for any $\mathbf{x}, \mathbf{y} \in V$ and any $\lambda, \mu \in \mathbb{R}$.

For Hilbert spaces V and W, a *Hilbert-space isomorphism* $\Lambda : V \to W$ is a linear isomorphism such that $(\Lambda\mathbf{x}, \Lambda\mathbf{y}) = (\mathbf{x}, \mathbf{y})$ for every $\mathbf{x}, \mathbf{y} \in V$.

Any Hilbert space H has an orthonormal basis $\{\mathbf{u}_a : a \in A\}$ for some set A, and H is Hilbert-space isomorphic to $l^2(A)$.[11]

It follows that $|A| = |B| \geq \aleph_0$ if and only if $l^2(A)$ is Hilbert-space isomorphic to $l^2(B)$. In general, the $l^2(A)$ spaces provide models for all abstract Hilbert spaces.

Finally, each Hilbert space isomorphism $\Lambda : V \to W$ is also a *metric space isometry*:

$$d_W(\Lambda\mathbf{x}, \Lambda\mathbf{y}) = \|\Lambda\mathbf{x} - \Lambda\mathbf{y}\| = (\Lambda\mathbf{x}, \Lambda\mathbf{y})^{\frac{1}{2}} = (\mathbf{x}, \mathbf{y})^{\frac{1}{2}} = \|\mathbf{x} - \mathbf{y}\| = d_V(\mathbf{x}, \mathbf{y}).$$

A7.7 Proposition *Let A be an infinite set, let $z \in A$, and let $A' = A \setminus \{z\}$. Then $l^2(A)$ is Hilbert-space isomorphic (and thus homeomorphic) to $l^2(A')$.*

PROOF. Select a bijection $a \mapsto \phi a$ of $A \to A'$; and for each $a \in A$, define $\mathbf{u}_{\phi a} = (u_{\phi b}^{\phi a})$ where $u_{\phi a}^{\phi a} = 1$ and $u_{\phi b}^{\phi a} = 0$ when $\phi a \neq \phi b$. Then $\{\mathbf{u}_{\phi a} : a \in A\}$ is the standard orthonormal basis of $l^2(A')$. Define a mapping $\mathbf{x} \mapsto \Phi\mathbf{x}$ of $l^2(A) \to l^2(A')$ by specifying that when $\mathbf{x} = (x_a)_{a \in A}$, then $\Phi\mathbf{x} = (x_{\phi a})_{\phi a \in A'}$ where $x_{\phi a} = x_a$ for each $a \in A$. That is, the "ath coordinate x_a of $\mathbf{x}$" equals the "(ϕa)th coordinate $x_{\phi a}$ of $\Phi\mathbf{x}$":

$$x_a = (\mathbf{x}, \mathbf{u}_a) = x_{\phi a} = (\Phi\mathbf{x}, \mathbf{u}_{\phi a}).$$

The map Φ is the desired Hilbert-space isomorphism because

$$(\Phi\mathbf{x}, \Phi\mathbf{y}) = \Sigma_{\phi a \in A'} x_{\phi a} y_{\phi a} = \Sigma_{a \in A} x_a y_a = (\mathbf{x}, \mathbf{y}).$$

Thus Φ is also a metric isometry, and hence a homeomorphism.   □

§A8 Convex Hulls and Closed Convex Hulls

For two points $\mathbf{v}$ and $\mathbf{w}$ in a vector space V, the set $[\mathbf{v}, \mathbf{w}] = \{\mathbf{y} : \mathbf{y} = t\mathbf{v} + (1-t)\mathbf{w}; 0 \leq t \leq 1\}$ is the *line segment* joining the *endpoints* $\mathbf{v}$ and $\mathbf{w}$. A non-empty set $C \subset V$ is *convex* if $\mathbf{v}, \mathbf{w} \in C$ implies $[\mathbf{v}, \mathbf{w}] \subset C$.

Every non-empty subset $K \subset V$ is contained in the convex set V. So the *convex hull* $H(K) = \cap\{C : K \subset C; C \text{ is convex}\}$ of K is the *smallest convex set containing K*. *Closed convex hulls* are similarly defined.

The set $H(K)$ may also be viewed as a union of sets: For the following proposition, we consider each non-empty finite set $F = \{\mathbf{v}_1, \ldots, \mathbf{v}_n\} \subset K$, and define $\sigma(F) = \{\Sigma_1^n \lambda_i \mathbf{v}_i : \Sigma_1^n \lambda_i = 1; 0 < \text{each } \lambda_i \leq 1\}$.

---

[11] See Rudin [1966, Section 4.19].

A8.1 Proposition *Let V be a linear space and let $\emptyset \neq K \subset V$. Then the convex hull $H(K) = \cup\{\sigma(F) : \emptyset \neq F \subset K \text{ is finite}\}$.*[12]

A8.2 Proposition *Let V be a linear topological space, and let $\emptyset \neq K \subset V$. Then the closed convex hull of K is the closure $\overline{H(K)}$ of $H(K)$.*

PROOF. Since $K \subset H(K)$ it is clear that $K \subset \overline{H(K)}$. To see that $\overline{H(K)}$ is convex, consider distinct vectors $\mathbf{x}, \mathbf{y} \in \overline{H(K)}$ and let sequences $\{\mathbf{x}_n\}$ and $\{\mathbf{y}_n\}$ in $H(K)$ converge, respectively, to $\mathbf{x}$ and $\mathbf{y}$. For each t, $0 \leq t \leq 1$, it follows (since $H(K)$ is convex and V is a linear topological space) that $t\mathbf{x}_n + (1-t)\mathbf{y}_n \in H(K)$ converges to $t\mathbf{x} + (1-t)\mathbf{y}$. Thus $t\mathbf{x} + (1-t)\mathbf{y} \in \overline{H(K)}$ for each such t. To see that $\overline{H(K)}$ is the smallest closed convex set containing K, suppose $K \subset C$ where C is a closed convex set. Then $H(K) \subset C$ by the previous proposition, and therefore $\overline{H(K)} \subset C$. $\qquad\square$

§A9 Standard Simplexes

A9.1 AFFINE TRANSFORMATIONS. An *affine transformation* T of $\mathbb{R}^{n+1}$ is a composition of a translation and nonsingular linear transformations. These mappings map each geometrically independent set of points onto a geometrically independent set of points. For example, let us suppose that $T(\mathbf{x}) = L(\mathbf{x} + \mathbf{q}) = L(\mathbf{x}) + \mathbf{p}$, where L is nonsingular linear, and that $\{\mathbf{v}_i\}$ is a geometrically independent set of points. If $\Sigma_i \lambda_i = 0$ and $\Sigma_i \lambda_i T(\mathbf{v}_i) = \mathbf{0}$, then $\Sigma_i \lambda_i T(\mathbf{v}_i) = \Sigma_i \lambda_i (L(\mathbf{v}_i) + \mathbf{p}) = \Sigma_i \lambda_i L(\mathbf{v}_i) + (\Sigma_i \lambda_i)\mathbf{p} = \Sigma_i \lambda_i L(\mathbf{v}_i)$. So

$$\Sigma_i \lambda_i T(\mathbf{v}_i) = L(\Sigma_i \lambda_i \mathbf{v}_i) = \mathbf{0} \Rightarrow \Sigma_i \lambda_i \mathbf{v}_i = \mathbf{0} \Rightarrow \text{ each } \lambda_i = 0.$$

For the standard basis $\{\mathbf{u}_1, \ldots, \mathbf{u}_{n+1}\}$ of $\mathbb{R}^{n+1}$ the translation $\mathbf{x} \mapsto (\mathbf{x} - \mathbf{u}_{n+1})$ maps the n-dimensional plane $P_u = \{\Sigma_1^n \lambda_i \mathbf{u_i} : \Sigma_1^n \lambda_i = 1; \lambda_i \in \mathbb{R}\}$ onto the vector subspace spanned by $\{\mathbf{u}_1 - \mathbf{u}_{n+1}, \ldots, \mathbf{u}_n - \mathbf{u}_{n+1}, \mathbf{0}\}$. This subspace has $\{\mathbf{u}_i - \mathbf{u}_{n+1}\}_1^n$ as a basis. Next, we follow this translation with a nonsingular linear transformation of $\mathbb{R}^{n+1}$ that maps, for $1 \leq i \leq n$, each $(\mathbf{u}_i - \mathbf{u}_{n+1}) \mapsto \mathbf{u}_i$. We thereby obtain an affine transformation S of $\mathbb{R}^{n+1}$ such that $S(\mathbf{u}_{n+1}) = \mathbf{0}$ and $S(\mathbf{u}_i) = \mathbf{u}_i$. Globally, S sends the plane P_u onto the plane $\mathbb{R}^n \times \{0\} \subset \mathbb{R}^{n+1}$ of the first n coordinates in $\mathbb{R}^{n+1}$.

A9.2 EXAMPLE. Consider $\mathbb{R}^2$, with $\mathbf{u}_1 = (1,0)$ and $\mathbf{u}_2 = (0,1)$. Using $\mathbf{x} \mapsto (\mathbf{x} - \mathbf{u}_2)$ to translate $\{\mathbf{u}_1, \mathbf{u}_2\}$ onto $\{\mathbf{u}_1 - \mathbf{u}_2, \mathbf{u}_2 - \mathbf{u}_2\} = \{(1,-1), (0,0)\}$, we send the line containing $(1,0)$ and $(0,1)$ onto the line containing $(1,-1)$ and $(0,0)$. Using

$$\mathbf{u}_1 = \begin{pmatrix} 1/\sqrt{2} & 0 \\ 0 & 1/\sqrt{2} \end{pmatrix} \begin{pmatrix} 1/\sqrt{2} & -1/\sqrt{2} \\ 1/\sqrt{2} & 1/\sqrt{2} \end{pmatrix} (\mathbf{u}_1 - \mathbf{u}_2),$$

---

[12]For a proof see Dugundji [1966, Appendix One, page 411].

we then rotate the line containing $(1, -1)$ and $(0, 0)$ counterclockwise by $45°$ and scale by $1/\sqrt{2}$. In short, we have an affine map that takes the line ℓ_1 containing $(1, 0)$ and $(0, 1)$ in $\mathbb{R}^2$ onto the line $\ell_2 \subset \mathbb{R}^1 \times \{0\}$. Moreover, the point $\lambda_1(1, 0) + \lambda_2(0, 1) \in \ell_1$ maps to the point $\lambda_1(1, 0) + \lambda_2(0, 0) \in \ell_2$.   □

A9.3 FINITE-DIMENSIONAL SIMPLEXES.[13]   Given any geometrically independent set of vectors $\mathbf{v}_0, \ldots, \mathbf{v}_n$ in $\mathbb{R}^k$ where $k \geq n$, the set

$$[\mathbf{v}_0, \ldots, \mathbf{v}_n] = \{ \Sigma_0^n \lambda_i \mathbf{v}_i : \Sigma_0^n \lambda_i = 1; \ 0 \leq \text{each } \lambda_i \leq 1 \}$$

is an *n-dimensional simplex* with *vertices* $\mathbf{v}_0, \ldots, \mathbf{v}_n$. A point $\mathbf{v} = \Sigma_0^n \lambda_i \mathbf{v}_i$ in $[\mathbf{v}_0, \ldots, \mathbf{v}_n]$ provides a *unique* set of coefficients $\lambda_0, \ldots, \lambda_n$ that are called the *barycentric coordinates of* $\mathbf{v}$.[14]

When $[\mathbf{v}_0, \ldots, \mathbf{v}_n] \subset \mathbb{R}^k$ has the induced topology, each *barycentric coordinate function* $\mathbf{v} \mapsto \lambda_i = \lambda_i(\mathbf{v})$ is continuous: Using the figure below where S_v is an affine mapping, we note that S_v sends the *n*-simplex $[\mathbf{v}_0, \ldots, \mathbf{v}_n]$ "barycentrically" onto the *n*-simplex $[\mathbf{u}_0 = \mathbf{0}, \mathbf{u}_1, \ldots \mathbf{u}_n]$, i.e., $\Sigma_0^n \lambda_i \mathbf{v}_i \mapsto \Sigma_0^n \lambda_i \mathbf{u}_i$ is continuous; π_i is the continuous Cartesian coordinate projection from $\mathbb{R}^n$ onto its *i*th factor $\mathbb{R}$; and $\lambda_i = \lambda_i(\mathbf{v}) = \pi_i \circ S_v(\mathbf{v})$ is the barycentric coordinate projection. Since the diagram is commutative, it is clear that $\mathbf{v} \mapsto \lambda_i(\mathbf{v})$ is continuous.

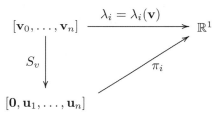

Any two *n*-dimensional simplexes $[\mathbf{v}_0, \ldots, \mathbf{v}_n]$ and $[\mathbf{w}_0, \ldots, \mathbf{w}_n]$ are homeomorphic under the *barycentric coordinate mapping* $\Sigma_0^n \lambda_i \mathbf{v}_i \mapsto \Sigma_0^n \lambda_i \mathbf{w}_i$.

The *standard n-dimensional simplex* $\Delta^n \subset \mathbb{R}^{n+1}$ is $[\mathbf{u}_1, \ldots, \mathbf{u}_{n+1}]$ where $\{\mathbf{u}_i\}$ is *the standard orthonormal basis of* $\mathbb{R}^{n+1}$, i.e.,

$$\mathbf{u}_1 = (1, 0, 0, \ldots, 0), \mathbf{u}_2 = (0, 1, 0, 0, \ldots, 0), \ldots, \mathbf{u}_{n+1} = (0, \ldots, 0, 1).$$

As discussed in the paragraph preceding Example A9.2, the standard *n*-simplex $\Delta^n = [\mathbf{u}_1, \ldots, \mathbf{u}_{n+1}]$ is (barycentrically) homeomorphic to $\Delta_n = [\mathbf{0}, \mathbf{u}_1, \ldots, \mathbf{u}_n] \subset \mathbb{R}^n$. The *subscript* n is a mnemonic that the dimension of the space containing Δ_n is "lower than" that of the space containing Δ^n.

Finally, using Proposition 8.2 and the definition of $\sigma(F)$, we may conclude that $H(\{\mathbf{u}_i\}) = \Delta^n$ and that $H(\{\mathbf{u}_i\})$ is closed in $\mathbb{R}^{n+1}$, i.e., Δ^n *is the closed convex hull of the standard orthonormal basis* $\{\mathbf{u}_i\}$ *of* $\mathbb{R}^{n+1}$.[15]

---

[13]See Munkres [1984, §1] and Kolmogorov and Fomin [1957, Theorem 3 page 76].
[14]If $\mathbf{v} = \Sigma_0^n \mu_i \mathbf{v}_i = \Sigma_0^n \tau_i \mathbf{v}_i$ where $\Sigma_i \mu_i = 1 = \Sigma_i \tau_i$, then $\Sigma_i(\mu_i - \tau_i) = 0$ and also $\Sigma_i(\mu_i - \tau_i)\mathbf{v}_i = 0$ yield $\mu_1 - \tau_1 = 0, \ldots, \mu_n - \tau_n = 0$.
[15]See Engelking and Siekulcki [1992, Section 2.1] and Alexandroff [1956].

A9.4 THE INFINITE-DIMENSIONAL STANDARD SIMPLEX Δ^A. Let A be an infinite set, and let $\{\mathbf{u}_a : a \in A\}$ be the standard orthonormal basis of $l^2(A)$. Then the *standard simplex* Δ^A is given by

$$\Delta^A = \{(x_a) \in l^2(A) : 0 \leq \Sigma_a x_a \leq 1; \, 0 \leq \text{each } x_a \leq 1\}.$$

A9.5 Proposition *Let A be an infinite set. Then Δ^A is the closed convex hull $\overline{H(\{\mathbf{u}_a\})}$ of the standard orthonormal basis $\{\mathbf{u}_a : a \in A\}$ of $l^2(A)$.*

PROOF. It suffices to prove inclusions (i) and (ii):

(i) $\overline{H(\{\mathbf{u}_a\})} \subset \Delta^A$: First, let $\mathbf{x} \in H(\{\mathbf{u}_a\})$. Then $\mathbf{x} = (x_a) \in \sigma(F)$ for some non-empty finite $F \subset \{\mathbf{u}_a\}$. So for the finite set $A_F = \{a \in A : \mathbf{u}_a \in F\}$, we have $\Sigma_a x_a = \Sigma_{a \in A_F} x_a = 1$ and $0 \leq \text{each } x_a \leq 1$, i.e., $\mathbf{x} \in \Delta^A$. Second, let $\mathbf{x} \in \overline{H(\{\mathbf{u}_a\})} \setminus H(\{\mathbf{u}_a\})$. Then select a sequence $\mathbf{x}_n \in H(\{\mathbf{u}_a\})$ such that $\mathbf{x}_n \to \mathbf{x}$, i.e., $(x_a^n) \to (x_a)$. Now for each n, we have $x_a^n \to x_a$ because $|x_a^n - x_a| \leq \|\mathbf{x}_n - \mathbf{x}\|$. And $x_a^n \to x_a$ coupled with $0 < \text{each } x_a^n \leq 1$, yields $0 \leq \text{each } x_a \leq 1$. Moreover, since addition of real numbers is continuous, for each non-empty finite set $A_F \subset A$, we see that $\Sigma_{a \in A_F} x_a^n \to \Sigma_{a \in A_F} x_a$. It follows, since each $\Sigma_{a \in A} x_a^n = 1$ that the value 1 is an upper bound of $\Sigma_a x_a$. Thus again $\mathbf{x} \in \Delta^A$.

(ii) $\Delta^A \subset \overline{H(\{\mathbf{u}_a\})} = \text{cl}(H)$: First, observe that the zero vector $\mathbf{u}_0 = \mathbf{0}$ of $l^2(A)$ is in both Δ^A and $\text{cl}(H)$. The former claim is clear, while the latter may be demonstrated by considering a countably infinite list $a_1, a_2, \ldots$ of members of A, defining the sequence $\mathbf{x}_n = (x_a^n)$ in $H = H(\{\mathbf{u}_a\})$ by

$$x_{a_1}^n = \cdots = x_{a_n}^n = 1/n \text{ and } x_{a_i}^i = 0 \text{ for } i \geq n+1$$

and noting that $\|\mathbf{x}_n - \mathbf{u}_0\| = 1/\sqrt{n}$. Since $\mathbf{u}_0$ is a point in the closed and convex hull $\text{cl}(H)$ of $\{\mathbf{u}_a : a \in A\}$, we see that $\text{cl}(H)$ is also the closed convex hull of $\{\mathbf{u}_a\}_A \cup \{\mathbf{u}_0\}$.

Now let $x = (x_a) \in \Delta^A$, and then order the elements in $\{a \in A : x_a \neq 0\}$ as $a_1, a_2, \ldots$. We define

$$\mathbf{x_n} = x_{a_1} \mathbf{u}_{a_1} + \cdots + x_{a_n} \mathbf{u}_{a_n} + (1 - \Sigma_1^n x_{a_i}) \mathbf{u}_0.$$

Then each $\mathbf{x}_n \in H(\{\mathbf{u}_a\})$ because $0 \leq \Sigma_1^n x_{a_i} \leq \Sigma_a x_a \leq 1$ and $0 \leq \text{each } x_a \leq 1$. Moreover, since $\|\mathbf{x}^n - \mathbf{x}\|^2 \leq \Sigma_{n+1}^\infty (x^{a_i})^2$, which goes to zero as $n \to \infty$, it follows that $\mathbf{x} \in \text{cl}(H)$. $\qquad\square$

A9.6 Proposition *Let A be infinite, let $z \in A$, and let $A' = A \setminus \{z\}$. Also let $\phi : A \to A'$ be a bijection, and define $\Phi : \Delta^A \to \Delta^{A'}$ by*

$$\mathbf{x} = (x_a)_{a \in A} \quad \mapsto \quad \Phi \mathbf{x} = (x_{\phi a})_{\phi a \in A'},$$

where $x_{\phi a} = x_a$. Then the ath coordinate of $\mathbf{x}$ is the (ϕa)th coordinate of $\Phi \mathbf{x}$ and Φ is a homeomorphism.

PROOF. The correspondence Φ is a metric isometry, hence homeomorphism $l^2(A) \rightarrow l^2(A')$, as detailed in the proof of Proposition 7.7. The only observation that is required is that of showing $\Phi(\Delta^A) = \Delta^{A'}$. This equality follows since $(x_a) \in \Delta^A$ is equivalent to $0 \leq \Sigma_A x_a \leq 1$ and $0 \leq$ each $x_a \leq 1$, which is equivalent to $0 \leq \Sigma_{A'} x_{\phi a} \leq 1$ and $0 \leq$ each $x_{\phi a} \leq 1$ because the coordinates of (x_a) are also the coordinates of $(x_{\phi a})$.   □

APPENDIX 3

Measures and Fractal Dimension

This appendix provides a convenient and concise basic development of Hausdorff measures that leads to a definition of fractal dimension — Hausdorff measures and dimension (§A10), The Lebesgue and Hausdorff m_ε^p-measures (§A11), Hausdorff p-measures (§A12), Hausdorff dimension (§A13), and fractal dimension (§A14).

§A10 Hausdorff Measures and Dimension

The mathematics known as measure theory evolved from constructions of sets, functions, and integrals that were outside of the classical calculus. Indeed, as stated in the Preface of Rogers [1970], "E. Borel in his 1894 thesis essentially introduced the Lebesgue outer measure as a means of estimating the size of certain sets, so that he could construct certain pathological functions, while Lebesgue [1904] applied measure theory to obtain his integral."

The seeds of Hausdorff measures were planted by Carathéodory [1914], who introduced "general (Carathéodory) outer measures" and showed how to construct "p-measures" for certain integer values of p. Subsequently, Hausdorff [1919] extended the range of values of p to all positive reals, and also showed that "in a certain sense" Cantor's set has "fractional dimension" $\log 2 / \log 3$.

The following few sections contain definitions and propositions that lead to Hausdorff measures and Hausdorff dimension. An in-depth and careful development may be found in Rogers [1970].

To begin the spadework, recall that unlike each *topological dimension* function $D_T = \mathrm{ind}$, Ind, or $\dim$, which was formulated as a topological invariant — X homeomorphic to Y implies $D_T(X) = D_T(Y)$ — a *measure* function typically requires a metric, i.e., the distance function d of a metric space (X, d), and consequently is not necessarily a topological invariant.

For relevant examples, the unit interval $I = [0, 1]$ as a subset of the real line $\mathbb{R}$ has Lebesgue measure unity, but its homeomorphic image $[0, 1/2]$ in $\mathbb{R}$ given by the imbedding $x \mapsto x/2$ has Lebesgue measure $1/2$. Similarly, the subspace of irrational numbers in the unit interval has Lebesgue measure unity, but its homeomorphic image in Cantor's space given by the imbedding $\Sigma_1^\infty a_n/2^n \to \Sigma_1^\infty (2a_n)/3^n$ has Lebesgue measure zero. In contrast, each "counting measure" is topologically invariant because homeomorphisms preserve cardinality.[1]

---

[1] Recall that a *counting measure* $\mu : 2^X \to [0, \infty]$ on the family 2^X of all subsets of a

§A11 Lebesgue and Hausdorff m_ε^p-Measures

The Lebesgue and Hausdorff m_ε^p-measures may be unified when viewed as separate applications of a general method of constructing outer measures (Taylor [1965, Chapter 4]; Rogers [1970, "Method II"]. Roughly, each application yields an outer measure $\mu^* : 2^X \to [0, \infty]$ for a given set X. In turn, then, μ^* induces a σ-algebra $S \subset 2^X$ such that μ^* restricted to S, i.e., $\mu = \mu^*|_S : S \to [0, \infty]$, is a measure.

Digging deeper, we recall that a *σ-algebra (σ-field)* is a non-empty family S of subsets of a non-empty set X that is closed under complementation and countable unions. It follows since

$$E \in S \neq \emptyset \;\Rightarrow\; X = E \cup E^\sim \in S \;\Rightarrow\; \emptyset = X^\sim \in S,$$

that both X and $\emptyset$ are members of S, and, since

$$\cap_n E_n = (\cup_n E_n^\sim)^\sim \in S \quad \text{and} \quad E - F = E \cap F^\sim \in S$$

that S is also closed under countable intersections and set difference. An example of a σ-algebra is the family 2^X of all subsets of X.

For a σ-algebra S, a *measure* $\mu : S \to [0, \infty]$ is a "countably additive" set function that maps the empty set to zero. In other words, $\mu(\emptyset) = 0$, and, for each countable list $E_1, E_2, \dots \in S$ of pairwise disjoint sets, $\mu(\cup_j E_j) = \Sigma_j \mu(E_j)$. In particular, the list $E_1 = A$, $E_2 = E - A$, $E_3 = \emptyset$, $E_4 = \emptyset$, $\dots$ coupled with the countable additivity of μ show that μ is also *monotone*, i.e., $A, E \in S$ and $A \subset E$ imply $\mu(A) \leq \mu(E)$.

An *outer measure* $\mu^* : 2^X \to [0, \infty]$ is a monotone and "countably *sub*additive" set function that maps the empty set to zero. In other words, $\mu^*(\emptyset) = 0$, $\mu^*(A) \leq \mu^*(E)$ when $A \subset E$, and, any countable list $E_1, E_2, \dots \in 2^X$ yields $\mu^*(\cup_j E_j) \leq \Sigma_j \mu^*(E_j)$.[2]

To construct the μ^*-induced S, we call $E \subset X$ μ^*-*measurable* if $\mu^*(T) = \mu^*(T \cap E) + \mu^*(T - E)$ for all "test sets" $T \subset X$ — E is μ^*-*measurable* if μ^* is additive on sets that are separated by E. The collection S of all μ^*-measurable sets is a σ-algebra, and μ^* restricted to S (denoted μ) is a measure on S.

When $X = (X, d)$ is a metric space, an outer measure μ^* on 2^X is a *metric outer measure* if $\mu^*(A \cup B) = \mu^*(A) + \mu^*(B)$ whenever the subsets A and B of X are positively separated, i.e. $d(A, B) = \inf \{d(x, y) : x \in A, y \in B\} > 0$. It turns out that whenever μ^* is a metric outer measure, then the μ^*-induced σ-algebra S contains all open (and hence Borel) sets in X.

---

set X is given by

$$\mu(E) = \begin{cases} k & \text{if } E \text{ is a finite set with } k \text{ elements,} \\ \infty & \text{otherwise.} \end{cases}$$

It turns out that the Hausdorff 0-measure m^0 is a counting measure.

[2]The countably subadditive property does not imply the monotone property, e.g., consider $X = \{1, 2\}$ and let $\mu^* : 2^X \to [0, \infty]$ be given by $\mu^*(\{2\}) = 1$ and $\mu^*(E) = 0$ otherwise.

With these concepts, we now describe a "general method" for constructing outer measures. Let $\mathcal{G}$ denote a family of subsets of X such that $\emptyset \in \mathcal{G}$ and, for each $E \subset X$, there is a countable subcollection $\{G_n\}$ of $\mathcal{G}$ whose union $\cup_n G_n$ is a superset of E. In this case, call the collection $\{G_n\}$ a $\mathcal{G}$-*covering* of E. Next, let $\lambda : \mathcal{G} \to [0, \infty]$ be any map such that $\lambda(\emptyset) = 0$, and, for $E \subset X$, define

$$\mu^*(E) = \inf \Sigma_n \lambda(G_n) \qquad \{G_n\} \subset \mathcal{G} \text{ varies over the } \mathcal{G}\text{-coverings of } E.$$

Then $\mu^* : 2^X \to [0, \infty]$ is an outer measure.

One application of this "general method" yields Lebesgue outer measures: For example, let $X = \mathbb{R}^1$, let $\mathcal{G}$ contain the empty set and all non-empty open intervals (a, b), and let $\lambda : \mathcal{G} \to [0, \infty]$ be the "length function," i.e., $\lambda(a, b) = |(a, b)| = b - a$ and $\lambda(\emptyset) = 0$. Then for $E \subset X$, the (Lebesgue) outer measure $\mu^* : 2^{\mathbb{R}} \to [0, \infty]$ is given by

$$\mu^*(E) = \inf \Sigma_n \lambda(G_n) \qquad \{G_n\} \subset \mathcal{G} \text{ varies over all } \mathcal{G}\text{-coverings of } E.$$

Again, because μ^* is a metric outer measure each open subset (and therefore each Borel subset) of $\mathbb{R}$ is Lebesgue measurable.

Another application of the "general method" yields the Hausdorff metric outer measures m_ε^p.

A11.1 Definition (m_ε^p metric outer measures)   Let $X = \mathbb{R}^n$ be Euclidean n-space with the usual metric d. Let $p \in [0, \infty)$, let $\varepsilon > 0$, and let $\mathcal{G} = \mathcal{G}_\varepsilon$ denote the collection of subsets G of $\mathbb{R}^n$ whose diameter $|G| = \sup \{d(x, y) : x, y \in G\} < \varepsilon$. (The empty set $\emptyset \in \mathcal{G}$ because $|\emptyset| = 0$ by definition.) Define $\lambda : \mathcal{G}_\varepsilon \to [0, \infty]$ by $\lambda(G) = |G|^p$. (When $p = 0$, define $|G|^p = |G|^0 = 0$ if $G = \emptyset$, and $|G|^0 = 1$ otherwise.) For each $E \subset X = \mathbb{R}^n$, define

$$m_\varepsilon^p(E) = \inf \Sigma_n |G_n|^p \qquad \{G_n\} \subset \mathcal{G}_\varepsilon \text{ varies over all } \mathcal{G}_\varepsilon\text{-coverings of } E.$$

Thus, for each $p \in [0, \infty)$ and each $\varepsilon > 0$, we have the *metric outer measure* $m_\varepsilon^p : 2^{\mathbb{R}^n} \to [0, \infty]$.

These metric outer measures m_ε^p are used in the following section to define the Hausdorff p-measures m^p.

§A12 Hausdorff p-Measures

For each fixed $p \in [0, \infty)$, the Hausdorff p-*measure* (a metric outer measure) is given by

$$m^p(E) = \sup_{\varepsilon > 0} m_\varepsilon^p(E).$$

A12.1 Proposition   Let $p \in [0, \infty)$ be fixed. Then $m^p(E) = \sup_{\varepsilon > 0} m_\varepsilon^p(E) = \lim_{\varepsilon \to 0} m_\varepsilon^p(E)$.

PROOF. Since $\varepsilon' < \varepsilon$ implies $\mathcal{G}_{\varepsilon'} \subset \mathcal{G}_{\varepsilon}$, we have $m_\varepsilon^p(E) \leq m_{\varepsilon'}^p(E)$, showing that $m_\varepsilon^p(E)$ is nondecreasing as $\varepsilon \to 0$. So $\lim_{\varepsilon \to 0} m_\varepsilon^p(E)$ exists in $[0, \infty]$ and equals $\sup_{\varepsilon > 0} m_\varepsilon^p(E)$. $\qquad\square$

A12.2 Proposition (properties of m^0) *Let (X, d) be a metric space, and let $p \in [0, \infty)$. Then*

   (i)   *the p-measure is monotone, i.e., $F \subset E$ implies $m^p(F) \leq m^p(E)$;*
   (ii)   *when $E = \emptyset$, $m^0(E) = 0$ and $m^q(E) = 0$ for each $q > 0$;*
   (iii)   *when $E = \{x_1, \ldots, x_k\}$ is a finite set with $k > 0$ elements,*
           $m^0(E) = k$ and $m^q(E) = 0$ for each $q > 0$; and
   (iv)   *when E is infinite, $m^0(E) = \infty$.*

PROOF. (i) Each cover of E is also a cover of F. (ii) The family $\{\emptyset\}$ covers E, $|\emptyset|^0 = 0$ by definition, and $0^q = 0$ when $q > 0$. (iii) For $\varepsilon = \min\{d(x_i, x_j) : i \neq j\}$, a cover $\mathcal{G}_\varepsilon$ has at least k members G_n such that $|G_n| < \varepsilon$. For $p = 0$, each $|G_n|^0 = 1$; and for $q > 0$, each $|G_n|^q < \varepsilon^q$. (iv) Select a tower $E_1 \subset E_2 \subset \cdots$ of subsets of E where E_k has size k and apply (i). $\qquad\square$

In passing, notice that (ii), (iii), and (iv) show that m^0 is a "counting measure."

12.3 Proposition (p-measure bifurcation) *Let (X, d) be a metric space and let $E \subset X$. If there exists a real number $p \leq \inf\{r : m^r(E) = 0\}$ such that $m^p(E)$ is finite, then*

   (v) $p = \inf\{q : m^q(E) = 0\}$ and $m^q(E) = 0$ if $q \in (p, \infty)$; and
   (vi)   *if $p > 0$ is positive, then the p-measure induces a bifurcation*

$$m^q(E) = \begin{cases} \infty & \text{if } q \in [0, p) \\ 0 & \text{if } q \in (p, \infty) \end{cases} \quad \text{and}$$

$$p = \inf\{q : m^q(E) = 0\} = \sup\{q : m^q(E) = \infty\}.$$

PROOF. If $p = 0$, then (v) is valid because $0 \leq m^0(E) < \infty$ and properties (ii) and (iii) of m^0 apply. So let $p > 0$ and $q \in [0, p) \cup (p, \infty)$. Then $q - p \neq 0$ and

$$\Sigma_n |G_n|^q = \Sigma_{G_n \neq \emptyset}(|G_n|^p |G_n|^{q-p}) + \Sigma_{G_n = \emptyset}|G_n|^q = \Sigma_{G_n \neq \emptyset}(|G_n|^p |G_n|^{q-p})$$

because $|\emptyset|^q = 0$ when $q = 0$. For $q \in (p, \infty)$ and each $G_n \in \mathcal{G}_\varepsilon$, we have $q - p > 0$ and $|G_n|^{q-p} \leq \varepsilon^{q-p}$, showing

$$\Sigma_{G_n \neq \emptyset}(|G_n|^p |G_n|^{q-p}) \leq \varepsilon^{q-p} \Sigma_n |G_n|^p.$$

It follows that $q \in (p, \infty)$ implies

$$m^q(E) = \lim_{\varepsilon \to 0} \inf \Sigma_n |G_n|^q \leq \left(\lim_{\varepsilon \to 0} \varepsilon^{q-p}\right) m^p(E) = 0$$

because $0 \leq m^p(E) < \infty$. Thus (v) is valid. On the other hand, for $q \in [0,p)$ we have $q - p < 0$ and each $|G_n|^{q-p} \geq \varepsilon^{q-p} = 1/\varepsilon^{p-q}$, showing

$$m^q(E) = \lim_{\varepsilon \to 0} \inf \Sigma_n |G_n|^q \geq \left(\lim_{\varepsilon \to 0} 1/\varepsilon^{p-q} \right) m^p(E) = \infty$$

when $0 < m^p(E) < \infty$. The lone remaining case is "$q \in [0,p)$ and $m^p(E) = 0$." We do, however, know:

(vii) $\quad \begin{cases} \text{if } 0 < m^s(E) < \infty \text{ and } s \leq \inf \{r : m^r(E) = 0\}, \\ \text{then } s = \inf \{r : m^r(E) = 0\}. \end{cases}$

Now suppose "$q \in [0,p)$ and $m^p(E) = 0$." Then $m^q(E) > 0$ — otherwise $p \leq \inf \{r : m^r(E) = 0\} \leq q$ while $q < p$. Since $m^q(E) > 0$, we only need to show that $m^q(E)$ is not finite. But $0 < m^q(E) < \infty$ and $q < p = \inf \{r : m^r(E) = 0\}$ provide a substitution of q for s in (vii) — so $p = q$ and $q < p$. It follows that $m^q(E) = \infty$ when $q \in [0,p)$ and $m^p(E) = 0$. Finally, the bifurcation in (vi) yields the "inf" and "sup" equalities. $\qquad \square$

§A13 Hausdorff Dimension

The following proposition shows that Hausdorff dimension (defined below) is well defined.

13.1 Proposition *Let (X,d) be a metric space and let $E \subset X$. Then there is a unique $p \in [0,\infty]$ such that*

$$m^q(E) = \begin{cases} \infty & \text{if } q \in [0,p) \\ 0 & \text{if } q \in (p,\infty). \end{cases}$$

Thus, for each subset E of X either $m^q(E) = \infty$ for every q-measure, or, $m^q(E)$ is finite for some q. In the latter case, there exists a unique $p = \inf \{q : m^q(E) = 0\}$. This correspondence $E \mapsto D(E)$ given by

$$D(E) = \begin{cases} p = \inf \{q : m^q(E) = 0\} & \text{if } m^q(E) \text{ is finite for some } q \\ \infty & \text{if } m^q(E) = \infty \text{ for every } q \end{cases}$$

is the *Hausdorff dimension function* $D : 2^X \to [0,\infty]$.

This bifurcation property applies to any infinite (and hence interesting) subset $E \subset \mathbb{R}^n$ for which there is a positive p such that $m^p(E) \in (0,\infty)$. Indeed, such a p must be unique, and the Hausdorff dimension $D(E) = p$ for our set E.

(Even though such a p may not be a positive integer, the idea of saying E is fundamentally "p-dimensional" is nevertheless analogous to the idea that a square $E \subset \mathbb{R}^2$ is fundamentally "2-dimensional," — when viewed in the context of Lebesgue measures in $\mathbb{R}^q$, for those integer values q where $q < 2$

and $q > 2$ in particular, we may (in the former case) view E as a "space-filling" 1-dimensional curve of infinite length, or (in the latter case) view E as a subset in 3-space with zero volume.)

We close this section by recalling that the Hausdorff dimension of the Cantor set $\mathcal{C}$ is $\ln(2)/\ln(3)$. (For a proof, see page 14 of Falconer [1985].)

§A14 Fractal Dimension

From Mandelbrot [1983], a subset E of $\mathbb{R}^n$ such that $D(E) \neq D_T(E)$ is a *fractal* where the dimension function "D" is the *Hausdorff dimension function* (sometimes called the *Hausdorff-Besicovitch dimension function*) defined in §A13, and D_T denotes the topological dimension, i.e., any of ind, Ind, or dim.

According to this definition of fractal, we find that the Cantor set $\mathcal{C}$ is a fractal with dimension (for a proof, see page 14 of Falconer [1985])

$$D(\mathcal{C}) = \ln(2)/\ln(3) = 0.639\ldots \neq 0 = \operatorname{ind} \mathcal{C} = D_T(\mathcal{C}).$$

Over the years, the term "fractal" has taken on various meanings in various contexts (see for example, Peitgen, Jürgens, and Saupe [1992]).

In this text, we use *fractal dimension* to mean *self-similarity dimension*. To illustrate the basic idea, let us, for the moment at least, follow the opening of Chapter 2 in Crownover [1995]: Suppose a line segment is divided into N equal pieces, each being thought of as a scaled copy of the whole segment. If the scaling ratio is r, then the relation between N and r is $Nr = 1$.

Similarly, if a square has its *sides scaled by the factor* r into N equal subsquares, then $Nr^2 = 1$, and for a cube, $Nr^3 = 1$.

With these examples, it is not difficult to notice that the dimension of the object being scaled shows up as the exponent of the scaling factor r, i.e.,

$$Nr^d = 1.$$

To consider non-integral values of the dimension d, suppose $r = 1/3$ and consider the Cantor set $\mathcal{C}$. Then since $\mathcal{C}$ may be partitioned into $N = 2$ sets $(\mathcal{C} \cap [0, 1/3]) \cup (\mathcal{C} \cap [2/3, 1])$, we have

$$Nr^d = (2)(1/3)^d = 1 \implies \ln 2 + d(\ln(1/3)) = 0 \implies d = \ln 2/\ln 3.$$

The value $\ln 2/\ln 3 = \ln N/\ln(1/r)$, where the scaling factor $r = 1/3$ and N is the number 2 of copies of $\mathcal{C}$, agrees with the Hausdorff dimension $D(\mathcal{C})$ of Cantor's set. The formula $\ln N/\ln(1/r)$ is fundamental in calculating the self-similarity dimension.

An introductory discussion of the self-similarity dimension may be found in Chapter 5 of Crownover [1995]. For our purposes, however, suppose there are similitudes $S_1, \ldots, S_N$, each with scale factor r such that a compact set $E \subset \mathbb{R}^n$ satisfies

$$E = S_1(E) \cup \cdots \cup S_N(E)$$

and the Hausdorff d-measure, where $d = \ln(N)/\ln(1/r)$, of the overlaps of the $S_i(E)$ sets are zero. Then the *self-similarity dimension* of E is $d = \ln(N)/\ln(1/r)$.

In particular, for the n-web ω^n, we see that the "overlaps" consist of a finite number of points, and that

$$\omega^n = w_0(\omega^n) \cup \cdots \cup w_n(\omega^n)$$

where the scale factor of each w_i is $1/2$. So the fractal dimension, i.e., the self-similarity dimension, of ω^n is $\ln(n+1)/\ln(2)$ (see §A10).

Bibliography

E. Akin

[1993]   *The General Topology of Dynamical Systems*, American Mathematical Society Graduate Studies in Mathematics, Vol. 1, Amererican Mathematical Society, Providence, RI.

C. C. Alexander

[1971]   Semidevelopable spaces and quotient images of metric spaces, Pacific J. Math., Vol. 37, No. 2, 277–293.

P. S. Alexandroff

[1932]   Dimensionstheorie, Math. Ann., Vol. 106, 161–238.

[1955]   The present status of the theory of dimension, Amer. Math. Soc. Transl. Ser. 2, Vol. 1, 1–26.

[1956]   *Combinatorial Topology*, Vol. 1, Graylock Press, Baltimore, MD.

[1957]   *Combinatorial Topology*, Vol. 2, Graylock Press, Rochester, NY.

[1960]   *Combinatorial Topology*, Vol. 3, Graylock Press, Albany, NY.

P. Alexandroff and H. Hopf

[1935]   *Topologie*, Springer, Berlin.

A. Arhangel'skii

[1963]   Some types of factor mappings, and the relations between classes of topological spaces, Dokl. Akad. Nauk SSSR, Vol. 153, 743–746 (Sov. Math. Dokl., Vol. 4, 1726–1729).

E. Artin and H. Braun

[1969]   *Introduction to Algebraic Topology*, Merrill Research and Lecture Series, Charles E. Merrill, Columbus, OH.

M. Barnsley

[1988]   *Fractals Everywhere*, Academic Press, Boston, MA.

M. Barnsley and A. Sloan

[1988]   A better way to compress images, BYTE Mag., January, 1988, 215–223.

R. G. Bartle

[1966]   *The Elements of Integration*, John Wiley and Sons, New York.

R. H. Bing

[1951]   Metrization of topological spaces, Canad. J. Math., Vol. 3, 175–186.

[1983]   *The Geometric Topology of 3-Manifolds*, American Mathematical Society Colloquium Publications, Vol. 40, American Mathematical Society, Providence, RI.

[1988]   Models for S^3, in *The Collected Papers of R. H. Bing*, Vol. I, Sukhjit Singh, Steve Armentrout, and Robert J. Daverman editors, American Mathematical Society, Providence, RI, 853–869.

K. Borsuk

[1967]   *Theory of Retracts*, Polska Akademia Nauk Monografie Matematiyczne, PWN–Polish Scientific Publishers, Warsaw.

[1975]   Remark on the Cartesian product of two 1-dimensional spaces, Bull. Acad. Polon. Sci. Sér. Sci. Math. Astron. Phys., Vol. 23, No. 9, 971–973.

N. Bourbaki

[1961]   *Topology Générale*, Hermann, Paris.

[1966]   *General Topology Part I*, Addison-Wesley, London.

L. E. J. Brouwer

[1913]   Über den natürlichen Dimensionsbegriff, J. Reine Angew. Math., Vol. 142, 146–152.

R. Brown

[1968]   *Elements of Modern Topology*, McGraw-Hill, London.

G. Cantor

[1883a] Grundlagen einer allgemeinen Mannigfaltigkeitslehre, Math. Ann., Vol. 21, 545–591.

[1883b] Fondaments d'une théorie générale des ensembles, Acta Math., Vol. 2, 381–408.

[1884]   De la puissance des ensembles parfait de points (On the power of perfect sets of points), extracted by the editors of the journal Acta Mathematica from a letter written to them by Cantor. English translation appears in Edgar [1993], 11–23.

C. Carathéodory

[1914]   Über das lineare Mass von Punktmengeneine Verallgemeinerung des
         Längenbegriffs, Nachr. Ges. Wiss. Göttingen, Band (Vol.) 1914, Heft
         (issue) 4, 404–426.

T. Crilly (with the assistance of Dale Johnson)

[1999]   The emergence of topological dimension theory, in *History of Topol-
         ogy*, I.M. James editor, North-Holland, Amsterdam, Chapter 1.

R. M. Crownover

[1995]   *Introduction to Fractals and Chaos*, Jones and Bartlett, Boston, MA.

J. Dieudonné

[1944]   Une généralisation des espaces compacts, J. Math. Pures Appl., Vol.
         23, 65–76.

C. H. Dowker

[1947]   *An imbedding theorem for paracompact metric spaces*, Duke Math.
         J., Vol. 14, 639–645.

J. Dugundji

[1966]   *Topology*, Allyn and Bacon, Boston, MA.

G. A. Edgar

[1993]   *Classics on Fractals*, Gerald A. Edgar editor, Addison-Wesley, Read-
         ing, MA.

R. Engelking

[1968]   *Outline of General Topology*, North-Holland, Amsterdam.

[1978]   *Dimension Theory*, North-Holland Mathematics Library, Vol. 8,
         North-Holland, New York.

R. Engelking and K. Sieklucki

[1992]   *Topology, a Geometric Approach*, Sigma Series in Pure Mathematics,
         Vol. 4, Heldermann Verlag, Berlin.

K. J. Falconer

[1985]   *The Geometry of Fractal Sets*, Cambridge Tracts in Mathematics,
         Vol. 85, Cambridge University Press, New York.

H. Federer

[1969]   *Geometric Measure Theory*, Springer-Verlag, New York.

V. V. Filippov

[1970]   Solution of a problem of P. S. Aleksandrov (a bicompact space with distinct inductive dimensions), Math. USSR Sb., Vol. 12, 41–57 (English Trans.).

W. Fleming

[1977]   *Functions of Severable Variables*, 2nd edition, Springer-Verlag, New York.

G. Flores

[1934]   Über n-dimensionale Komplexe die im R_{2n+1} absolute selbstverschungen sind, Ergeb. Math. Kolloq., Vol. 6, 4–7.

D. Gulick

[1992]   *Encounters with Chaos*, McGraw-Hill, New York.

O. Hájek

[1966]   Notes on quotient maps, Comment. Math. Univ. Carolin., Vol. 7, 319–323.

K. Hannabuss

[1996]   Forgotten fractals, Math. Intelligencer, Springer-Verlag, N.Y., Vol. 18, No. 3, 28–31.

Y. Hattori

[1989]   A note on universal spaces for finite dimensional complete metric spaces, Glas. Mat. Ser. III, Vol. 24, No. 44, 139–147.

F. Hausdorff

[1919]   Dimension und äusseres Mass, Math. Ann., Vol. 79, 157–179.

A. Hinz, S. Klavžar, U. Milutinović, D. Parisse, and C. Petr

[2005]   Metric properties of the Tower of Hanoi graphs and Stern's diatomic sequence, European J. Combin., Vol. 26, 693–708.

J. G. Hocking and G. S. Young

[1988]   *Topology*, Dover reprint of 1961 original (Addison-Wesley, Reading, MA), Dover Publications, New York.

W. Hurewicz

[1927]   Über das Verhältnis separabler Räume zu kompakten Räumen, Proc.
         K. Ned. Akad. Wet., Vol. 30, 425–430.

[1930]   Ein theorem der dimensionstheorie I, Ann. Math., Vol. 31, 176–180.

[1931]   Dimensionstheorie and Cartesische Räume, Proc. Akad., Amster-
         dam, Vol. 34, 399–400.

W. Hurewicz and H. Wallman

[1948]   *Dimension Theory* revised edition, Princeton University Press,
         Princeton, NJ (original copyright 1941).

J. E. Hutchinson

[1981]   Fractals and self similarity, Indiana Univ. Math. J., Vol. 30, No. 5.,
         713–747.

I. Ivanšić and U. Milutinović

[2002]   A universal separable metric space based on the triangular Sierpiński
         curve, Topol. Appl., Vol. 120, 237–271.

[2003]   Relative embeddability into Lipscomb's 0-dimensional universal
         space, Houston J. Math., Vol. 29, No. 4, 1001–1012.

[2005]   The pointed version of Lipscomb's embedding theorem, Houston J.
         Math., Vol. 31, No. 1, 173–192.

[2007]   Closed embeddings into Lipscomb's universal space, Glas. Mat. Ser.
         III, Vol. 42, No. 1, 95–108.

I. M. James

[1999]   *History of Topology*, I.M. James editor, Elsevier Science, Amsterdam.

D. M. Johnson

[1977]   Prelude to dimension theory: the geometrical investigations of
         Bernard Bolzano, Arch. Hist. Exact Sci., Vol. 17, 261–295.

M. Katětov

[1952]   On the dimension of non-separable spaces I, Czech. Math. J., Vol. 2,
         No. 77, 333–368 (in Russian).

J. L. Kelly

[1955]   *General Topology*, American Book Company, New York

J. L. Kelly and I. Namioka

[1963]   *Linear Topological Spaces*, Van Nostrand, Princeton, NJ.

S. Klavžar and U. Milutinović

[1997]   Graphs $S(n, k)$ and a variant of the Tower of Hanoi problem, Czech. Math. J., 47, No. 122, 95–104.

S. Klavžar, U. Milutinović, and C. Petr

[2002]   1-perfect codes in Sierpiński graphs, Bull. Aust. Math. Soc., Vol. 66, No. 3, 369–384.

S. Klavžar and B. Mohar

[2005]   Crossing numbers of Sierpiński-like graphs, J. Graph Theory, Vol. 50, 186-198.

A. N. Kolmogorov and S. V. Fomin

[1957]   *Elements of the Theory of Functions and Functional Analysis, Vol. 1 Metric and normed Spaces*, translated from the first (1954) Russian edition, Graylock Press, Rochester, NY.

[1961]   *Elements of the Theory of Functions and Functional Analysis, Vol. 2 Measure, the Lebesgue integral, Hilbert Space*, translated from the first (1960) Russian edition, Graylock Press, Albany, NY.

H. J. Kowalsky

[1957]   Einbettung metrischea Räume, Arch. Math., Vol. 8, 336–339.

W. Kulpa

[1970]   On uniform universal spaces, Fundam. Math., Vol. 69, 243–251.

E. Kummer

[1852]   Über Ergänzungssätze zu den allgemeinen Reziprozitätsgesetzen, J. Reine Angew. Math., Vol. 44, 93–146.

K. Kuratowski

[1930]   Sur le problème des courbes gauches en Topologie, Fundam. Math., Vol. 15, 271–283.

[1932]   Sur l'application des espaces fonctionnels à la théorie de la dimension, Fundam. Math., Vol. 18, 285–292.

[1937]   Sur les théorème de "plongement" dans la théorie de la dimension, Fundam. Math., Vol. 28, 336–342.

[1966]   *Topology Volume I* (translated from French), Academic Press, New York.

[1968]   *Topology Volume II* (translated from French), Academic Press, New York.

N. S. Lašnev

[1966]   Closed images of metric spaces, Dokl. Akad. Nauk SSSR, Vol. 170, 505–507 (Sov. Math. Dokl., Vol. 7, 1219–1221).

H. Lebesgue

[1904]   *Leçons sur l'integration et la recherche des fonctions primitives*, Gauthier-Villars, Paris.

S. Lefschetz

[1931]   On compact spaces, Math. Ann., Vol. 32, 521–538.

S. L. Lipscomb

[1973]   *Imbedding one dimensional metric spaces* (University of Virginia dissertation), University Microfilms, Ann Arbor, MI.

[1974]   A universal one-dimensional metric space, Lecture Notes in Mathematics, Vol. 378 *TOPO 72 General Topology and Its Applications*, Springer–Verlag, Berlin, 248–257.

[1975]   On imbedding finite-dimensional metric spaces, Trans. Amer. Math. Soc., Vol. 211, 143–160.

[1976]   An imbedding theorem for metric spaces, Proc. Amer. Math. Soc., Vol. 55, 165–169.

[2005]   A minimal extension of the iterated function system for Sierpiński's gasket to one whose attractor is the 2-simplex, Houston J. Math., Vol. 31, No. 4, 1065–1083.

[2007]   The Sierpiński-cheese iterated function system extended to a 3-simplex system, Houston J. Math., Vol. 33, No. 1, 169–207.

S. L. Lipscomb and J. C. Perry

[1992]   Lipscomb's $L(A)$ space fractalized in Hilbert's $l^2(A)$ space, Proc. Amer. Math. Soc., Vol. 115, 1157–1165.

B. B. Mandelbrot

[1975]   *Les objects fractals: forme, hasard et dimension*, Flammarion, Paris.

[1983] *The Fractal Geometry of Nature*, revised edition of 1977 version, W.H. Freeman and Company, New York.

J. C. Mayer, L. G. Oversteegen, and E. D. Tymchatyn

[1986] *The Menger Curve*, Dissertationes Mathematicae, PWN–Polish Scientific Publishers, Warsaw.

K. Menger

[1926a] Allgemeine Räume und Cartesische Räume, Proc. Akad. Wetensch. Amst., Vol. 29, 476–482.

[1926b] Über umfassendste n-dimensionale Mengen, Proc. Akad. Wetensch. Amst., Vol. 29, 1125–1128.

[1928] *Dimensionstheorie*, B. G. Teubner, Leipzig.

E. Michael

[1968] Bi-quotient maps and cartesian products of quotient maps, Ann. Inst. Fourier (Grenoble), Vol. 18, No. 2, 287–302.

[1972] A quintuple quotient quest, Gen. Topol. Appl., Vol. 2, 91–138.

[1974] Some classes of quotient maps, Topol. Structures, Math. Centre Tracts, Vol. 52, 55–58.

R. Miculescu and A. Mihail

[2008] Lipscomb's space ω^A is the attractor of an IFS containing affine transformations of $l^2(A)$, Proc. Amer. Math. Soc., Vol. 136, 587–592.

U. Milutinović

[1992] Completeness of the Lipscomb universal spaces, Glas. Math. Ser. III, Vol. 27, No. 47, 343–364.

[1993] *Contributions to the theory of universal spaces*, PhD thesis, University of Zagreb, Zagreb, (in Croatian).

[2006] Approximation of maps into Lipscomb's space by embeddings, Houston J. Math., Vol. 32, No. 1, 143–159.

K. Morita

[1954] Normal families and dimension theory for metric spaces, Math. Ann., Vol. 128, 350–362.

[1955] A condition for the metrizability of topological spaces and for n-dimensionality, Sci. Rep. Tokyo Kyoiku Daigaku Sect. A, Vol. 5, No. 114, 33–36.

K. Morita and S. Hanai

[1956]   Closed mappings and metric spaces, Proc. Jpn. Acad., Vol. 32, 10–14.

J. R. Munkres

[1984]   *Elements of Algebraic Topology*, Benjamin/Cummings, Reading, MA.

S. B. Nadler, Jr.

[1978]   *Hyperspaces of Sets*, Pure and Applied Mathematics, Vol. 49, Marcel Dekker, New York.

[1992]   *Continuum Theory, an Introduction*, Pure and Applied Mathematics, Vol. 158, Marcel Dekker, New York.

K. Nagami

[1957]   Some theorems in dimension theory for non-separable spaces, Proc. Jpn. Acad., Vol. 39, 80–92.

[1961]   A note on Hausdorff space with the star-finite property I, II, III, Proc. Jpn. Acad., Vol. 37, 131–134, 189–192, 356–357.

[1970]   *Dimension Theory*, Academic Press, New York.

J. Nagata

[1950]   On a necessary and sufficient condition of metrizability, J. Inst. Polytech., Osaka City Univ., Series A, Vol. 1, No. 2, 93–100.

[1958]   Note on dimension theory for metric spaces, Fundam. Math., Vol. 45, 143–181.

[1960]   On a universal n-dimensional set for metric spaces, Crelle J., Vol. 204, 132–138.

[1963]   A remark on general imbedding theorems in dimension theory, Proc. Jpn. Acad., Vol. 39, 197–199.

[1965]   *Modern Dimension Theory*, Bibliotheca Mathematica, Vol. 6, Interscience Publishers (John Wiley and Sons), New York.

[1967]   A survey of dimension theory, *General Topology and Its Relations to Modern Analysis and Algebra II*, Proceedings of the Second Prague Symposium, J. Novák editor, 1966, Academia, Prague, 259–270.

[1968]   *Modern General Topology*, Bibliotheca Mathematica, Vol. 7, North-Holland, Amsterdam.

[1983]   *Modern Dimension Theory (revised and extended edition)*, Sigma Series in Pure Mathematics, Vol. 2, Heldermann Verlag, Berlin.

A. Nagórko

[2006]   *Characterization and topological rigidity of Nöbeling manifolds*, Ph.D. thesis, Warsaw University.

A. G. Nemets

[1975]   On metric dimension, Dokl. Akad. Nauk SSSR, Vol. 225, 44–47. (English trans. in Sov. Math. Dokl., Vol. 16, 1975.)

G. Nöbeling

[1931]   Über eine n-dimensionale Universalmenge im R_{2n+1}, Math. Ann., Vol. 104, 71–80.

W. Olszewski and L. Piątkiewicz

[1992]   Closed embeddings of completely metrizable spaces into universal spaces, Glas. Math. J., Vol. 27, No. 47, 175–181.

P. Ostrand

[1971]   Covering dimension in general spaces, Gen. Topol. Appl., Vol. 1, No. 3, 209–221.

A. R. Pears

[1975]   *Dimension Theory of General Spaces*, Cambridge University Press, Cambridge.

H. Peitgen, H. Jürgens, and D. Saupe

[1992]   *Chaos and Fractals (New Frontiers of Science)*, Springer-Verlag, New York.

J. C. Perry

[1996]   Lipscomb's universal space is the attractor of an infinite iterated function system, Proc. Amer. Math. Soc., Vol. 124, 2479–2489.

J. C. Perry and S. L. Lipscomb

[2003]   The generalization of Sierpiński's triangle that lives in 4-space, Houston J. Math., Vol. 29, No. 3, 691–710.

W. J. Pervin

[1964]   *Foundations of General Topology*, Academic Press, New York.

H. Poincaré

[1912]   Rev. de Métaphy. Morale, Vol. 20, 486.

V. Ponomarev

[1960]   Normal spaces as images of zero-dimensional spaces, Dokl. Akad.
         Nauk SSSR, Vol. 132, 1269–1272.

L. S. Pontryagin

[1952]   *Foundations of Combinatorial Topology*, Graylock Press, Rochester,
         New York.

L. S. Pontryagin and G. Tolstowa

[1931]   Beweis des Mengerschen Einbettungssatzes, Math. Ann., Vol. 105,
         734–747.

F. Riesz and B. Sz.-Nagy

[1955]   *Functional Analysis*, (translated from the 2nd French edition of
         Leçons D'analyse Fonctionelle), Frederick Ungar, New York.

C. A. Rogers

[1970]   *Hausdorff Measures*, Cambridge University Press, Cambridge.

P. Roy

[1962]   Failure of equivalence of dimension concepts for metric spaces, Bull.
         Amer. Math. Soc., Vol. 68, 609–613.

[1968]   Nonequality of dimensions for metric spaces, Trans. Amer. Math.
         Soc., Vol. 134, 117–132.

W. Rudin

[1966]   *Real and Complex Analysis*, McGraw-Hill, New York.

N. A. Secelean

[2001]   Countable iterated function systems, Far East J. Dyn. Syst., Vol. 3,
         No. 2, 149–167.

W. Sierpiński

[1915]   Sur une courbe dont tout point est un point de ramification, C. R.
         Acad. Sci. Paris, Vol. 160, 302–305.

[1916]   Sur une courbe cantorienne qui contient une image biunivoquet et
         continue detoute courbe donnée, C. R. Acad. Sci. Paris, Vol. 162,
         629–632.

[1922]   Sur une propriété des ensembles frontières, Fundam. Math., Vol. 3, 7–13.

G. F. Simmons

[1963]   *Introduction to Topology and Modern Analysis*, McGraw-Hill, New York.

Y. M. Smirnov

[1951]   A necessary and sufficient condition for metrizability of a topological space, Dokl. Akad. Nauk SSSR. (N.S.), Vol. 77, 197–200.

[1956]   On strongly paracompact spaces, Izv. Akad. Nauk SSSR, Vol. 20, 252–274.

H. J. S. Smith

[1875]   On the integration of discontinuous functions, Proc. London Math. Soc., Vol. 6, 140–153; Collected Mathematical Papers, No. 25.

D. M. Y. Sommerville

[1958]   *An Introduction to the Geometry of N Dimensions*, Dover Publications, Inc., New York.

E. H. Spanier

[1966]   *Algebraic Topology*, McGraw-Hill, New York.

A. H. Stone

[1948]   Paracompactness and product spaces, Bull. Amer. Math. Soc., Vol. 54, 977–982.

[1956]   Metrizability of Decomposition Spaces, Proc. Amer. Math. Soc., Vol. 7, 690–700.

A. E. Taylor

[1965]   *General Theory of Functions and Integration*, Blaisdell, New York.

K. Tsuda

[1985a] A note on closed embeddings of finite dimensional metric spaces, Bull. London Math. Soc., Vol. 17, 275–278.

[1985b] A note on closed embeddings of finite dimensional metric spaces II, Bull. Polish Acad. Sci. Math., Vol. 33, 541–546.

P. Urysohn

[1925a] Zum Metrisationsproblem, Math. Ann., Vol. 94, 309–315.

[1925b] Mémoire sur les multiplicités Cantoriennes, Fundam. Math., Vol. 7, 30–137.

[1926] Mémoire sur les multiplicités Cantoriennes (suite), Fundam. Math., Vol. 8, 225–359.

L. Vietoris

[1927] Über den höheren Zusammenhang kompakter Räume und eine Klasse von zusammenhangstreuen Abbildungen, Math. Ann., Vol. 97, 454–472.

P. Vopěnka

[1958] On the dimension of compact spaces, Czech. Math. J., Vol. 8, 319–327 (in Russian).

A. Waśko

[1986] Spaces universal under closed embeddings of finite-dimensional complete metric spaces, Bull. London Math. Soc., Vol. 33, 541–546.

G. T. Whyburn

[1958] Topological characterizations of the Sierpiński curve, Fundam. Math., Vol. 45, 320–324.

[1964] *Topological Analysis* (revised edition), Princeton University Press, Princeton, NJ.

R. L. Wilder

[1938] The sphere in topology, in American Mathematical Society Semicentennial Publications, Vol. II, Semicentennial Addresses of the AMS, American Mathematical Society, Providence, RI, 136–184.

Index